W0055353

Fortschritte der Chemie organischer Naturstoffe

Progress in the Chemistry of Organic Natural Products

62

Founded by L. Zechmeister
Edited by W. Herz, G. W. Kirby, R. E. Moore,
W. Steglich, and Ch. Tamm

Authors:
S. V. Bhat, L. Minale, R. Riccio, F. Zollo

Springer-Verlag
Wien New York 1993

Prof. W. Herz, Department of Chemistry,
The Florida State University, Tallahassee, Florida, U.S.A.

Prof. G. W. Kirby, Chemistry Department,
The University, Glasgow, Scotland

Prof. R. E. Moore, Department of Chemistry,
University of Hawaii at Manoa, Honolulu, Hawaii, U.S.A.

Prof. Dr. W. Steglich, Institut für Organische Chemie der Universität
München, München, Federal Republic of Germany

Prof. Dr. Ch. Tamm, Institut für Organische Chemie der Universität Basel,
Basel, Switzerland

This work is subject to copyright.
All rights are reserved, whether the whole or part of the material is concerned, specifically those
of translation, reprinting, re-use of illustrations, broadcasting, reproduction by photocopying
machines or similar means, and storage in data banks.

© 1993 by Springer-Verlag/Wien
Softcover reprint of the hardcover 1st edition 1993

Library of Congress Catalog Card Number AC 39-1015

Typesetting: Macmillan India Ltd., Bangalore-25

Printed on acid free paper

With 52 Figures

ISSN 0071-7886
ISBN-13: 978-3-7091-9252-8   e-ISBN-13: 978-3-7091-9250-4
DOI: 10.1007/978-3-7091-9250-4

Contents

Steroidal Oligoglycosides and Polyhydroxysteroids from Echinoderms
By L. MINALE, R. RICCIO and F. ZOLLO 75

List of Contributors

BHAT, Prof. S.V., Department of Chemistry, Indian Institute of Technology, Powai, Bombay - 400 076, India

MINALE, Prof. L., Dipartimento di Chimica delle Sostanze Naturali, Università degli Studi di Napoli Frederico II, Via Domenico Monicsano 49, I-80131 Napoli, Italy.

RICCIO, Dr. R., Dipartimento di Chimica delle Sostanze Naturali, Università degli Studi di Napoli Frederico II, Via Domenico Monicsano 49, I-80131 Napoli, Italy.

ZOLLO, Dr. F., Dipartimento di Chimica delle Sostanze Naturali, Università degli Studi di Napoli Frederico II, Via Domenico Monicsano 49, I-80131 Napoli, Italy.

Forskolin and Congeners

SUJATA V. BHAT,
Department of Chemistry,
Indian Institute of Technology,
Powai, Bombay, India

Contents

* Dedicated to Professor Werner Herz on the occasion of his seventysecond birthday.

I. Introduction

Plants and plant products have been used as drugs for thousands of years and in recent history have provided a definite stimulus for the development of natural products chemistry. After the first report on forskolin (**1**) (*1*) in 1977, the last decade has witnessed an increasing amount of research on the chemistry, synthesis, biochemistry, pharmacology and various other aspects of the substance in many academic as well as industrial research laboratories all over the world, leading to a large number of publications, patents and dissertations.

A number of reviews summarising the chemical, biological, medicinal potential (*2–10*) and synthesis (*11, 12*) of forskolin have been reported. This review will provide a comprehensive survey of the information on forskolin and congeners available at this time. Forskolin was discovered as a result of screening programs directed towards the discovery of new leads from Indian medicinal plants at the Hoechst Research Centre,

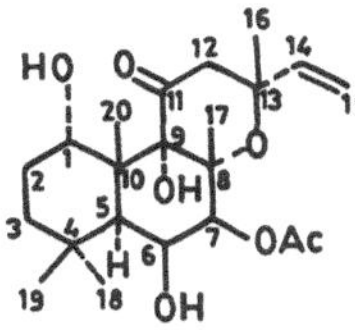

(1)

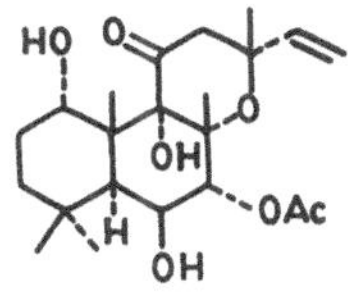

(2)

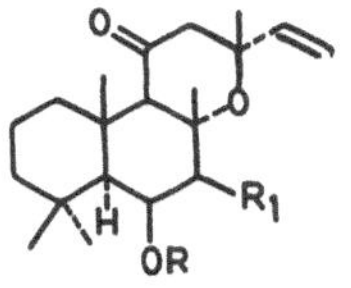

(3) R = H, R1 = OAc
(4) R = H, R1 = OH
(5) R = R1 = H

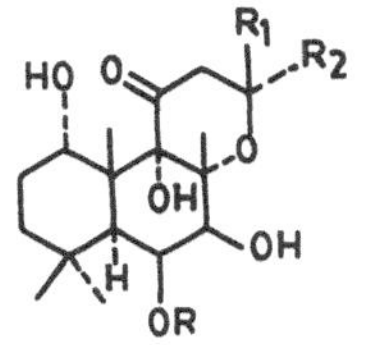

(6) R = H, R1 = Me,
R2 = –CH=CH2
(7) R = Ac, R1 = Me,
R2 = –CH=CH2
(8) R = Ac, R1 = –CH=CH2,
R2 = Me

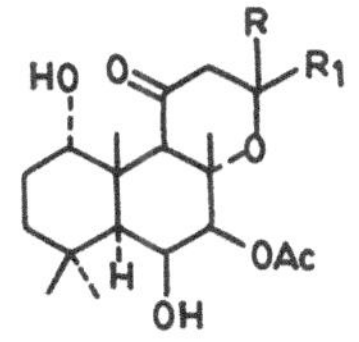

(9) R = Me, R1 = CH=CH2
(10) R = CH=CH2, R1 = Me

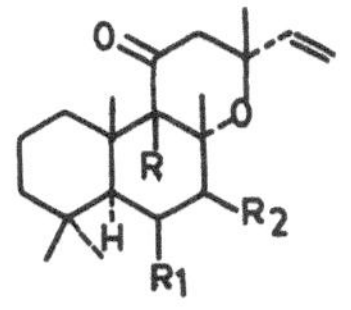

(11) R = OH, R1 = R2 = H
(12) R = R1 = OH, R2 = OAc
(13) R = R1 = R2 = OH
(14) R = R1 = R2 = H
(15) R = R1 = OH, R2 = H

Fig. 1. Diterpenoids of *Coleus forskohlii* (*1, 16–26*)

Bombay, India and the Central Drug Research Institute, (CDRI), Lucknow, India.

The Ayurvedic Sanskrit literature describes (13) the use of a plant known as 'Pashanbhedi' for the treatment of various ailments that can be associated with the cardiovascular, central nervous, bronchopulmonary and renal systems. One of the plants equated with 'Pashanbhedi' was *Coleus amboinicus*. Although this plant did not show any promising pharmacological activity, the methanolic extract from the roots of the related plant *Coleus forskohlii* Briq. displayed marked hypotensive and antispasmolytic activities (14). The active principle was identified as forskolin (1) (1, 15, 16) or coleonol (2) (17–20) by research groups at Hoechst, India and CDRI, India respectively. Subsequently the identity of coleonol with forskolin was established (21–23), and is represented by structure (1).

II. Isolation and Structure Determination

The methanolic or chloroform extract of dried and powdered roots of *Coleus forskohlii*, on column chromatography over silica gel, yielded the following labdane diterpenoids (Fig. 1); forskolin (1), 1,9-dideoxyforskolin (3), its 7-deacetyl (4) and 7-deacetoxy analogues (5), 7-deacetylforskolin (6), 6-acetyl-7-deacetylforskolin (7) and its 13-epimer (8), 9-deoxyforskolin (9) and its 13-epimer (10), 11-ketomanoyl oxide and its hydroxy derivatives (11–15). Physical properties of forskolin and congeners are given in Table 1 (1, 17–20, 24–27).

The mass spectrum of forskolin, $C_{22}H_{34}O_7$, showed the molecular ion at m/z 410 and major fragment ions at m/z 392 (M^+-H_2O), 364 (M^+-H_2O-CO), 332 ($M^+-H_2O-ACOH$) and 324 ($M^+-H_2O-C_5H_8$). The IR spectrum of (1) indicated the presence of hydroxyl (3430, 3230 cm^{-1}), carbonyl (1710 cm^{-1}) and vinyl (1650, 995, 913 cm^{-1}) groups.

The diterpene nature of the compound was evident from its ^{1}H- & ^{13}C-NMR spectra (Fig. 2 and Table 2). Spin decoupling NMR experiments indicated the presence of the $C_{5\alpha}H$-$C_{6\beta}OH$-$C_{7\beta}OAC$ sequence.

The NMR data supported by the chemical conversions shown in Fig. 3 indicated the presence of five tertiary C-methyls, a vinyl group attached to a quaternary carbon, an acetate, two secondary hydroxyls, a hindered tertiary hydroxyl and a carbonyl group, thus leading to the complete structure of forskolin as

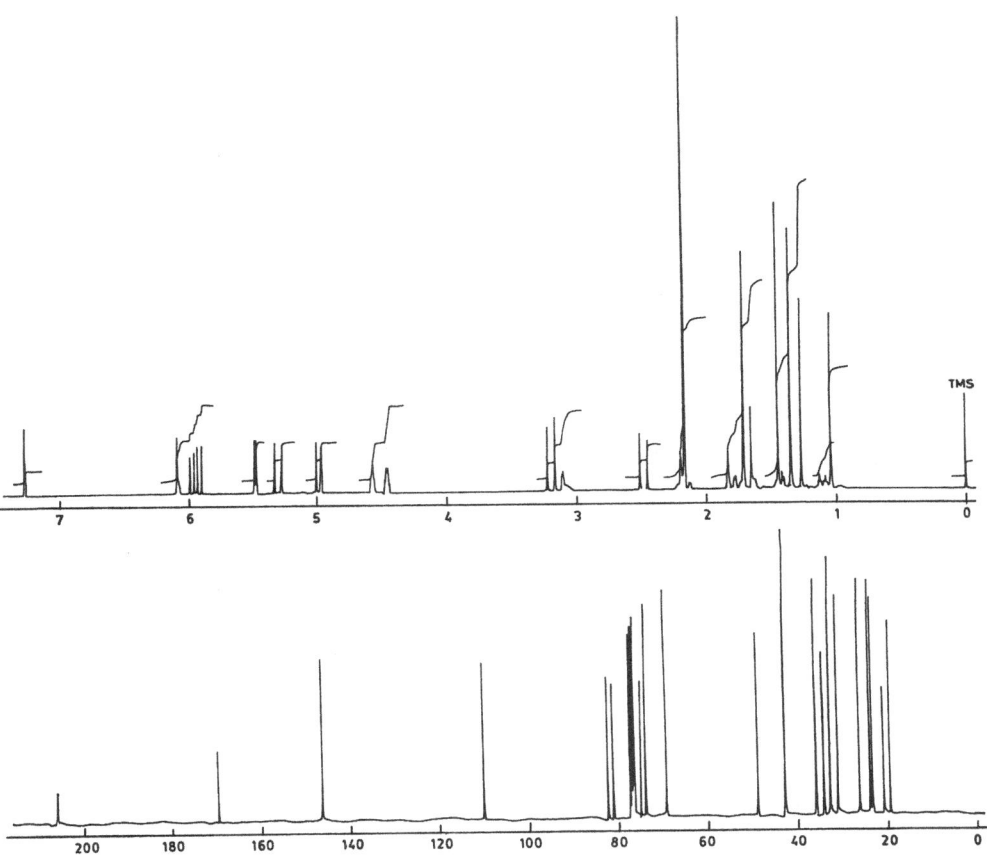

Fig. 2. ^1H (300 MHz) & ^{13}C (75.4 MHz decoupled) NMR spectra of (1) in $CDCl_3$

7-β-acetoxy-8,13-epoxy-1α,6β,9α-trihydroxy-labd-14-en-11-one   (1),
[1H-Naphtho[2,1-b]pyran-1-one,5-(acetyloxy)-3-ethenyl-dodecahydro-
6,10,10b-trihydroxy-3,4a,7,7,10a-pentamethyl-[3R-
(3α,4aβ,5β,6β,6aα,10α,10aβ,10bα)]].

Confirmation for the presence of the $C_{1α}$–OH, $C_{9α}$–OH and $C_{13α}$-
vinyl groups came from the facile formation of the 1,9-cyclic esters (20),
(25) and the 9,14-hemiacetal (18), respectively. The *trans* stereochemistry
of the A/B ring junction and the β-configuration at C_{10} was established
from the CD curves of (1), ΔE_{312} − 0.75, 14,15-dihydroforskolin (26),
$\Delta E_{312} = − 0.71$ and its 1-oxoderivative (27), $\Delta E_{312} = − 1.76$. The mass

Table. 1. *Physical Properties of Forskolin and Congeners*

Compound	MP°C (Solvent)	$[\alpha]_D^a$	Ref
(1)	230–232 (A)	− 26.19 (c 1.68)	*(1)*, *(17)*
(3)	162–165 (A)	− 89.4 (c 2.45)	*(1)*, *18c*
(4)	150–151 (B)	− 127.2 (c 0.1)	(1)
(5)	120–121 (C)	− 133.0 (c 0.1)	*(24)*
(6)	176–178 (A)	− 21.25 (c, 0.8)	*(1)*
(7)	208–210 (D)	+ 2.92 (c 1.0)	*{1)*, *(18a)*
(8)	205–6 (E)	− 6.7 (c 1.0)	*(18a)*
(9)	210–212 (F)	—	*(16)*, *(18c)*
(10)	180	− 77.7 (c 1.0)	*(27)*
(11)	85	+ 7.05 (c 1.7)	*(18b)*
(12)	135–136 (B)	− 23.1 (c 0.1)	*(24)*
(13)	148–149 (G)	− 25.9 (c 0.1)	*(24)*
(14)	96–97 (G)	− 103.2 (c 0.2)	*(24)*
(15)	192	− 134 (c 1.0)	*(25)*

a   in $CHCl_3$ at 23–25°C
A,   Ethyl acetate: petroleum ether
B   n-Hexane
C   Ethyl acetate: cyclohexane
D $CHCl_3$: petroleum ether
E   Acetone: hexane
F   Ethyl acetate: benzene
G   2-Propanol

spectrum of the 1-oxoderivative (**27**) had the base peak at m/z $(C_9H_{15}O)$ (Fig. 4) a fragmentation typical of 1-oxo-steroids (*15*). The application of Mills rule to the Δ^5-compound (**23**), $[M]_D + 347.17°$, derived by thionyl chloride/pyridine treatment of the 1-methylether of forskolin and the corresponding deacetyl derivative (**24**), $[M]_D − 32.03°$, established the β-configuration of the 7-OAc substituent (*1*).

Finally the absolute configuration of (**1**) was confirmed by X-ray analysis of forskolin and its 1-benzyl-7-deacetyl-7-α-bromo-isobutyryl derivative (**28**) (*28, 29*).

Recently support for the absolute stereochemistry of (**1**) was provided (*30*) by applying the exciton chirality circular dichroism method to 6,7-dibenzoyl-7-deacetyl-1-oxo forskolin (**29**) (*30*) and by 2D NMR studies (*22, 31*). In the solid state forskolin exists in the all chair conformation (X-ray, *29*) whereas in solution the C ring of (**1**) is present in the boat conformation (high resolution NMR *31*) (Fig. 5).

Table 2. ^1H and ^{13}C Chemical Shifts and Multiplicity in NMR Spectra of Forskolin in $CDCl_3$

Atom	δH	Multiplicity	δC	Multiplicity[a]
1	4.56	dd, $J_{1.2}$ = 3.4 & 2.5	74.5	d
2 eq	1.44	m	26.6	t
2 ax	2.13	m		
3 eq	1.11	m		
3 ax	1.77	m	36.1	t
4			34.4	s
5	2.19	d, $J_{5.6}$ = 1.9	42.9	d
6	4.46	dd, $J_{5.6}$ = 1.9, $J_{6.7}$ = 3.9	69.9	d
7	5.48	d, $J_{6.7}$ = 3.9	76.7	d
8			81.4	s
9			82.7	s
10			42.95	s
11			205.3	s
12 eq	2.47	d, J_{gem} = 17.1	48.8	t
12 ax	3.20			
13			75.0	s
14	5.94	dd, $J_{14, 15}$ = 17.1, 10.2	146.4	d
15	4.98	dd, $J_{14.15}$ = 10.2, 1.0 gem	110.7	t
	5.30	dd, $J_{14.15}$ = 17.1, 1.0 gem		
16	1.34	s	31.5	q
17	1.71	s	23.6	q
18	1.03	s	33.0	q
19	1.26	s	24.3	q
20	1.44	s	19.8	q
21			169.6	s
22	2.17	s	21.0	q

[a] from single frequency off resonance decoupling

III. Botanical and Biogenetic Considerations

The genus *Coleus* belongs to the family Labiatae popularly known as the mint family. *Coleus* is a large genus of herbs and shrubs comprising nearly 200 species which are widely distributed in the tropical and subtropical regions of Asia, Africa, Australia and the Pacific Islands. In India about 8 species are recorded. Many species are cultivated for ornamental purposes and some for edible tubers and leaves, and for use as pickles and salads.

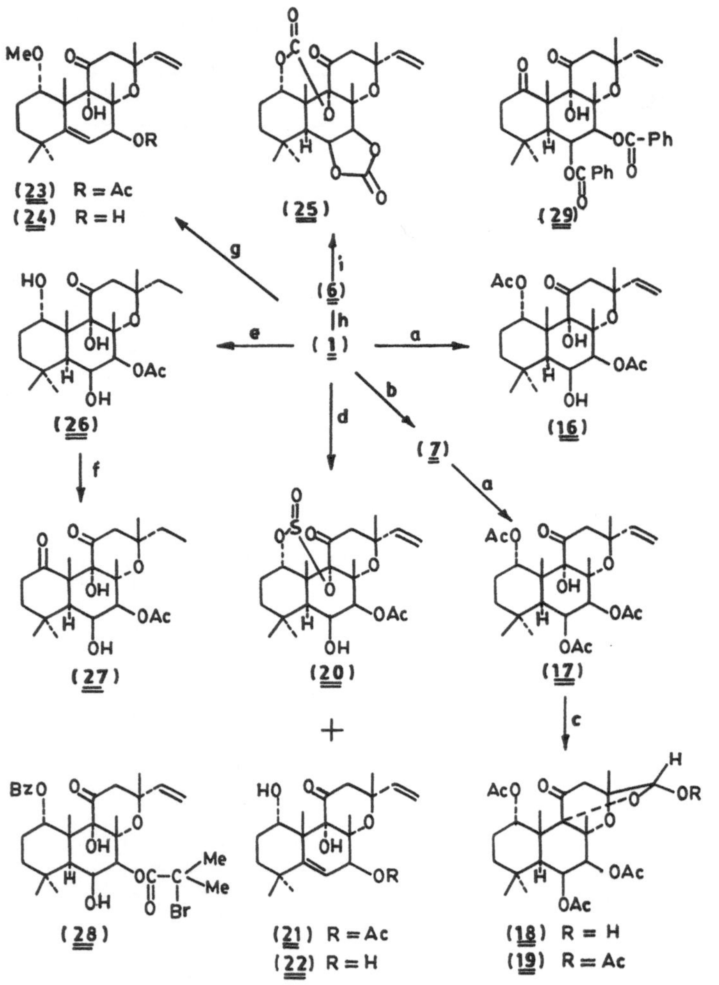

Fig. 3. Chemical conversions used in structure elucidation of (1), **a** Ac$_2$O, pyridine, **b** Al$_2$O$_3$, benzene, **c** O$_3$, CHCl$_3$, **d** SOCl$_2$, pyridine, **e** H$_2$, Pd/C, **f** Jones reagent, **g**(i) Me$_2$SO$_4$, K$_2$CO$_3$. **g**(ii) SOCl$_2$, pyridine, **h** KOH, MeOH, H$_2$O, **i** COCl$_2$, pyridine

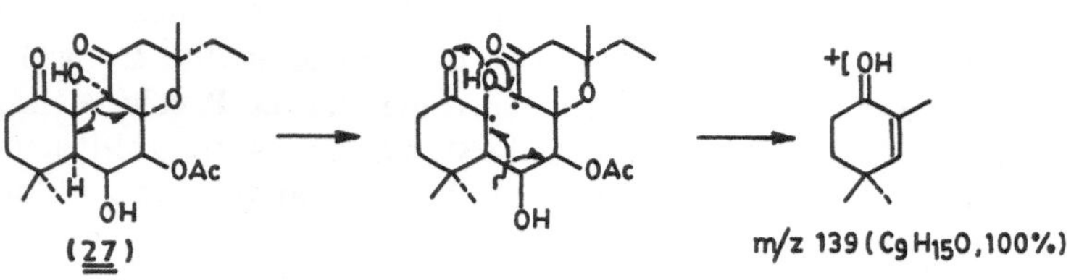

Fig. 4. Base peak in mass spectrum of (27) *(15)*

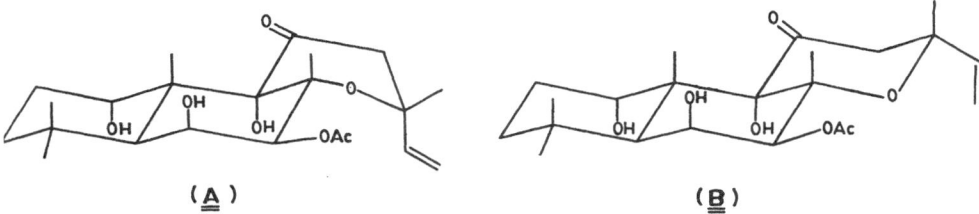

Fig. 5. Conformations of forskolin (A & B) derived from NMR studies (*31*) and X-ray crystal structure (*29*) respectively

Coleus forskohlii Briq. which grows perenially over the tropical and subtropical regions of India, Pakistan, Sri Lanka, Tropical East Africa and Brazil, is cultivated in some parts of India for its use in pickles. Its ethnomedical uses for relief of cough, eczema, skin infections, tumors and boils have been recorded (*10*).

Recently in a systematic study of the constituents of *Coleus forskohlii* a wide range of variation of forskolin content (0.01–0.44%) in dry roots was reported by VISHWAKARMA *et al.* (*32*).

The production of forskolin by suspension culture of *Coleus forskohlii* in an airlift bioreactor which produces a maximum of 730 mg/kg after 19 days of fermentation has been standardised by MERSINGER *et al.* (*33*). *In vitro* clonal multiplication of *Coleus forskohlii* and propagation under field conditions has also been reported (*34*). Dry roots of such plants produced 0.1% of forskolin.

Thin layer, GLC and HPLC chromatographic methods have been developed for the assay of forskolin (*35, 36*). When such assay methods were applied to crude plant extracts forskolin could not be detected in six other *Coleus* species and also in six taxonomically related *Plectranthus* species (*37*). BASLAS-PRADEEPKUMAR (*38*) reported phytochemical study of the genes *Coleus* describing the chemical composition and medicinal importance of prominent species.

Many interesting diterpenoids have been isolated from other *Coleus* species. Some of these are listed in Fig. 6, e.g. coleon C (**30**) and coleon D (**31**) from *Coleus aquaticus*; coleon E (**32**), coleon F (**33**), barbatusin (**34**), and cyclobarbatusin (**35**), from *Coleus barbatus*, coleon B (**36**) and coleon A (**37**) from *Coleus ligniaricus*, coleon L (**38**), coleon I (**39**) etc. from *Coleus somaliensis* due to EUGSTER and coworkers (*39*) and WANG *et al.* (*40*). It is noteworthy that many of these terpenoids are abietanes with oxygenation at C_3 or C_2 rather than at C_1 in the labdane skeleton of (**1**).

Since it elaborates ca. 14 labdanes which possess various types of oxygen functions, biosynthesis and sequential introduction of oxygen into

the 11-oxomanoyl oxide skeleton of *Coleus forskohlii* is of considerable
interest. In order to find out the mode of introduction of oxygen in the
diterpenes (1)–(15), MVA labelled with ^{3}H and ^{14}C at C_2, C_4 or C_5 was
fed to roots of *Coleus forskohlii* in different sets of experiments by AKHILA
and coworkers (26). It is very well established that C_2 of MVA gives rise
to C_1, C_7, C_{12} and C_{18} of the diterpenes. From the isotopic and atomic
(^{14}C/^{3}H) ratio in forskolin (1) and the congeners (3), (5), (9), (11), (12), and
(14) isolated after the plant was fed with labelled MVA, these authors
have suggested the possible biogenetic route to forskolin and congeners
shown in Figure 7. In the postulated biosynthetic pathway 8,13-epoxy-
labd-14-en-11-one (14) is the first monooxygenated labdane-type diterpene; subsequent sequential addition of oxygen gives 1,9-dideoxy-, 9-
deoxy-, 1-deoxy- and forskolin along with the other terpenes listed in
Figure 1.

IV. Chemistry of Forskolin

Molecule (1) has been subjected to various chemical modifications,
initially in the course of structure elucidation, subsequently in construction of analogues for derivation of structure-activity relationships as well
as for improvement of biological profile and more recently as the result of
synthetic studies based on retrosynthetic analysis and transformations
using microbes and newer reagents. The influence of functional groups
on activity was studied through i) acylation and alkylation reactions,
ii) oxidation and reduction reactions, iii) biological oxidations, and
iv) isomerisation, fragmentation and rearrangement reactions

i) Acylation and alkylation Reactions

On acylation (Fig. 8) with acyl anhydrides or halides in the presence
of base forskolin gave 1-acyl derivatives (40). Similarly, 7-deacetylfor-
skolin (6) on controlled acylation gave 7-monoacyl derivatives (41) (*15,
16, 41*) which can also be obtained from (1) through protection of 1-
hydroxy group as in compound (42), selective hydrolysis of 7-acetate, and
acylation of the 7-hydroxy group followed by deprotection of 1-OH
group. Reaction of 7-deacetyl-forskolin (6) with *p*-toluenesulfonyl chloride or methanesulfonyl chloride led to the corresponding 7-sulfonate
derivatives (41)f and (41)g. Recently preferential acylation of the C_1-
hydroxyl group of 7-deacetylforskolin (6) with bromoacetyl bromide was

References, pp. 51–74

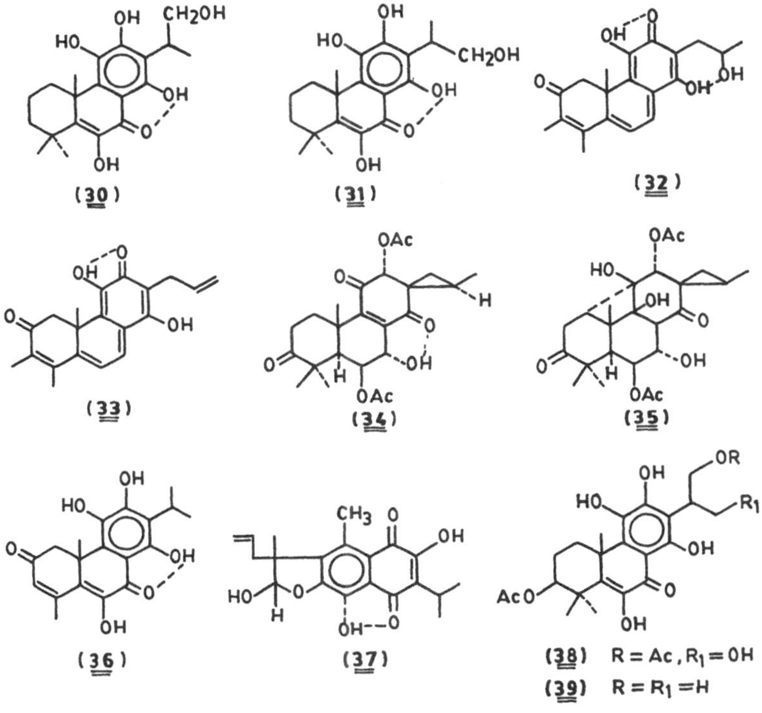

Fig. 6. Some diterpenoids of other *Coleus* species (*38, 39*)

reported (*42*). A large number of C_7-acyl derivatives of (**6**) have been prepared (*1, 14, 15, 19, 41–44*) in an effort to improve the pharmacological profile of (**1**). However access to esters of the 6-hydroxy group in high yield remains difficult because of the congested environment imposed by the three 1,3-diaxially juxtaposed methyl groups of C_4, C_8 and C_{10}. The base-catalysed acyl migration of the 7-acyl group of (**1**) or (**41**), with alumina (*1, 15*), piperidine (*41*), NaOH (*41*) NaOMe (*41*) or Li N(TMS)$_2$ (*42*), was utilised extensively to obtain 6-acyl derivatives (**43**). The hydrolysis of the 6,7-orthoester (**44**) was also used to obtain 6-acetyl-7-deacetylforskolin (**7**) (*21*). Similarly, the 6-benzoyl ester was synthesised through hydrolysis of the corresponding 6,7-orthoester (*30*). 6,7-Diacyl derivatives (**45**) were easily obtained through controlled acylation of 6-acyl derivatives (**43**). The 6,7-diacyl derivatives (**45**) or 1,6-diacyl derivatives (**46**) on further acylation yielded 1,6,7-triacyl derivatives (**47**).

In alkylation reactions of 7-deacetylforskolin (**6**), preferential reactivity of the 7-hydroxy group does not generally obtain. Thus partial

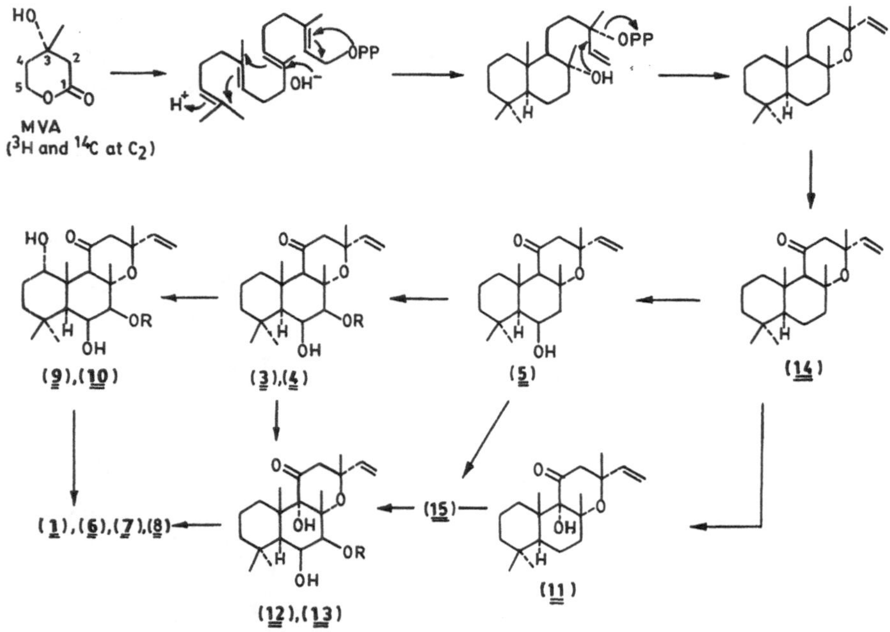

Fig. 7. Possible biogenetic route to forskolin and congeners from MVA (26)

alkylation of 7-deacetylforskolin (6) with alkyl halides in the presence of anhydrous K_2CO_3 gave 1-alkyl derivatives (48) (15, 16). Recently the synthesis of 7-(2-deoxy-α-D-arabino-hexopyranosyl)- derivative (49) was reported by SEQUIN et al. (43). In this glycosidation reaction in the presence of p-toluene-sulphonic acid the 7-OH group of (6) had reacted preferentially.

The carbamate (50) (Fig. 9) was obtained from (1) by treatment with phenyl isocyanate (16) while reaction of (6) with dialkylcarbamoyl chloride under controlled conditions yielded carbamate (51). The pyrrolidene carbamates (55) and (57) were obtained by O'MALLEY et al. (44) from 1,9-dimethylaminomethylidene acetal (52) through the sequence of reactions shown in Figure 9. The acetal (52) on selective hydrolysis and reaction with 1,1-carbonyldiimidazole (CDI) in the presence of $HNEt_2$ yielded 6,7-carbonate (53) which on reaction with pyrrolidine followed by hydrolysis of the acetal yielded 6-carbamates (54) and (55) respectively. The 7-carbamate (57) was similarly synthesised through carbamate (56) using slight modifications in the procedure.

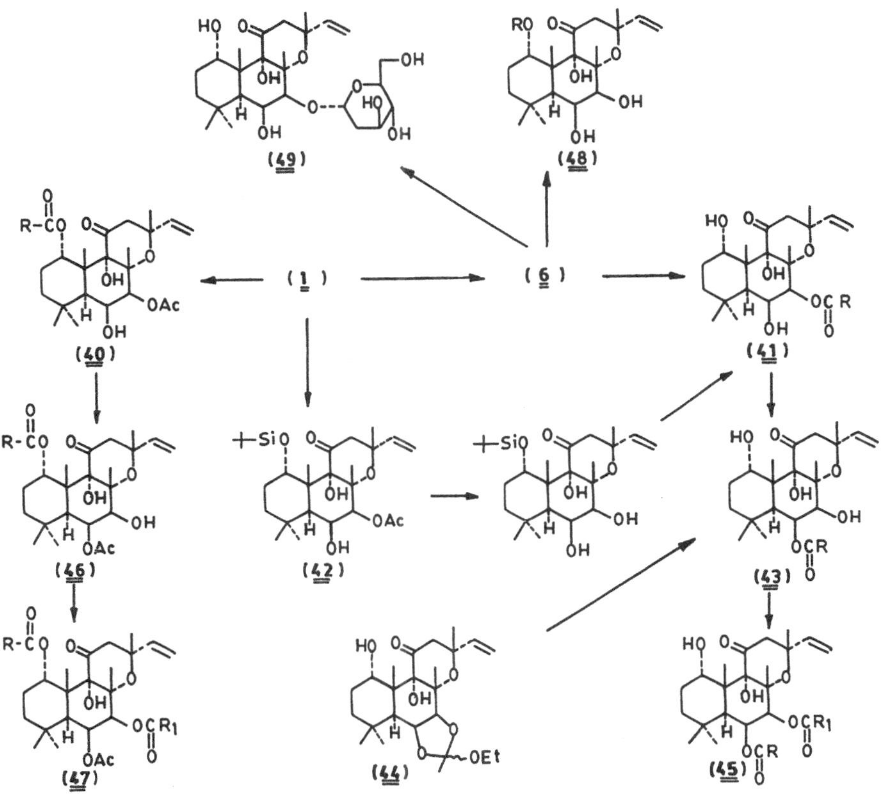

Fig. 8. Acyl, alkyl derivatives of (1) (*15, 19, 21, 41–43*)

Forskolin is a lipophilic molecule with poor solubility in water, hence water soluble derivatives were made available through introduction of amino groups or amino bearing moieties at the C_1-, C_6-, C_7- or C_{15}-positions (Fig. 10) and conversion of the resulting derivatives to salts. The C_7-aminoacyl derivatives (59) (Figure 10) were synthesised (*41*) either through nucleophilic displacement of halogen with amines in the haloacyl derivatives (58) or through condensation of the C_7 hydroxy group of (6) with aminoalkyl acids in the presence of DCC. The Michael addition of amines to 7-acryloyl-7-deacetylforskolin was also used to produce 7-aminoacyl derivatives (*41*). The base-induced acyl migration of (59) yielded 6-aminoacyl-7-deacetylforskolin (60). Table 3 describes some 6- and 7-acyl, aminoacyl and carbamate derivatives of (1). The C_1-aminoacyl derivatives (61) were also synthesised (*42*). The forskolin 14,

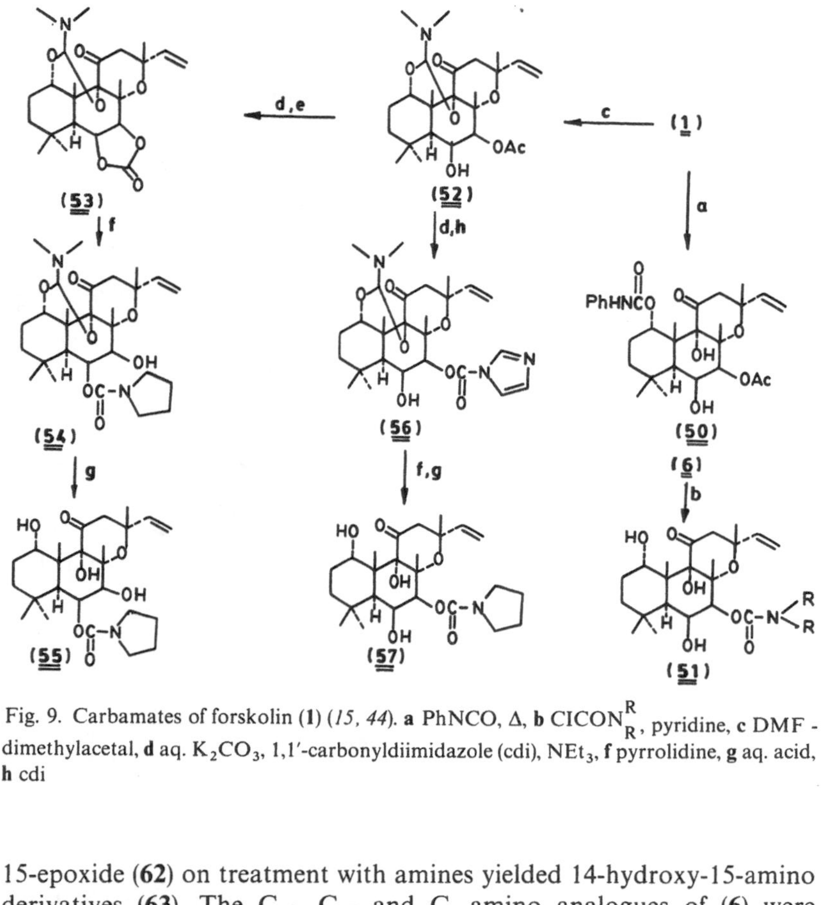

Fig. 9. Carbamates of forskolin (1) (15, 44). a PhNCO, Δ, b CICON$_R^R$, pyridine, c DMF - dimethylacetal, d aq. K$_2$CO$_3$, 1,1'-carbonyldiimidazole (cdi), NEt$_3$, f pyrrolidine, g aq. acid, h cdi

15-epoxide (62) on treatment with amines yielded 14-hydroxy-15-amino derivatives (63). The C$_1$-, C$_6$- and C$_7$-amino analogues of (6) were obtained through reduction of the oximes of C$_1$-, C$_6$- and C$_7$-keto derivatives (15).

Derivatization of the hindered 9-α-OH group could be achieved through 1,9-cyclic carbonate (25), sulfite (20) (15), phosphite (BHAT et al., unpublished results), acetonide (101) dimethylformamide acetal (52) (42) or through 9,14-hemiacetal (18), 9,6-hemiacetal (79) derivatives (15).

Interesting photoaffinity labels (64), (65), (66), (68) and (69) containing ^{125}I were obtained from 6,7-dicarbonate (53), 7-chloroacetyl-7-deacetylforskolin (58) and 7-hemisuccinyl-7-deacetylforskolin (67) respectively (45–48) (Fig. 11).

14,15-Ditritioforskolin ([3H]-dihydro-forskolin) was produced by catalytic reduction with tritium gas. The [3H] forskolin (49, 50) was prepared

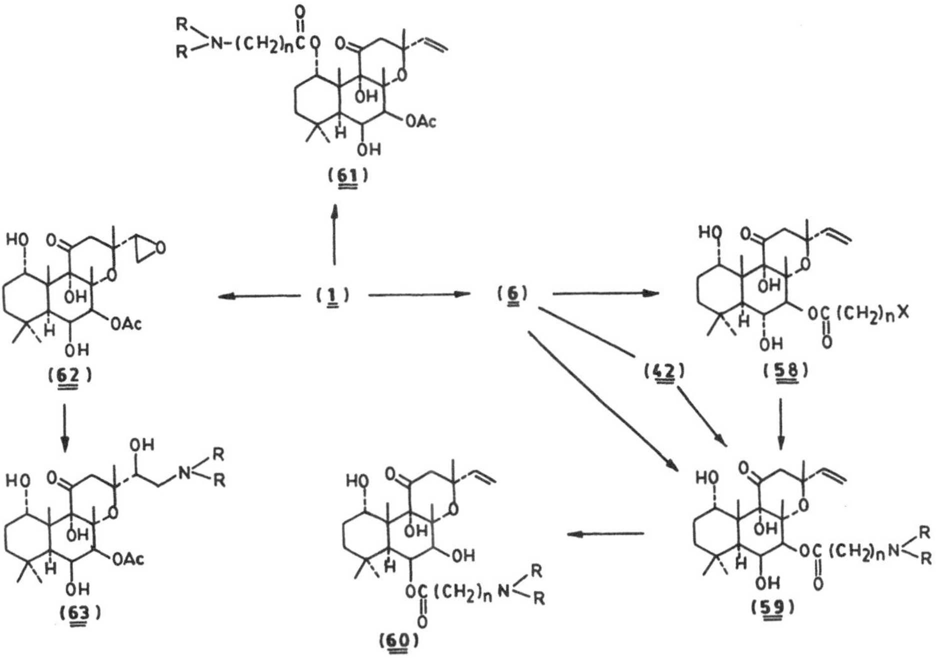

Fig. 10. Water soluble derivatives of forskolin (41, 42)

through base catalysed exchange of hydrogen at the 12-position with tritiated water. Both [^{3}H]-dihydroforskolin and [^{3}H]-forskolin are stable and relatively easy to purify. Radiolabelled forskolin containing tritium in the acetate group at C_7 was also synthesised; however its easy deacetylation in the bovine brain membrane was observed (51). ^{14}C-Forskolin with ^{14}C-label at the C_{15}-position was obtained by Wittig reaction of C_{14}-aldehyde (71) with (^{14}C)-methylene-triphenyl phosphorane.

ii) Oxidations and Reductions

The 15-norderivatives (71–73) were obtained through ozonolysis or oxidative cleavage of the 14,15-double bond of (1) (Fig. 12).

The $C_{1\alpha}$–OH group was selectively oxidised using either Collins or Jones reagent to yield the 1-oxo derivative (74). The C_6- and C_7-oxo derivatives (75 and 76) were obtained by Jones oxidation of derivatives in which the other hydroxyl groups were protected (15, 53). An interesting

SUJATA V. BHAT

Table 3. *6- & 7-Acyl, Haloacyl, Aminoacyl and Carbamate Analogues of (1) (Fig. 6,7,8)**

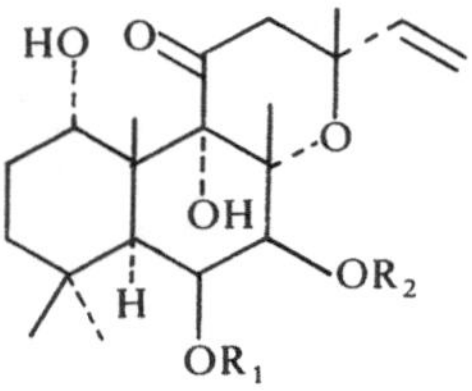

Comp No.	R_1	R_2	M.P°C (Solvent)	Ref
(41)a	H	COH	225–227 (A)	
(41)b	H	COEt	175–176 (A)	
(41)c	H	COnPr	176–177 (A)	
(41)d	H	COOEt	166–167 (A)	(16)
(41)e	H	CO(3′,4′,5′OMe₃)-Ph	273–275 (A)	
(41)f	H	SO₂Me	161–162 (A)	
(41)g	H	SO₂(4′ Me)-Ph	187–189 (A)	
(41)h	H	COCHOHCH₂OH	165 (B)	
(41)i	H	COCHC(Me)₂	148 (C)	(43)
(58)	H	COCH₂Cl	200–202 (A)	(53)
(67)	H	COCH₂CH₂COOH		(48)
(59)a	H	COCH₂NEt₂ . HCl H₂O	170–172 (A)	
(59)b	H	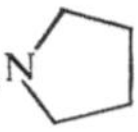 COCH₂N⟨ ⟩ . HCl, 1.5 H₂O	189–92 (A)	
(59)c	H	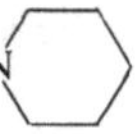 COCH₂N⟨ ⟩ . HCl	194–97 (D)	
(59)d	H	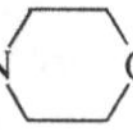 COCH₂N⟨ ⟩O . HCl	174–178 (D)	(41)
(59)e	H	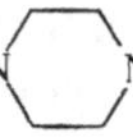 COCH₂N⟨ ⟩N–Me . 2HCl . 2H₂O	193–194 (D)	
(59)f	H	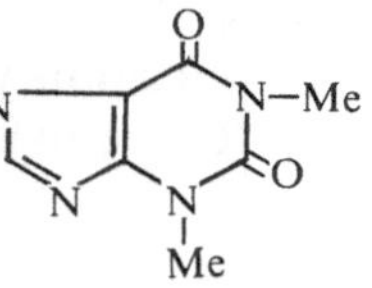 COCH₂N⟨purine⟩N–Me	198 (D)	
(59)g	H	COCH(Me)N⟨  ⟩O . 1.25 . HCl . H₂O	185–187 (D)	
(59)h	H	COC(Me)₂N⟨  ⟩O . HCl	274 (D)	
(59)i	H	CO(CH₂)₃N⟨ 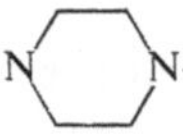 ⟩N–Me . HCl	203–207	
(51)	H	CONEt₂	216–219 (D)	(16)

References, pp. 51–74

Table 3. (*Continued*)

Comp No.	R_1	R_2		M.P°C (Solvent)	Ref
(57)	H	CON⟨cyclopentyl⟩ 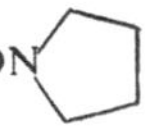		230–233	(44)

Comp No.	R_1		R_2	M.P°C (Solvent)	Ref
(7)	COMe		H	208–210 (F)	(16)
(43)	COEt		H	195–200 (A)	
(55)	CON⟨cyclopentyl⟩ 		H	135–140 (A)	(44)
(60)a	COCH$_2$N⟨pyrrolidine⟩ HCl O. 2H$_2$O		H	137–139 (D)	
(60)b	COCH$_2$N⟨thiomorpholine⟩S. HCl, 1.5 H$_2$O		H	173–176 (D)	(41)
(60)c	COCH$_2$N⟨piperidine⟩ .HCl . H$_2$O		H	204–205 (D)	
(60)d	COCH$_2$N⟨morpholine⟩O HCl, O. 5H$_2$O		H	152–155 (D)	
(60)e	CO(CH$_2$)$_2$N⟨piperidine⟩ . HCl, 1.5 H$_2$O		H	195–197 (D)	
(60)f	CO(CH$_2$)$_2$N⟨piperazine⟩N–Me . 2.5HCl 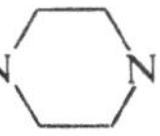		H	221–222 (D)	
(60)g	CO(CH$_2$)$_3$N⟨piperazine⟩N–Me . 2HCl 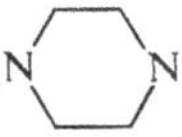		H	249–250 (D)	
(45)a	COMe		COMe	299–301 (E)	
(45)b	COMe		COEt	229–231 (A)	
(45)c	COMe		COnPr	196–198 (G)	(16)
(45)d	COMe		COOEt	177–179 (A)	
(45)e		CO		218–222 (E)	

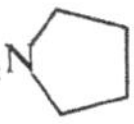

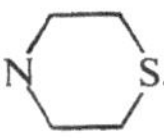

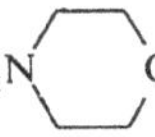

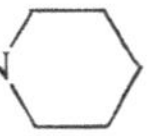

* Numerous 6- and 7-aminoacyl analogues of (1) have been synthesised, hence only representative derivatives are given.

A   Ethyl acetate: petroleum ether
B   CHCl$_3$
C   MeOH

D   MeOH: Et$_2$O
E   Ethyl acetate: benzene
F   CHCl$_3$: petroleum ether
G   C$_6$H$_6$: petroleum ether

rearrangement shown in more detail in (Fig. 16, *vide infra*) was observed during formation of the oxo derivative (76). Dihydroforskolin (26) on oxidation with Sarett's reagent gave the 1,6-diketo derivative (77). The 1,7-diketo derivative (78) was obtained through a similar oxidation of dihydro-6-acetyl-7-deacetyl-forskolin.

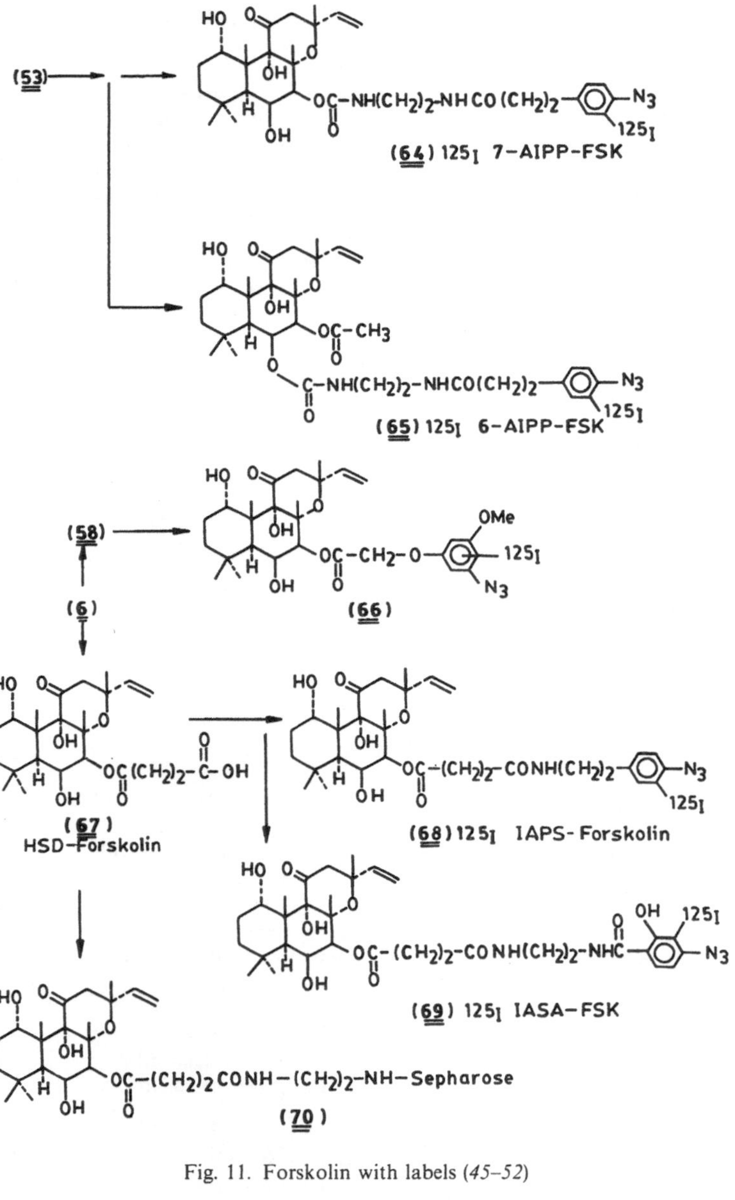

Fig. 11. Forskolin with labels (45–52)

Periodate oxidation of 1-benzyl-7-deacetyl-forskolin gave γ-lactol (**79**) which on further oxidation yielded γ-lactone (**80**) (*15*).

The dihydroxy analogue of 7-deacetylforskolin (**81**) was obtained by basic hydrolysis of forskolin 14,15-epoxide (**62**). Recently DELPECH and LETT obtained 14,15-dehydroforskolin (**82**) *via* condensation of C$_{14}$-aldehyde (**71**) with dimethyl diazomethylphosphonate anion (*54*).

The 12-oxoforskolin (**83**) was obtained from (**1**) through oxidation with selenium dioxide (*55*). Conversion of 9-deoxy-forskolin (**9**) to forskolin was achieved through selective oxidation of the 9,11-enolmethyl ether by HIRB (*56*). Using similar reaction conditions WELZEL and coworkers (*57*) obtained 12-α-silyloxy-1,9-dideoxyforskolin-6-TMS ether while syntheses of 12-fluoro (**84**), 12-chloro (**85**) and 12-bromo (**86**) derivatives was achieved by SHUTSKE (*58*).

Reduction of forskolin with LiAlH$_4$ yielded the 11-β-hydroxy-7-deacetyl analogue (**87**) while the isomeric 11-α-hydroxy-7-deacetyl deriv-

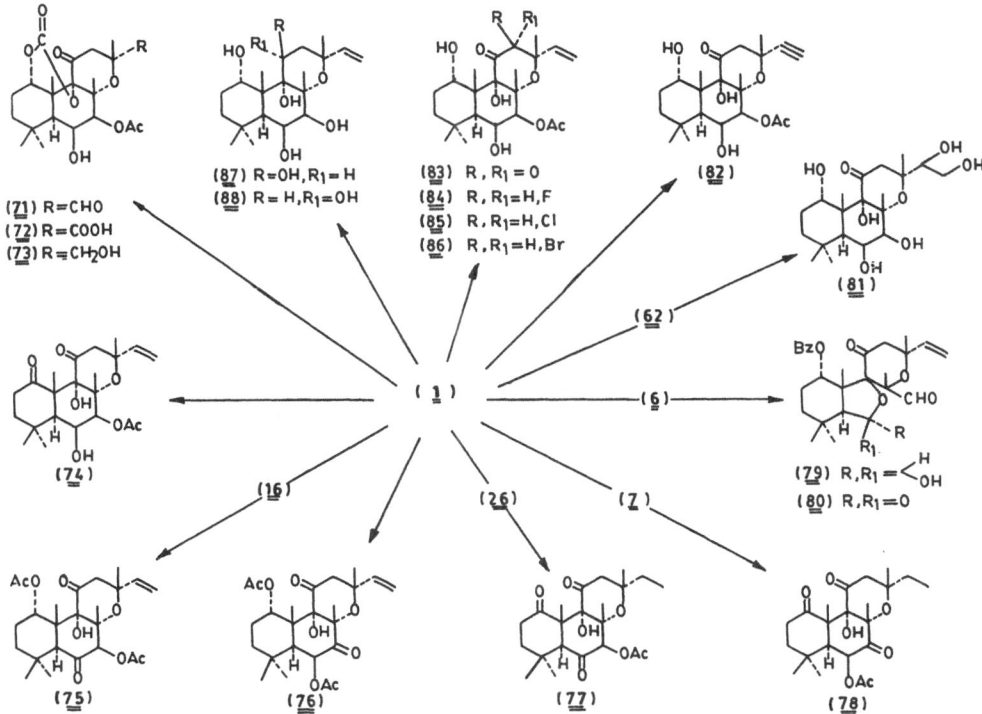

Fig. 12. Oxidation reduction reactions of (**1**)

ative (**88**) was obtained through reduction of (**1**) with Na–EtOH, which also gave the 9-deoxy analogue of (**88**). Lithium tri-*t*-butoxy aluminium hydride reduction of (**78**) yielded the 1-*β* isomer of dihydro-7-deacetylforskolin (*15*).

iii) Biological Oxidations

An extensive study of the microbial hydroxylation of 1,9-dideoxyforskolin (**3**) which invariably cooccur with forskolin in *Coleus forskohlii* in amounts equal to or sometimes even in excess of forskolin, was of practical interest (*59–63*). As shown in Fig. 13 the fungal strain, *Scopuloriopsis sp* (H/134), was found to convert 7-deacetyl-1, 9-dideoxyforskolin (**4**) to 7-deacetylforskolin (**6**), albeit in low yield (0.76%), and the 2*β*-hydroxy derivative (**89**). 1,9-Dideoxyforskolin (**3**) on similar treatment yielded 6-acetyl-7-deacetyl-1,9-dideoxyforskolin (**90a**) and the 2*β*-hydroxy derivative (**90b**). Incubation of (**3**) with *Neurospora crassa* (ATCC 10336) for 120 h gave 9-deoxyforskolin (**9**) as well as the 2-α-hydroxy- (**91**) and 3-α-hydroxy derivatives (**92**). The shake culture of (**3**) with *Mortierella isabellina* (ATCC 160074) for 120 h yielded 9-deoxyforskolin (**9**), 2*β*-hydroxy- (**90b**), 12-hydroxy- (**93**), 18-hydroxy- (**94**) and 3-keto- (**95**) derivatives. Incubation of (**3**) with *Aspergillus niger* (DCM 3210) yielded 3*β*-hydroxy derivatives (**96**) and (**97**). Incubation of 7-deacetylforskolin (**6**) with FF 406 (Aspergillus sp.) yielded the 3*β*-acetoxy- (**98**), 3*β*-hydroxy- (**99a**) and 18-hydroxy (**99b**) derivatives. Acetylation of (**99a**) gave acetyl derivatives (**98**) and (**100**). Incidentally the 3*β*-hydroxy analogues (**99a**), (**100**) and (**101**) are the main metabolites of forskolin in rats and dogs, identified after administration of 15^{14}C-labelled forskolin (*64*). The transformation using FF 406 and with (**3**) as substrate yielded the 3-*β*-hydroxy derivative (**97**).

The microbial oxidation of *ent*-13-epimanoyl oxide to obtain *ent*-forskolin analogues has also been reported (*65*).

iv) Isomerisation, Fragmentation and Rearrangement Reactions

Recently the unnatural 1*β*-isomer of forskolin (**102**) (Figure 14) 7α-analogue (**103**) and forskolin 6,7-epoxide (**104**) were obtained by VISHWAKARMA and TANDON through application of MITSUNOBU's reaction (*66*).

Forskolin under Lewis acid catalysis rearranged to spiroforskolin (**105**) which reverted back to forskolin after several days in methanol (*67*).

References, pp. 51–74

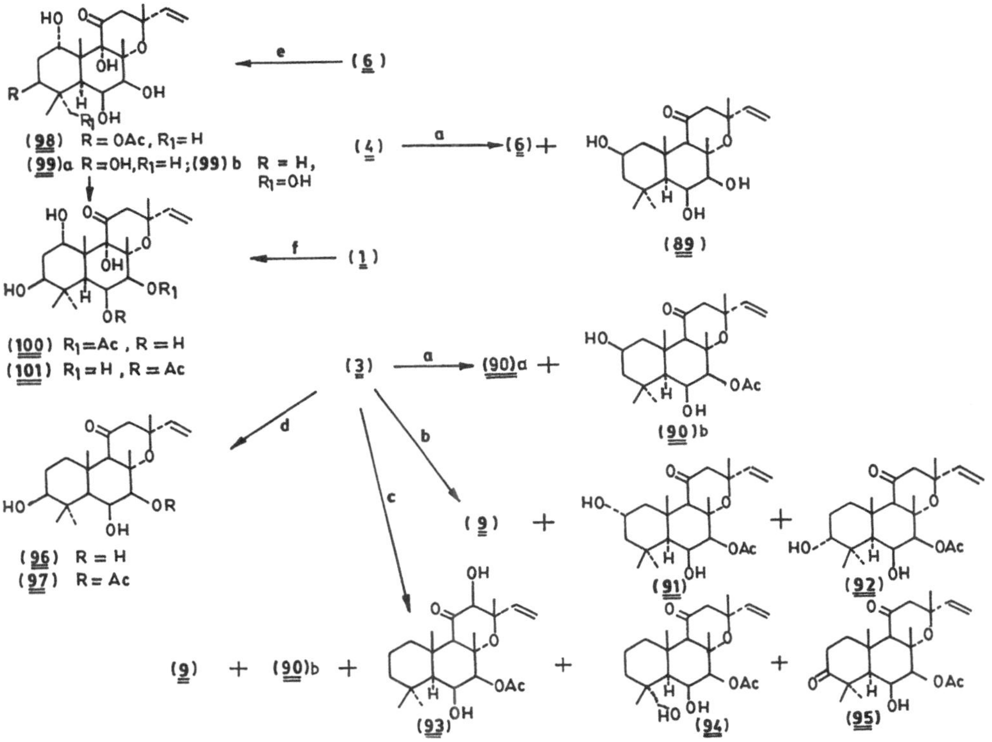

Fig. 13. Biological hydroxylation/oxidation of (1) and congeners (58–64). a scopuloriopsis sp [H/134], b Neurospora crassa (ATCC 10336), c Mortierella isabellina (ATCC 160074), d Aspergillus niger (DCM no 3210), e Aspergillus sp (FF 406), f metabolism in cats and dogs

SAKSENA et al. have reported (68) a convenient degradation of forskolin 1,9-carbonate by ozonolysis, conversion of the resulting aldehyde to an aldoxime (106) and attempted dehydration of the latter with 1,1'-carbonyldiimidazole which unexpectedly resulted in β-elimination to give the 14,15-dinor derivative (107). This substance later served as an important relay in synthetic routes to forskolin. However, preliminary experiments at reconstructing the forskolin skeleton by conjugate addition of $(CH_2 = CH)_2CuLi_2$ to (107) met with failure (68).

Subsequently DELPECH and LETT (69) examined several conditions for conjugate addition to (107). These reactions again failed; the only products which could be isolated were produced by hydrolysis of the 1,9-carbonate and/or the 7β-OAc, and by a 7β- to 6β-acetate transposition,

(> 75%). Therefore the 1,9; 6,7-bis-isopropylidene derivative (108) was degraded to the dinor analogue (109); however many attempts at achieving conjugate addition to (109) with methyl, ethyl or vinyl copper reagents, had failed. Success was finally achieved with dimethyl cuprate in the presence of BF_3–OEt_2 which reacted with (109) to give 1,4-adduct (110). Finally Conjugate addition of vinyl cuprate to (109) in the presence of BF_3–OEt_2 gave (108) as the major product (59%) along with C_{13} epimer (111) (14%) (Fig. 14).

IKEGAMI et al. (70) also reported the successful conjugate addition of a vinyl group to synthetic-9-hydroxydihydropyranone using $(CH_2=CH)_2CuCNLi_2$ to yield (±)-1,6,7-trideoxyforskolin, whereas ZIEGLER et al. (71) developed a completely different method for adding

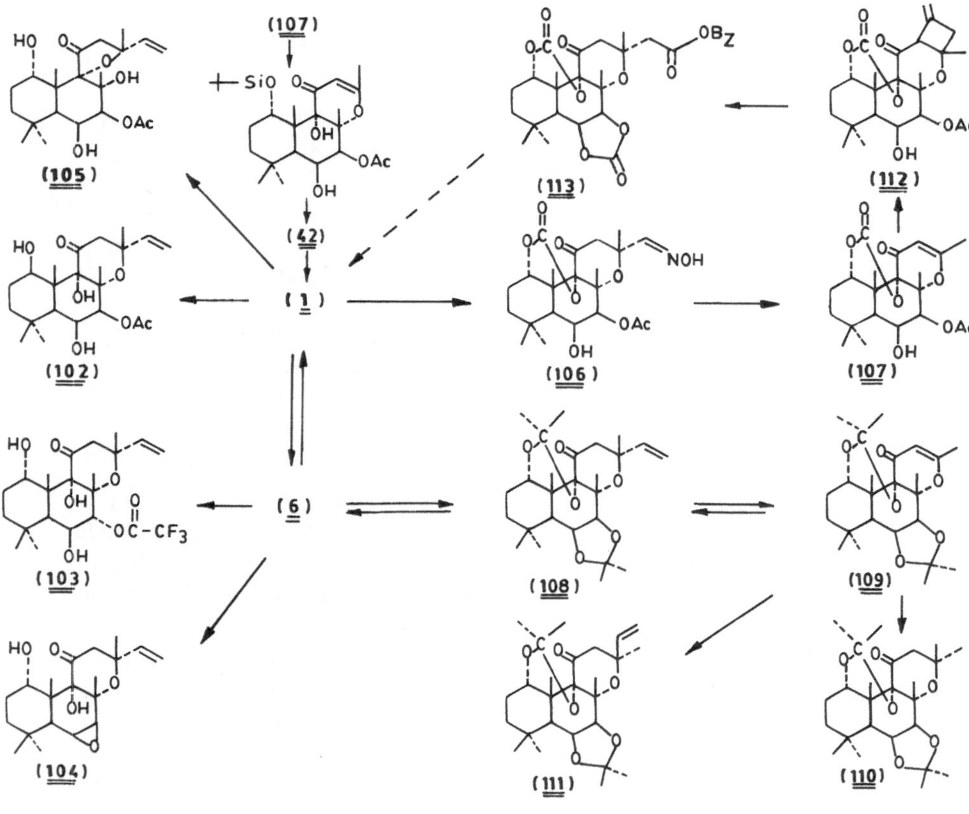

Fig. 14. Isomerization and degradation reactions of (1) (66–77)

the vinyl group to the C ring of dihydropyranone (**107**). Irradiation of (**107**) in the presence of allene yielded a single photo adduct (**112**) which after ozonolysis of its 1,9,6,7-dicarbonate followed by solvolysis yielded benzyl ester (**113**). Reductive elimination applied to (**113**) followed by hydrolysis and partial acetylation yielded forskolin. Subsequently these authors were successful in synthesising (±) forskolin through conjugate addition of vinyl cuprate to synthetic (±) dihydropyranone (*72*).

WELZEL *et al.* (*73*) in an attempt to prepare the 9(11)-silylenol ether of naturally-occurring 1,9-dideoxyforskolin (**3**) (Fig. 15) found that treatment of the latter with trimethylsilyl iodide and hexamethyldisilazane (HMDS) followed by selective cleavage of the C-8 silyl ether afforded bicyclic products (**114**) and (**115**) in 45% and 18% yield respectively. However, attempts of reconstruction of ring C from (**114**) and (**115**) were

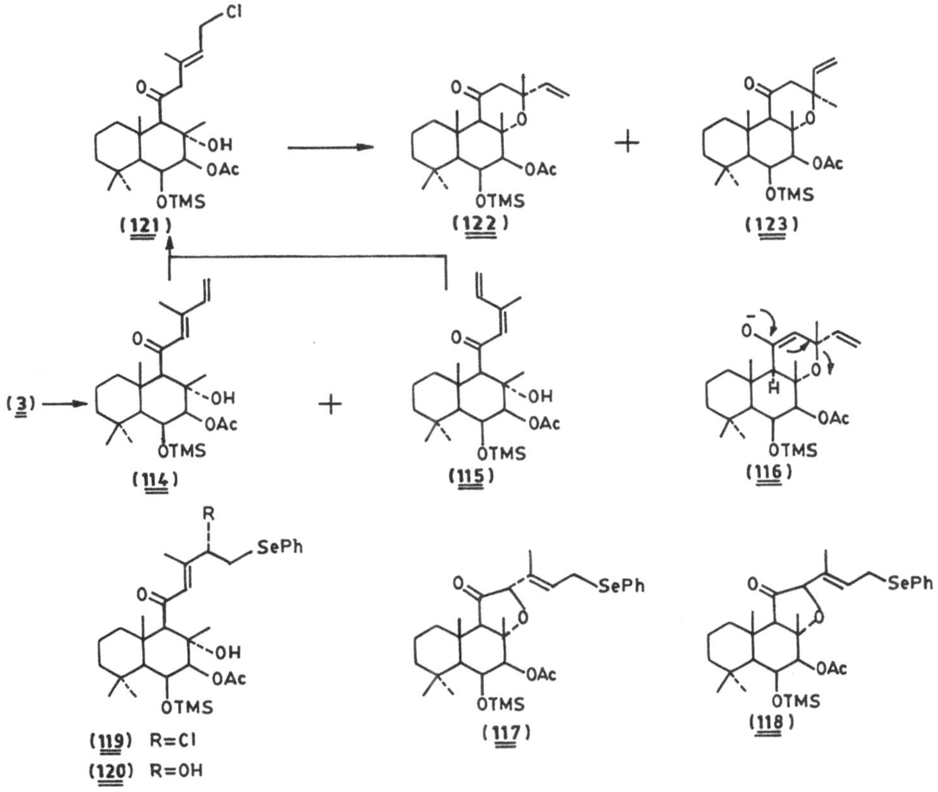

Fig. 15. Fragmentation and reconstruction of ring C in (**3**), (*73*)

unsuccessful. Conjugate addition of the C8-OH group to the enone unit under acidic or basic conditions failed. Apparently the enolates of type (116) revert easily to the bicyclic starting material.

In view of these results, organoselenium mediated cyclization reactions were studied, since it was assumed that under such conditions, the electrophilic selenium species would attack C_{12} to trap intermediate (116). However, treatment of (114) with N-phenyl-selenophthalimide (N-PSP) in the presence of camphorsulphonic acid (CSA) led exclusively to the tetrahydrofurans (117) and (118) in 21% and 7% yield respectively, and when (114) was treated with phenylselenyl chloride, only the bicyclic product (120), presumably formed from (119) during chromatograph of the reaction mixture, was isolated. The formation of (120) led the authors to suggest that the reaction with N-PSP proceeds through an intermediate of type (119) from which (117) and (118) are formed by SN_2 displacement reaction involving nucleophilic attack of C 8-hydroxyl group at C-12. Subsequently these authors observed that hydrogen chloride at low temperature with mixture of (114) and (115) gave addition products from which (121) was obtained in low yield. The cyclization of

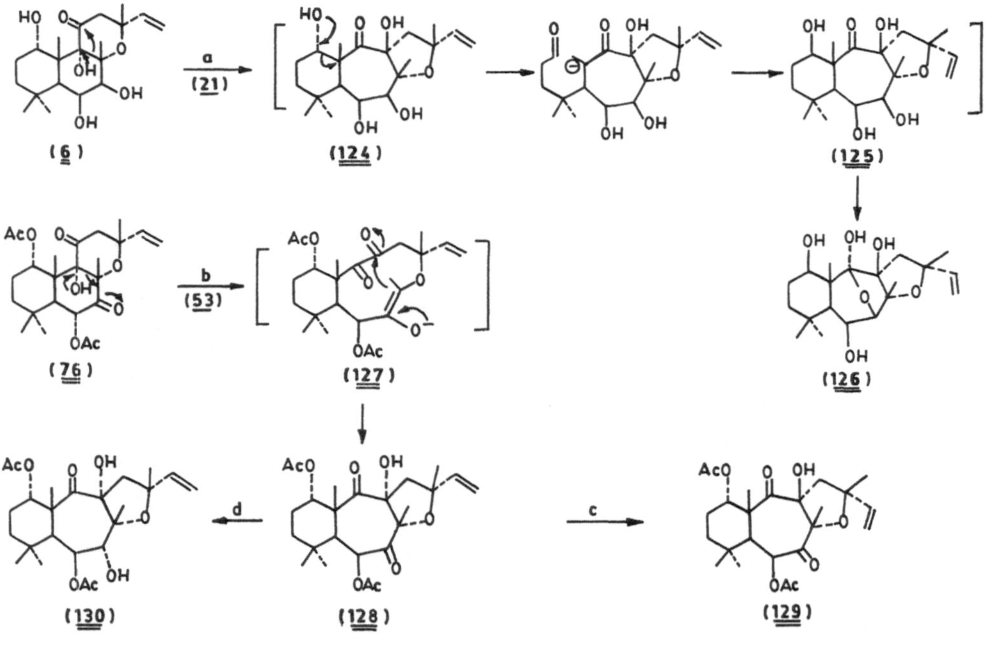

Fig. 16. Rearrangement reactions of forskolin (21, 53). a ethanolic KOH, reflux, b Jones oxidation, c pyridine, d NaBH₄, pyridine

References, pp. 51–74

(121) in the presence of N-PSP and $SnCl_4$ followed by reductive elimination with Zn–Cu couple yielded 6-trimethylsilyl ether (122) (41%) and its 13-epimer (123) (12%).

A remarkable rearrangement (Fig. 16) of 7-deacetyl-forskolin (6) on vigorous alkaline hydrolysis to yield tetracyclic product (126) was reported by SAKSENA et al. (21). A possible mode of formation of (126) suggested by these authors involves an α-ketol rearrangement with migration of 8,9-bond to yield (124). This is followed by a retroaldol-aldol process that accounts for the inversion at C_1 and finally by the formation of the hemiketal involving the $C_7\beta$-hydroxy and C_9-keto groups of the resulting tetrol (125) KHANDELWAL et al. (53) observed a similar rearrangement during Jones oxidation of 1,6-diacetyl-7-deacetylforskolin. This yielded not only the expected 7-oxo derivative (76) but also rearranged diketone (128) which on treatment with pyridine gave epimeric diketone (129). The formation of (128) from (76) was similarly postulated to proceed via a retroaldol-aldol process involving the intermediate (127). Reduction of (128) with $NaBH_4$ in pyridine yielded the epimeric diketone (129) and the reduced product (130).

V. Synthesis of Forskolin

The unique structural features and the biological properties of forskolin have aroused the interest of leading organic chemists thus resulting in enormous activity directed towards synthesis of this substance. Since reviews (11, 12) detailing various approaches towards total synthesis of (1) and intermediates have been published recently, the present report will describe only salient features of various efforts towards this goal.

The discussion is organised into three primary sections
i)   Approaches towards A,B ring system
ii) Approaches towards A,B,C ring system
iii) Total synthesis of (1)

i) Approaches to A, B Ring System of Forskolin

For the construction of A, B rings of decalin system present in forskolin (1) various strategies have been employed.

Intramolecular Diels Alder (IMDA) Strategy
JENKINS and coworkers (74) reported IMDA reaction of maleate ester (131) (Figure 17) which yielded tricyclic lactones (132) and (133). In

subsequent work (75), the original stereochemical assignment of adducts
(132) and (133) (as the methyl esters) was confirmed by x-ray crystallography; similarly the IMDA reaction of fumarate ester (134) was shown to
give adduct (135). The use of dideuteromaleate ester (136) under the same
conditions indicated that the formation of (132) was due to epimerisation
at C-9 of the highly strained lactone (137) which is the kinetically
controlled product generated *via* an *exo*-transition state. Using similar
strategy NICOLAOU et al. (76) synthesised tricyclic lactone (139) from
trimethylsilylacetylenic ester (138).

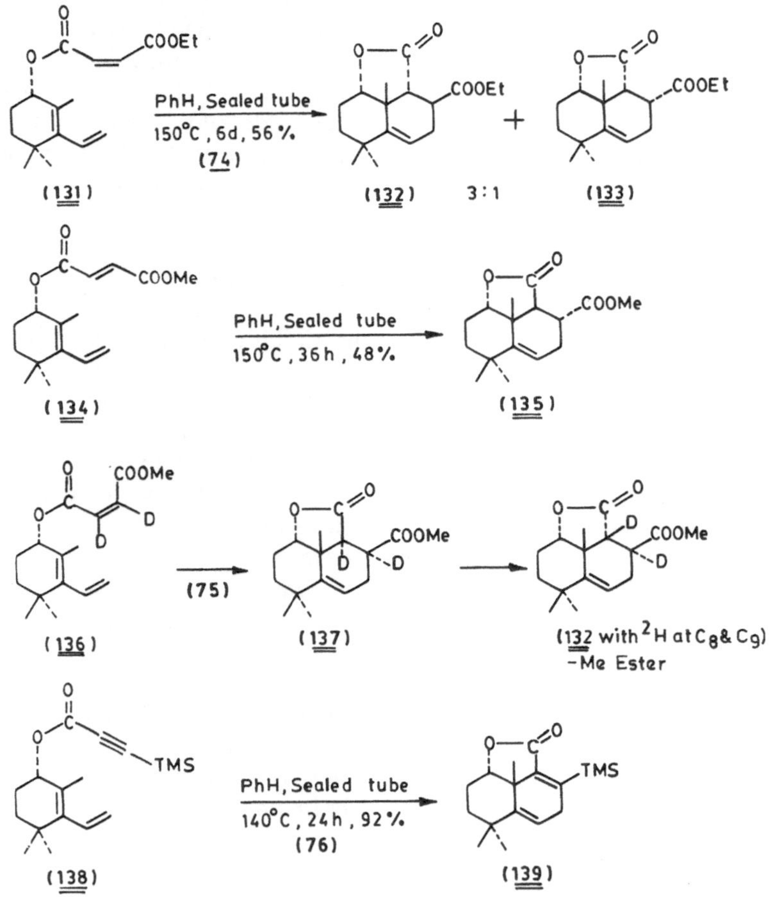

Fig. 17. IMDA reaction of esters towards synthesis of AB rings of (1) (74–76)

Liu *et al.* (*77*) reported that cyclization of acetylenic ether (**140**) to yield the expected adduct (**141**) (Fig. 18) which was further oxidized to afford lactone (**142**) in excellent yield. Recently KANEMATSU *et al.* (*78, 79*) reported a synthesis of the ZIEGLER intermediate (**148**) by using the IMDA reaction of allenic ether generated *in situ* from acetylenic ether (**143**) which yielded adduct (**144**). Acid catalysed addition of methanol to (**144**) yielded (**145**); this was followed by introduction of the keto group at C_6 and equilibration in basic medium to give *trans*-decalin (**146**) which was converted to enone (**147**) by conjugate addition of lithium dimethyl cuprate followed by regeneration of the double bond and oxidation of the acetal moiety with *m*-chloroperbenzoic acid in the presence of boron trifluoride etherate to afford intermediate (**148**). More recently these

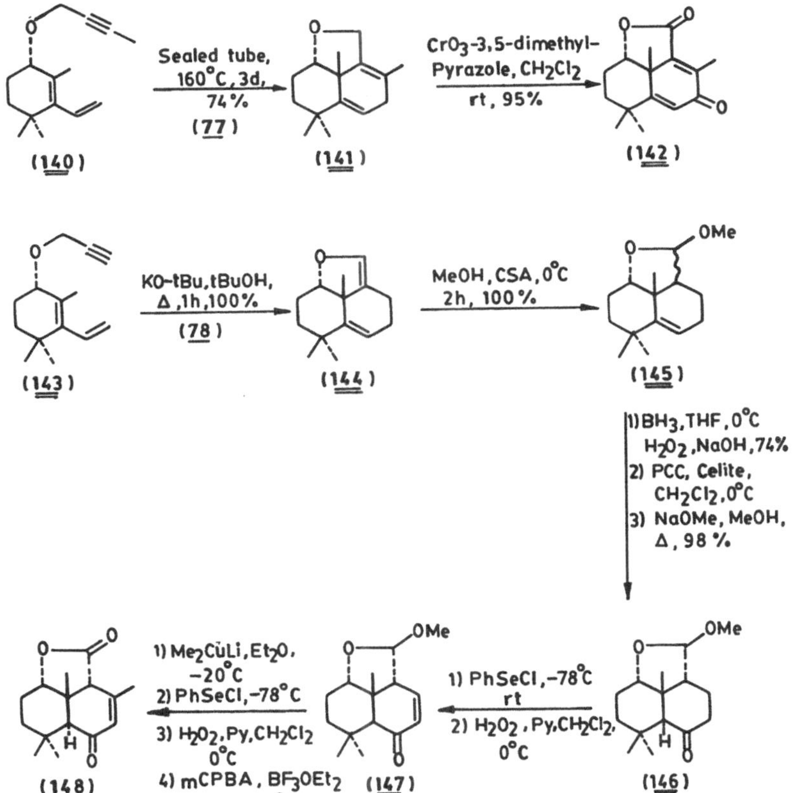

Fig. 18. IMDA reaction of acetylenic ethers towards AB rings of (**1**) (*77, 78*)

authors (79) have synthesised chiral (148) starting from optically active
ester (143) as shown in Figure 30 (vide infra).

TROST et al. (80) reported a completely different approach towards the
synthesis of A–B ring system of (1) (Fig. 19). The acetylenic esters
(149 a–f) were cyclised thermally or under Lewis acid catalyzed conditions to yield the adducts (150), (151) or (152) depending on the substrate
and the reaction conditions used.

Recently TSANG and FRASER REID (81) synthesised chiral tricyclic
lactone (155) through a IMDA reaction of diene (153) obtained by a
series of reactions starting from 2,3,4,6-O-benzyl-glucono-1,5-lactone.
Adduct (154) on further oxidation and methanolysis gave lactone (155).

The Intermolecular Diels-Alder Strategy

SNIDER et al. (82) described the intermolecular Diels-Alder reaction of
cyclopropyl diene (156) with 2,6-dimethylbenzoquinone (157) to yield
adduct (158) (Fig. 20). Using a similar approach SIH and coworkers (83)
reported the Diels-Alder reaction of diene (159) with quinone (157) to
give adducts (160) and (161). Adduct (161) was converted to (±) dione

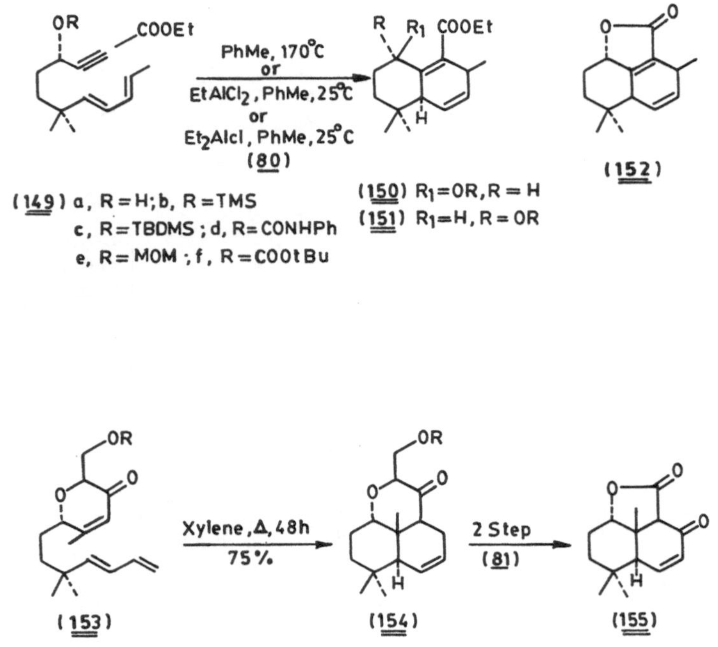

Fig. 19. Alternate IMDA reaction towards AB rings of (1) (80, 81)

(162) through a series of reactions and finally the chloroacetyl ester of (±) (162) was kinetically resolved using porcine pancreatic lipase to provide (+) (162).

BHAKUNI and coworkers (84) utilised the Diels-Alder reaction of the diene (163) and analogues derived from D-glucose with the *in situ* generated methoxycarbonyl-*p*-quinone (164) to yield adduct (165) and

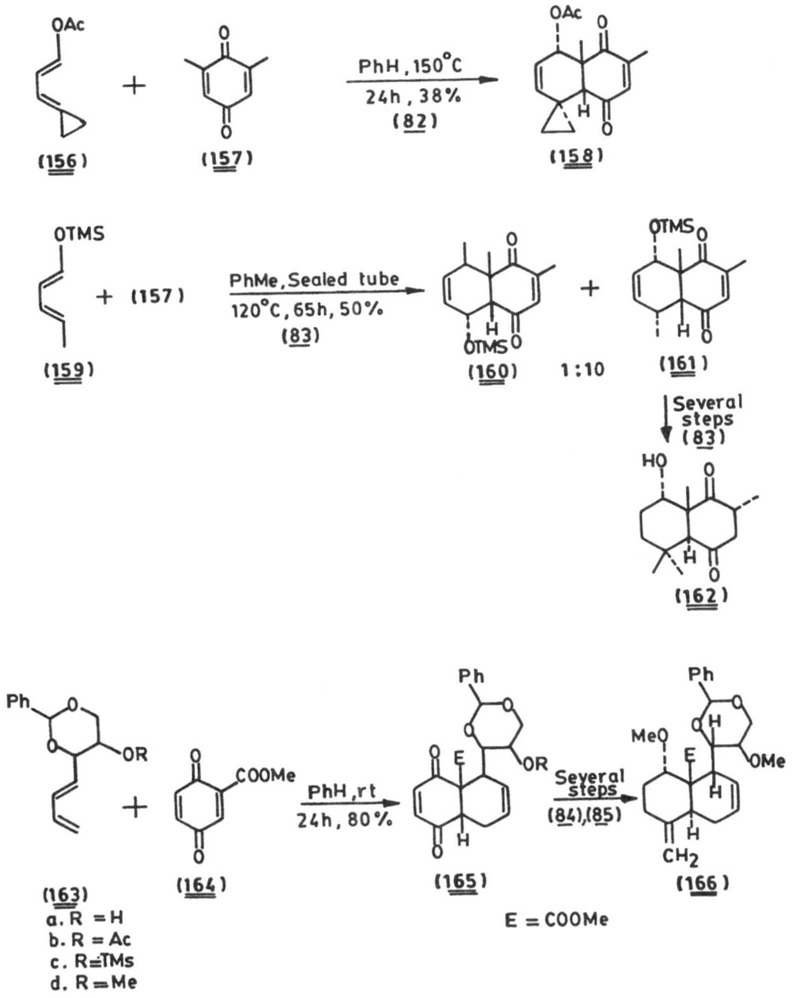

Fig. 20. Intermolecular Diels-Alder approach towards AB rings of (1) (82–85)

analogues, which on further elaboration through series of reaction led to *trans*-bicyclic derivative (**166**) (*85*).

Intramolecular Nitrile Oxide Cycloaddition Approach

BARCO, POLLINI and coworkers (*86*) applied an intramolecular nitrile oxide [3 + 2] cycloaddition (INOC) reaction for the synthesis of isooxazoline (**169**) (Figure 21). The nitrile oxide generated *in situ* from nitro derivative (**167**) underwent INOC reaction on heating to yield tricyclic derivative (**168**). The isooxazoline (**169**) was obtained from (**168**) on treatment with *p*-toluene-sulfonic acid in ethylene glycol followed by hydrolysis of the resulting ketal.

Subsequently the same authors (*87*) reported an improvement in the yield of the cycloaddition step. Acetate (**170**) gave adduct (**171**) in excellent yield, which was converted to acetate (**172**). Subsequently acetate (**172**) was used for synthesis of the A, B, C rings of forskolin as described in Figure 26 (*vide infra*).

KOZIKOWSKI *et al.* (*88*) utilised (4 + 2) and (3 + 2) cycloaddition strategy to obtain derivatives (**177**) (Fig. 21). The isomeric mixture of dienes (**173**) on reaction with dimethyl acetylenedicarboxylate yielded adducts (**174**) which were converted to isomeric aldehydes (**175**) through a series of reactions. The nitrile oxide obtained from aldehydes (**175**) through oxidation of oxime with sodium hypochloride underwent cycloaddition to yield adducts (**176**). The structure of the major product (**177**) obtained after hydrolysis of (**176**) was unambigously established by x-ray crystallography.

Intramolecular Radical Mediated Cyclization Followed
by Mukaiyama Aldolisation

PATTENDEN and coworkers (*89* to *91*) achieved synthesis of ZIEGLER intermediate (**148**) (Fig. 22) through radical cyclization followed by MUKAIYAMA's version of the aldol reaction. The bromo derivative (**178**) on treatment with vitamin B_{12} and $LiClO_4$ in methanol yielded a diastereomeric mixture of *trans*-bicyclic products (**179**) which was converted in three steps to silyl enol ether (**180**). MUKUYAMA aldolization of (**180**) gave tricyclic lactones (**181**) and (**182**) which on elimination gave either (**183**) or (**184**) depending on the reaction conditions. The ZIEGLER intermediate (**148**) and the isomeric ketolactone (**148**)a were obtained from (**184**) on oxidation with PDC and tBuOOH (*91*).

The Intermolecular Michael-Aldol Condensation Strategy

KOFT *et al.* (*92*) found that the acetoacetate ester (**185**) on treatment with cesium carbonate underwent tandem Michael-aldol reactions fol-

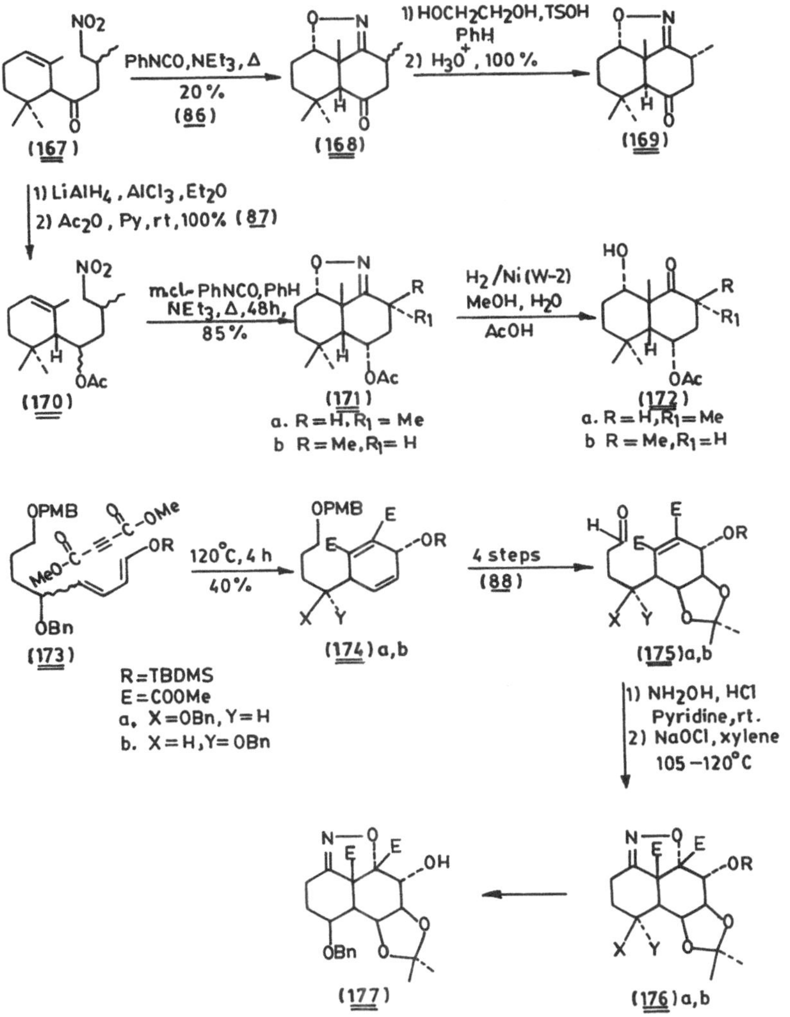

Fig. 21. Intramolecular nitrile oxide cyclization towards AB rings of (1) (*86–88*)

lowed by deacetylation and double bond migration to give lactone (**186**) (Fig. 23).

By following a similar approach LI and WU (*93*) reported synthesis of the tricyclic lactones (**189**), (**190**). Treatment of acetoacetate ester (**187**) with sodium hydride followed by cyclization of the resulting ketone (**188**)

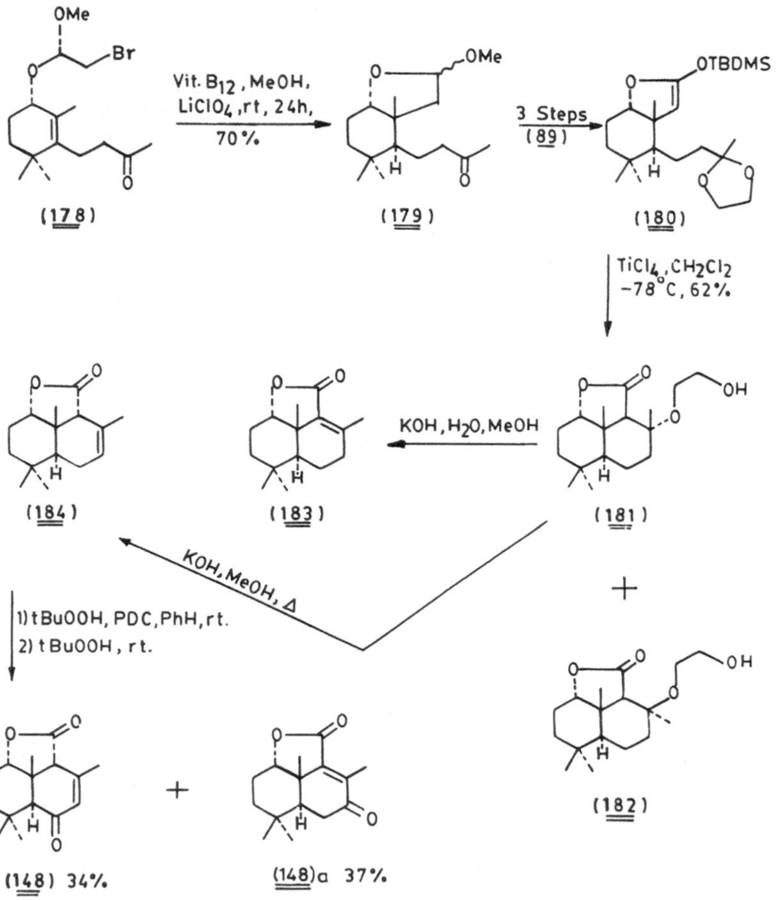

Fig. 22. Radical cyclization followed by Mukaiyama's adolization towards AB rings of (1) (89–91)

under acid conditions gave a mixture from which lactone (189) was isolated in good yield after crystallization. The mother liquors after chromatography gave the minor lactones (190) and (191). Recently RUVEDA and coworkers (94, 95) standardised conditions to improve the yield of tricyclic lactone (189). Refluxing of a dilute solution of (188) in 1,2-dichloroethane in the presence of an equimolecular amount of p-toluenesulfonic acid monohydrate afforded mainly lactone (189) with negligible formation of lactones (190) and (191). Subsequently these

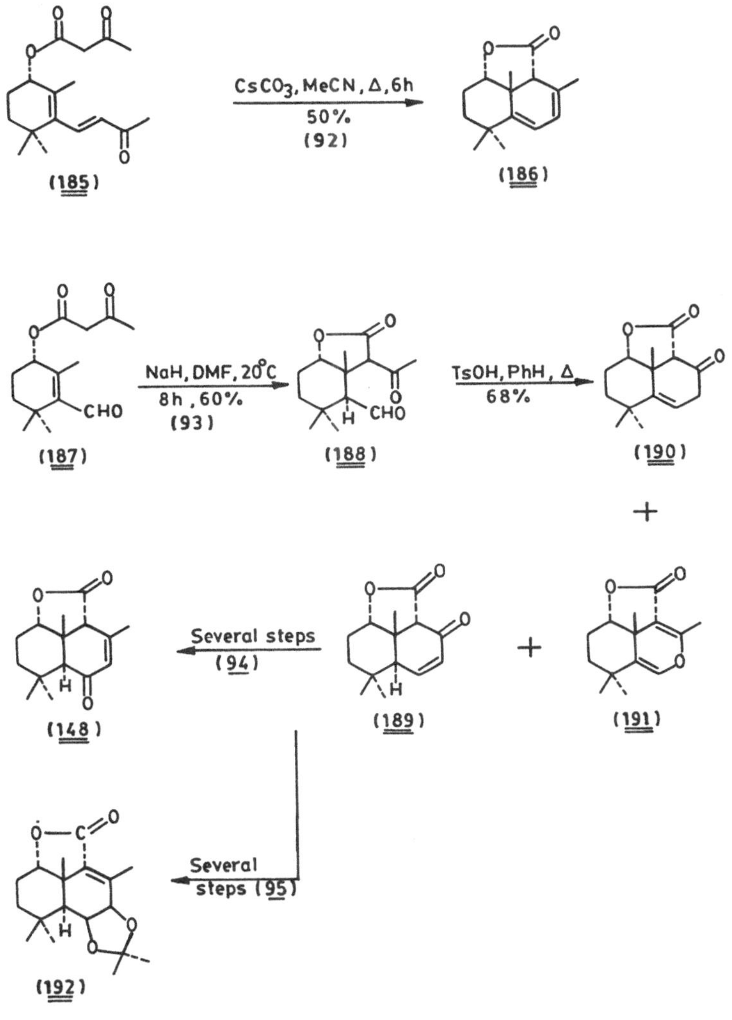

Fig. 23. Intramolecular Michael-Aldol strategy towards AB rings of forskolin (92–95)

authors demonstrated the utility of lactone (189) by converting it through several steps to ZIEGLER intermediates (148) and (192) (94, 95).

Intramolecular Olefin Cyclization

By using electrocyclization CHA et al. (96) reported the synthesis of bicyclic derivative (194) from triene ester (193). The bicyclic ester (194) was converted to intermediate (148) in several steps (Fig. 24).

Subsequently LECLAIRE and LALLEMAND (97) found that thermolysis
or photolysis of (193) led to a complex mixture of products from which
the tricyclic lactone (186) was obtained in low yield. Therefore these
authors thermolysed the unprotected triene carboxylic ester (195). Under
these conditions (195) afforded a mixture of lactone (186) and hydroxy
ester (196).

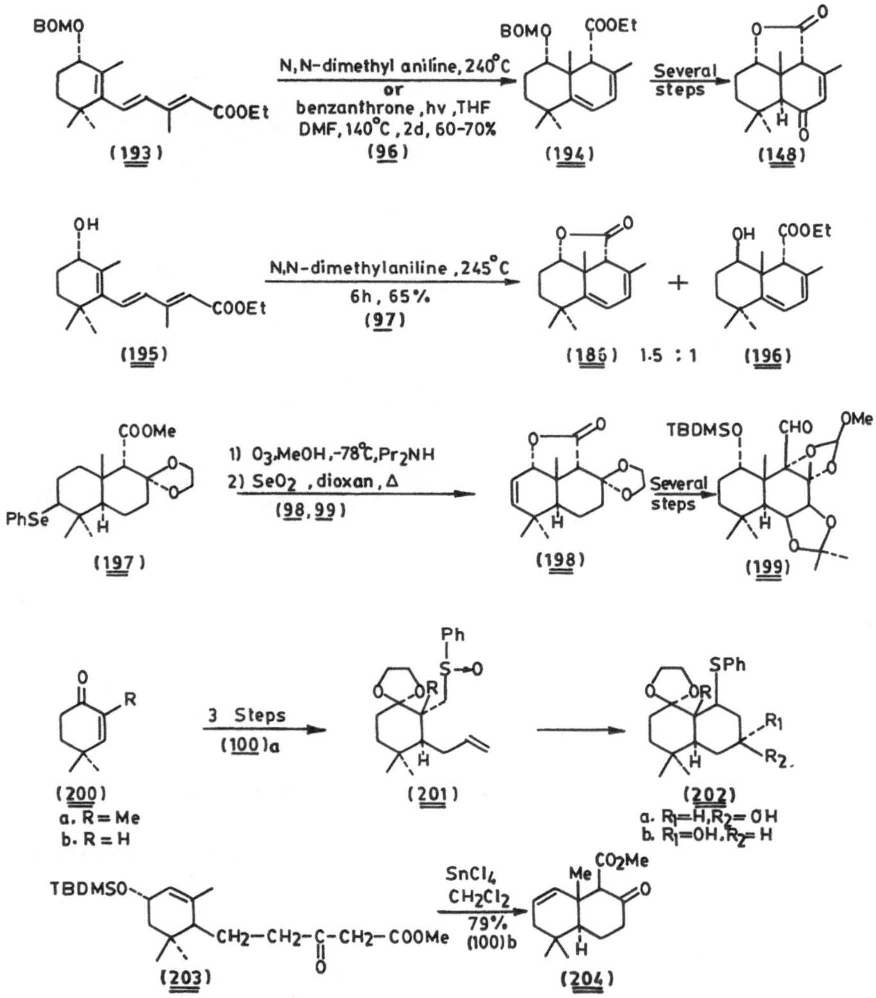

Fig. 24. Olefin cyclization approach towards AB rings of forskolin (96–101)

NICOLAOU *et al.* adopted (*98*) a completely different synthetic route to aldehyde (**199**) by starting from bicyclic selenide (**197**) prepared by the known cation-mediated polyene cyclization (*99*). Oxidative elimination followed by selenium dioxide oxidation of (**197**) gave tricyclic lactone (**198**) which was converted in several steps to aldehyde (**199**).

WELZEL and coworkers (*100a*) have described unsuccessful attempts at effecting a Diels-Alder reaction of 4,4-dimethyl-cyclohexenone (**200**) with silyloxydienes. Later they converted cyclohexenone (**200**) to olefenic sulfoxide (**201**) which by a Pummerer type arrangement yielded bicyclic derivative (**202**). LU *et al.* (*100b*) reported synthesis of the bicyclic derivative (**204**) from the olefinic β-Keto ester (**203**) by treatment with tin tetrachloride.

ii) Approaches to ABC Ring System of Forskolin

PAQUETTE *et al.* (*101*) obtained the model tricyclic compound (**207**) (Fig. 25) containing the A, B, C ring system of forskolin by using an anionic oxy-Cope rearrangement as the key step. Heating the potassium salt of enol methyl ether (**205**) at 70°C for 20 minutes followed by trapping of enolate with phenyl selenyl chloride gave the rearranged product (**206**). Oxidative elimination, methylation followed by hydrogenation of (**206**) gave ketone (**207**).

In an attempt to expand this approach to the synthesis of more advanced tricyclic intermediates to forskolin, PAQUETTE *et al.* (*102*) prepared several highly functionalized analogues of (**205**) including some in enantiomerically pure form. However these derivatives did not undergo the oxy-Cope rearrangement probably because steric and/or electronic effects of the substituents in the dihydropyran ring, prevented the [3, 3] sigmatropic rearrangement.

WELZEL and coworkers (*103*) reacted bicyclofarnesol sulfone (**208**) with both racemic and (R)-methyloxirane to yield the diastereomeric

Fig. 25. Anionic oxycope rearrangement strategy for ABC rings of (**1**) (*101, 102*)

alcohols (209)a and (209)b in racemic and optically active form respectively. Selenium induced cyclization followed by reductive elimination
with tri-butyl-tin hydride or cyclization with mercuric trifluoroacetate
followed by reduction with sodium borohydride gave sulfones (210)a and
(210)b. However because of the incorrect relative stereochemistry at C_8
and C_{10} these compounds were not suitable for further transformation
into more advanced intermediates towards forskolin.

BARCO, POLLINI and coworkers (87) synthesised the tricyclic model
compound (212) from cis-decalin (172) (see Fig. 21). Protection of the C_1-
hydroxyl group in (172) followed by addition of 4-lithio-2-methyl-3-
butyn-2-O-THP ether provided acetate (211) which was converted to
tricyclic derivative (212) in several steps.

iii) Total Synthesis of Forskolin

Ziegler's Approach

In 1985, ZIEGLER and coworkers reported the synthesis of tricyclic
lactone (192), (104) (Fig. 27). Thermolysis of diene (213) gave adduct (214)
which on hydroboration, oxidation and epimerisation followed by oxidative elimination with lead tetraacetate gave lactone (148). Reduction of
(148), followed by epoxidation, treatment of the resulting epoxide (215)
with $LiNEt_2$ and protection of the vicinal diol gave tricyclic lactone (192).
In the subsequent publication these authors (105) reported conversion of
lactone (192) to tricyclic pyranone (220). Reduction of (192) with lithium
aluminium hydride, followed by acetylation and oxidation with osmium
tetroxide gave triol (216). Protection of the hydroxy groups of (216)
followed by selective hydrolysis of acetate and Collins oxidation yielded
aldehyde (217). Addition of 1-lithiopropyne to (217) followed by second
oxidation, partial hydrolysis of the orthoformate protecting group at the
C_8- and C_9-hydroxyl groups and treatment with base led to tricyclic
dihydropyranone (218) and E-olefin (219) which was smoothly converted
to (218).

The synthesis of dihydropyranone (218) together with the photo-
chemical and vinyl cuprate addition reactions discussed earlier in Fig. 14
constitute the first formal synthesis of forskolin. The acetonide (220) was
hydrolysed and the resulting (±) (6) was reacetylated under controlled
condition to give (±) (1).

Ikegami's Approach

The second total synthesis of (±) (1) was achieved by IKEGAMI et al.
(106) (Fig. 28). The intramolecular Diels-Alder reaction of diene lactone

(221) yielded adduct (222) which on oxidation with osmium tetroxide followed by treatment with Parikh-modified Moffat reagent gave diketone (223) with a methyl thiomethyl group at C-8. Reductive elimination of the methoxy group of C-8 followed by reduction of carbonyl groups of (223) with the bulky t-butylamino borane led to the desired 6β, 7β diol which was converted to acetonide (224). Thermolysis of the sulfoxide obtained by oxidation of (224) followed by base catalyzed isomerization of the resultant exocyclic double bond gave the conjugated lactone (192).

Starting with (192) and following a sequence similar to that previously reported by the same authors (70) for the synthesis of ($\pm$) 1,6,7-trideoxy forskolin, the total synthesis of forskolin was completed. Lactone (192) was reduced with lithium aluminium hydride and the primary hydroxyl group was protected followed by hydroxylation of the C_8, C_9-double bond with osmium tetroxide to yield triol (225). Protection of the C_1-hydroxy group, hydrolysis and acetylation of the C_{11}-hydroxy group and protection of the 8,9-diol led to orthoformate (226). Reduction of (226) with lithium aluminium hydride, subsequent oxidation with modified Moffat reagent and addition of 4-lithio-3-butyn-1-O-TBDPS ether followed by manganese dioxide oxidation of the resulting alcohol gave alkynone (227). The conjugate addition of dimethyl cuprate followed hydrolysis of the orthoformate, selenium mediated cyclization and reduction yielded tetrahydropyranone (228). Deprotection of the 15-hydroxy group, elimination through a selenide gave (229), and the deprotection of 1,6,7 hydroxyl groups in (229) followed by selective acetylation of the C_7-hydroxyl groups gave ($\pm$) forskolin (1).

Corey's Approach

The third total synthesis of forskolin was reported by COREY and coworkers (107) (Fig. 29). Intramolecular Diels-Alder reaction of the acetylenic acid with diene alcohol (230) yielded adduct (139) which had previously been obtained by NICOLAOU et al. (see Fig. 17). Treatment of lactone (139) with lithium dimethyl cuprate in the presence of boron trifluoride etherate resulted in replacement of tosyl group by methyl. Isomerisation of the double bond in a basic medium led to tricyclic lactone (186) which had been prepared previously by KOFT et al. (see Fig. 23). Cycloaddition of singlet oxygen to (186) gave endoperoxide (231) which on reduction followed by benzoylation gave monobenzoate (232). 1,3-Oxidative rearrangement of (232) followed by reduction and esterification with diazomethane led to methyl ester (233). Deprotection of the C_1-hydroxy group, simultaneous lactonization and reduction of C_6-Keto group followed by epoxidation of 7,8-double bond gave ZIEGLER intermediate (215) (Fig. 27) which was converted to lactone (192) by using procedures similar to ZIEGLER et al. (104).

Direct ethynylation of lactone (192) followed by protection of the C_1-hydroxyl group with dimethylcarbamoyl chloride afforded ynone (234). Conjugate addition of a hydroxyl group to ynone (234), resilylation of the primary hydroxy group and acetylation yielded β-acetoxyenone which on treatment with singlet oxygen yielded endoperoxide (236) apparently through intermediate (235). By treatment of (236) with sodium ethoxide in the presence of tributylphosphine followed by acetic acid–acetic anhydride and subsequently with methyl copper, tributylphosphine complex in the presence of boron trifluoride etherate, β-conjugate addition product (237) was obtained in excellent yield. By a series of functional group manipulations similar to IKEGAMI's approach (Fig. 28) the synthesis of (±) forskolin was completed.

Subsequently COREY et al. (108, 109) synthesised (S)-alcohol (230) by reduction of enone (238) with (R)-oxazaborolidine (239) as chiral catalyst

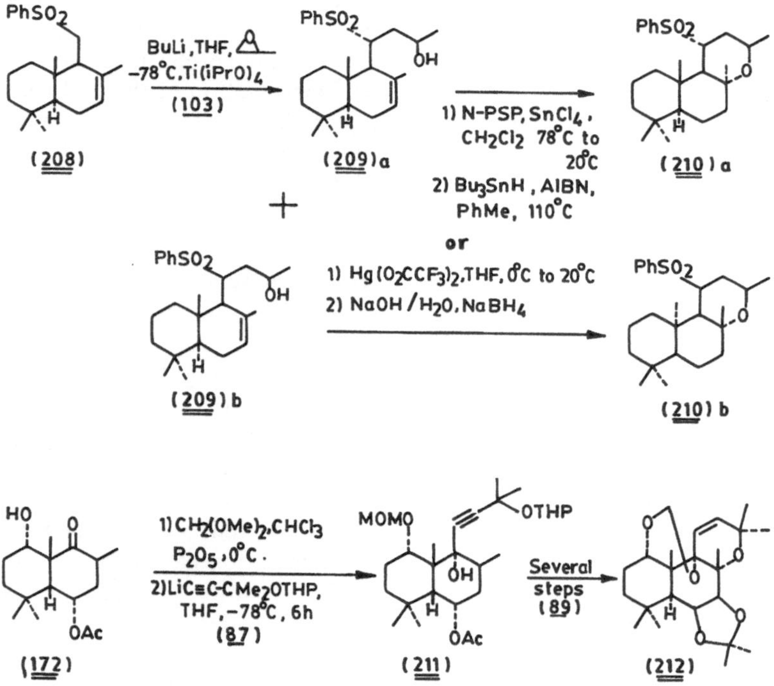

Fig. 26. Condensation of Drimane type of intermediates for the ABC rings of (1)

References, pp. 51–74

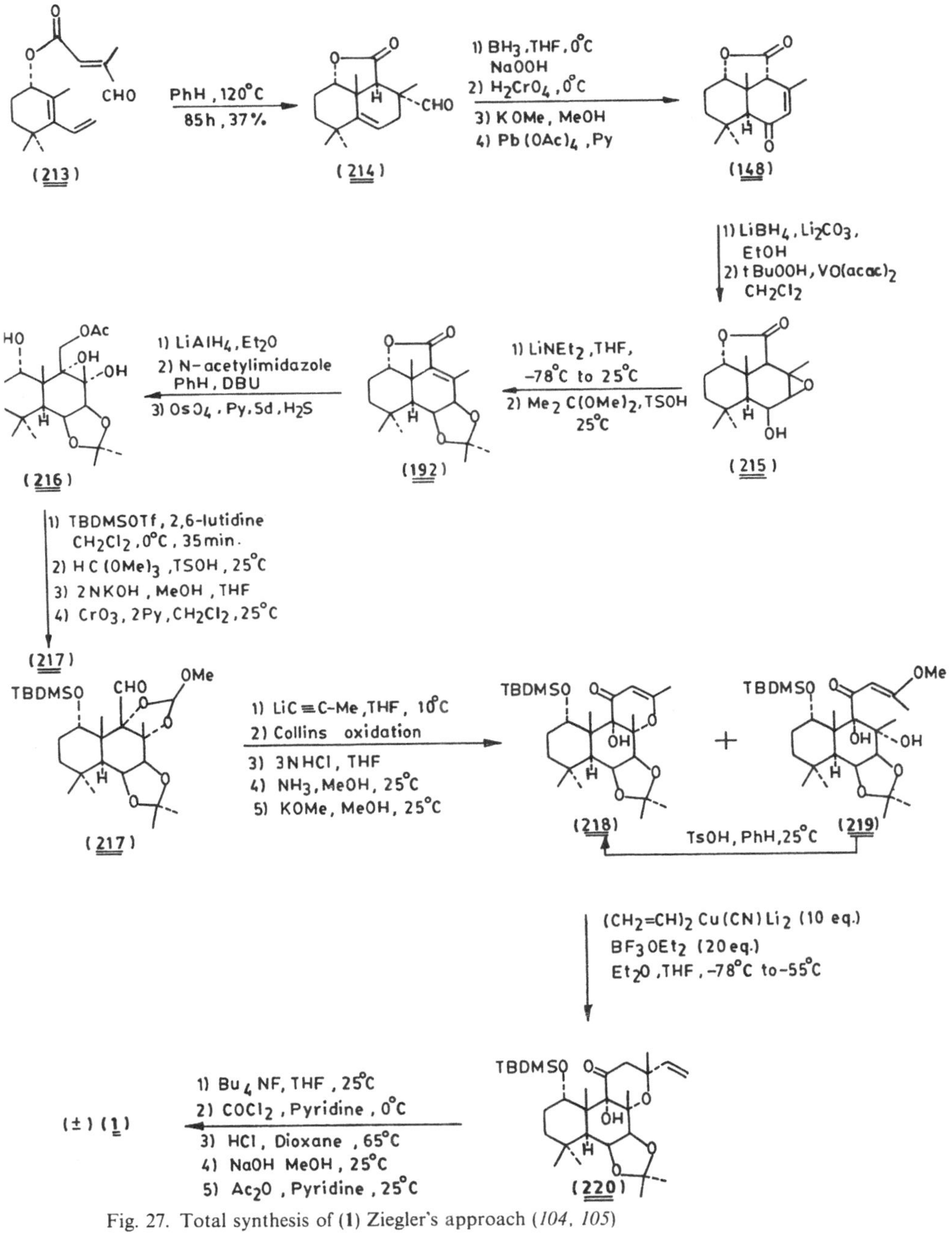

Fig. 27. Total synthesis of (1) Ziegler's approach (104, 105)

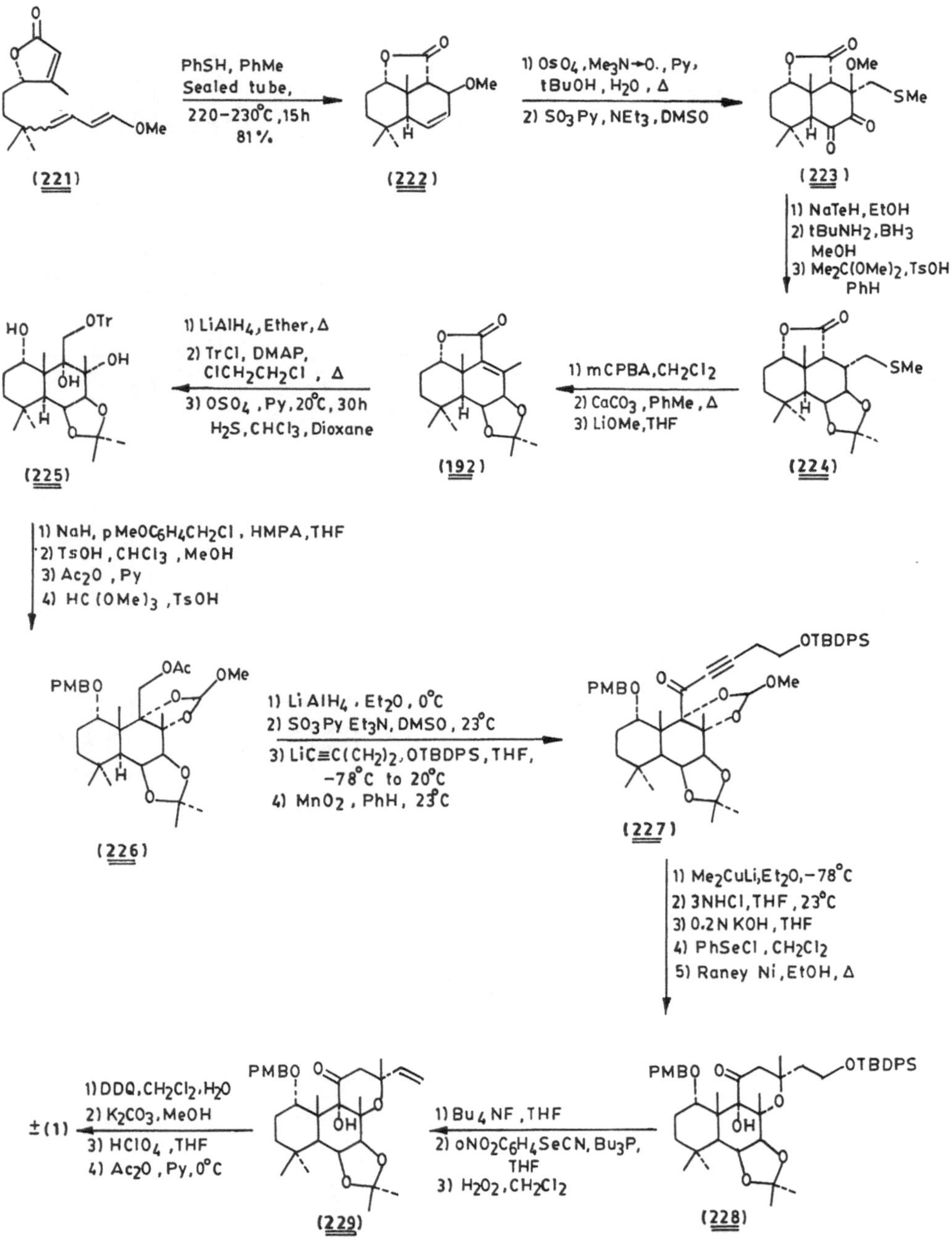

Fig. 28. Total synthesis of (1), Ikegami's approach (106)

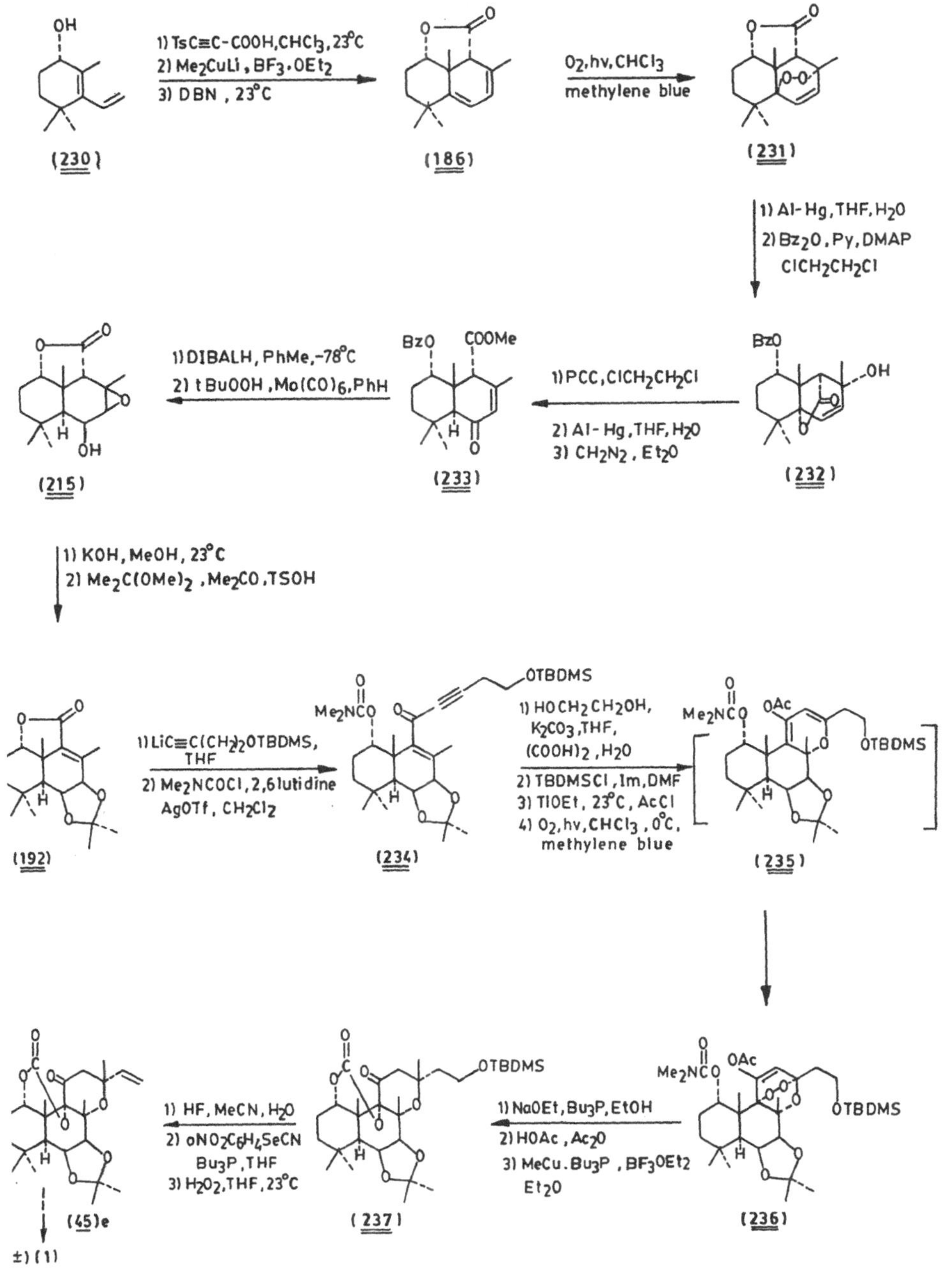

Fig. 29. Total synthesis of (1) Corey's approach (107)

and borane as reducing agent (109) (Fig. 30). Using essentially the same sequence previously described by the same authors (107) the tricyclic lactone (240) was synthesised in optically active form.

In a subsequent report Corey et al. (110) described a new route to the Ziegler intermediate (−) (148) as shown in Fig. 30. Reduction of enone (241) with borane in the presence of (R) (239) yielded optically active alcohol (−) (242) which was transformed through an IMDA reaction into the tricyclic lactone (+) (243) in good yield. Displacement of the tosyl group with methyl and a hydrolysis of the enol ester followed by isomerization of the double bond led to optically active Ziegler intermediate (−) (148). Recently Kanematsu et al. also synthesised optically active (−) (148) starting from optically active ether (−) (143) obtained from (−) (230) and using same sequence described in Fig. 18 (79).

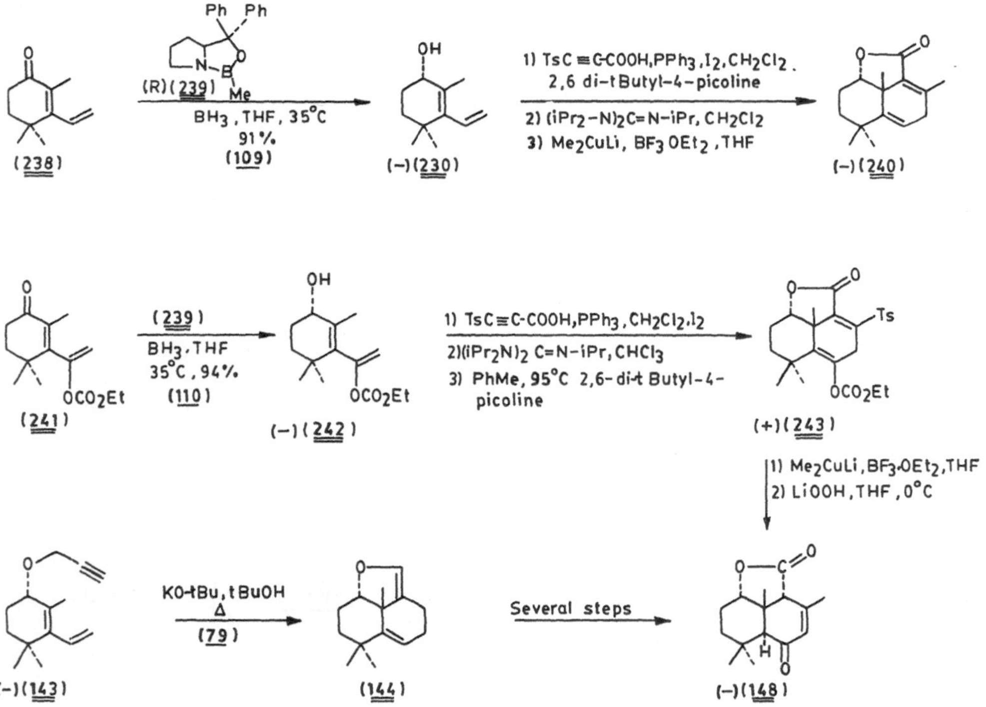

Fig. 30. Synthesis of optically pure intermediates through asymmetric reduction with chiral oxazaborolidine

VI. Biological Activities

Forskolin has unique cardiotonic, antihypertensive, platelet aggregation inhibitory, intraocular pressure lowering, bronchodilatory, antiinflammatory and adenylate cyclase activating properties. The molecule is relatively nontoxic with LD_{50} values of 3100 mg kg^{-1} p.os and 105 mg kg^{-1} i.p. in mice and 2550 mg kg^{-1} p.os and 92 mg/kg i.p. in rats (*10*).

i) Activation of Adenylate Cyclase

The ability of forskolin to activate adenylate cyclase was first reported in 1981 (*111, 112*). Forskolin stimulated adenylate cyclase rapidly and reversibly *in vivo* and *in vitro*, with EC_{50} values in the 5–15 μM range, resulting in increased intracellular cAMP. The unique nature of the adenylate cyclase activating property of forskolin has been described in detail in several reviews (*2–10*). Forskolin activated hormone sensitive adenylate cyclase even in the absence of a functional stimulating guanine nucleotide regulatory protein (Ns protein). This distinguished the action of forskolin from those of hormones and other activators of adenylate cyclase, such as NaF, guanine nucleotides and cholera toxin, and suggested that forskolin was acting directly at the catalytic subunit of adenylate cyclase or at a protein closely related to it. It soon became clear that the ability of forskolin to stimulate adenylate cyclase is markedly affected by the nucleotide regulatory proteins that mediate hormones and other effectors. Thus hormonal agonists that act through the stimulatory guanine nucleotide regulatory protein (Ns protein) markedly potentiate the activity of forskolin and vice versa. Conversely stimulation of adenylate cyclase by forskolin is markedly reduced by hormonal agonists that act through the inhibitory guanine nucleotide regulatory protein (Ni protein). At very low doses (in the picomolar to nanomolar range) forskolin has inhibitory action on adenylate cyclase and generation of cAMP. (*113, 114*) (Fig. 31).

Its specific enzyme activating property makes forskolin a useful tool in the study of the pathophysiology of a variety of disorders related to modulation of adenylate cyclase. Thus forskolin has become a very potent tool for studying the role of cAMP in regulating ion channels, metabolic pathways, cell growth, muscle contractions, secretary process and for studying defects in the expression of a hormonal signal at the level of activated protein kinases or phosphorylated proteins. These correlations between adenylate cyclase activation and various pathophysiological effects exhibited by forskolin have been discussed in detail

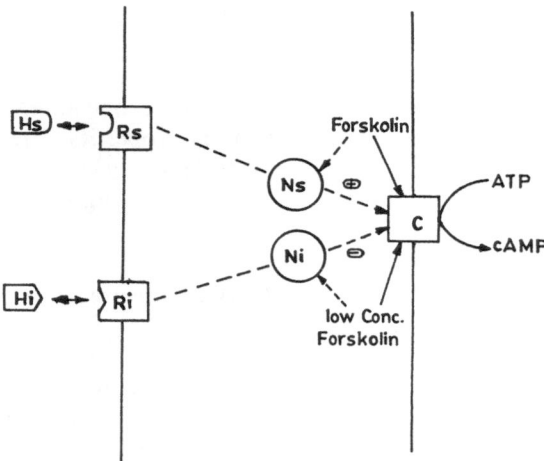

Fig. 31. Schematic representation of the hormone sensitive adenylate cyclase system and possible site of action of (**1**)

by Seamon and Daly in a recent review (*7*). Additional important references after 1985 are listed at the end of this report.

ii) Pharmacological Activities of Forskolin with Therapeutic Significance

The hemodynamic profile of forskolin was studied in anaesthetised normotensive dogs and cats. Doses ranging from $5–100\,\mu g\,Kg^{-1}$ i.v. resulted in dose dependent fall in systemic blood pressure. Forskolin at $10\,mg\,kg^{-1}$ given orally for 5 days to spontaneously hypertensive rats lowered the systolic blood pressure by 30%. Similar results were obtained in renal hypertensive and Doca-saline hypertensive animal models (*10*). The fall in blood pressure is attributed to the peripheral vasodilatory property of forskolin (*115–117*).

In very small doses forskolin exhibited potent positive inotropic activity on heart of guinea pig, dog, cat, rat and rhesus monkey. In the guinea pig atrium model EC_{50} for augmentation of force of contraction by forskolin was found to be 2.48×10^{-8} M. These effects were dose dependent and reversible (*10, 115*). The increase in contractile force elicited by forskolin requires the presence of calcium and is inhibited by calcium channel blocker verapamil (*115, 117*).

Forskolin inhibited human platelet aggregation induced by epinephrine, ADP or collagen with an IC_{50} of approximately 6 μM (*118–121*), whereas the IC_{50} for inhibition of arachidonic acid induced aggregation was much lower, at about 0.6 μM (*120*). The ability of forskolin to inhibit platelet aggregation and to produce positive inotropic activity was correlated with its ability to raise intracellular levels of cAMP (*120, 122*). In addition to its direct effects, a low concentration of forskolin was found to markedly augment the efficacy of potency of prostaglandins in inhibiting platelet aggregation (*119*).

Forskolin applied topically resulted in a decrease in intraocular pressure in rabbit, human, cat and monkey eyes (*123–130*). The decrease in intraocular pressure produced by topical application of a 1% suspension of forskolin persisted approximately 6 hr (*125*). Forskolin was also found to increase ciliary blood flow after topical application (*126*). The decrease in intraocular pressure produced by forskolin was correlated to its ability to increase cAMP in the ciliary epithelia, leading to reduced aqueous flow (*125*).

Forskolin is capable of producing smooth muscle relaxation in lung and parenchymal tissue previously contracted by a calcium ionophore or by ovalbumin (*131–135*). Forskolin did reduce airway resistance in human asthmatics challenged by methacholine (*133*), when given as an aerosol with a dose of 1 mg/puff per patient. The reduction in airway resistance occurred with in 5 min. However, in these experiments forskolin was ineffective when given orally.

In the carrageenin induced acute rat paw edema, forskolin, when administered intraperitoneally, exhibited potent inhibition of oedema in

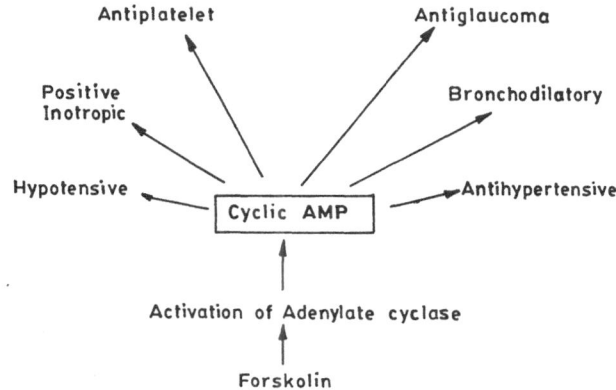

Fig. 32. Pharmacological activities of forskolin

a dose dependent manner with the ED_{50} value of ~ 5 mg/kg. Forskolin when applied topically on croton oil-induced local inflammation showed potent activity with ED_{50} value of 1.15 mg ear^{-1}. In adjuvant induced palyarthritis in rats, a significant activity was observed after i.p. administration of forskolin at 2.5 mg/kg dose (10).

In conclusion, therefore, forskolin provides a spectrum of interesting activities which are of therapeutic importance to mankind (Fig. 32). The broad action of forskolin in stimulating pharmacological phenomena associated with increase in cAMP in most tissues may, however limit its clinical usefulness. Forskolin and some derivatives are in preclinical/clinical trials. However, formulations & pharmacodynamic studies deserve further investigation to produce tissue specific activities with desired duration of action.

iii) Modulation of Membrane Transport Proteins

More recently forskolin has been shown to inhibit a number of membrane transport proteins and channel proteins through both cAMP dependent and independent mechanisms (9). Such proteins include glucose transporter, nicotinic acetyl choline receptor, voltage dependent K^+ channels, $GABA_A$ receptors, p-glycoprotein multidrug transporter etc.

Forskolin inhibits glucose transport in a number of different cells including adipocytes, erythrocyte, platelets, cardiomyocytes, bone cells (136–141), and in vesicle preparation from erythrocytes and adipocytes (137–140). Since these vesicle preparations do not produce cAMP, the inhibitory effects of forskolin cannot therefore be attributed to activation of adenylate cyclase. 1,9-Dideoxyforskolin also inhibites glucose transport in rat adipocytes. ^{3}H-forskolin (49) and an $[^{125}I]$-aryl-azido derivative of forskolin (43) are covalently incorporated into a 18 KDa fragment the erythrocyte glucose transporter when photolysed in the presence of erythrocyte membranes (49). Incorporation of both analogues is inhibited by cytochalasin B and by D glucose.

cAMP dependent and independent effects of forskolin have been observed on acetyl choline induced desensitization of the nicotinic acetyl choline receptor (141–146). At a concentration of 1 μM forskolin (but not its 1,9-dideoxy analogue which does not activate adenylate cyclase) enhance nicotinic receptor depolarization after 30 min in rat muscle, consistent with activity through cAMP. Both forskolin and 1,9-dideoxyforskolin at concentrations of 20 μM, rapidly inhibited nicotinic receptors in rat muscle cultures, which is consistent with a cyclic AMP independent effect (142).

Forskolin rapidly and reversibly inhibited the carbachol stimulated influx of $^{86}Rb^+$ through nicotinic receptors of PC12 cells and chick myotubes (144, 145). The inhibitory effect was also observed with 1,9-dideoxyforskolin with slightly higher potency (145). Thus forskolin inhibits agonist induced $^{86}Rb^+$ influx through the nicotinic receptor by a mechanism that is independent of adenylate cyclase and cAMP. Forskolin also modulates acetyl choline release in the hippocampus independently of adenylate cyclase activation (147).

Forskolin modulates voltage dependent K^+ conductances through cAMP dependent and cAMP independent mechanisms (148–156). Forskolin inhibited delayed rectifying K^+ channels in pancreatic β-cells, as well as voltage dependent K^+ channels in *Helix* neurons, nudibranch neurons, human T cells and PC 12-cells in a cAMP independent manner (154, 156). However forskolin did not affect Ca^{2+} dependent K^+ channels (156). Therefore, it seems unlikely that the effects of forskolin are due to a general perturbation of the lipid bilayer. Forskolin increased the active transport of Na^+ and also enhanced the permeability of urea of isolated frog skin (157, 158). The calcium channels were also affected by forskolin (159–165). Interestingly forskolin and 1,9-dideoxyforskolin inhibit muscimol stimulated Cl^- fluxes through the $GABA_A$ receptor in brain synaptoneurosomes (166). Forskolin also binds to p-glyco-protein multidrug transporter (47).

iv) Effect on Cell Growth

Forskolin stimulated cultured human and animal epithelial cell proliferation (167). The cultured cells may be useful as autologous skin transplants for treatment of extensive burns. Biphasic action of forskolin on growth hormone and prolactin secretion by rat anterior pitutary cells *in vitro* was observed (168). In the primary monolayer cultures of rat adenohypophyseal cells prepared from both immature (12 days old) and adult (6 weeks to 4 month old) male and female rats, the dose-related stimulation of growth hormone release by forskolin was biphasic. At increasing forskolin concentrations from 0.03–3.16 µM, growth hormone release increased progressively to maximal values of 442% and 303% of basal release in cells from immature and adult rats respectively. Forskolin concentrations above 3.16 µM induced progressively diminished growth hormone responses. Forskolin stimulated prolactin release to a lesser degree than it did growth hormone release and prolactin response to forskolin was also biphasic (168). Forskolin was found to promote growth of live stock (169).

v) Other Utility

Forskolin treated antithrombogenic surgical implants have been evaluated (*170*). The utility of the preparations containing forskolin for hair growth and hair graying suppression was demonstrated (*171, 172*).

VII. Structure Activity Correlations

More than 250 semisynthetic derivatives of forskolin have so far been prepared bearing single or multiple changes at various functional groups in the molecule using procedures discussed earlier in section 4.

The antihypertensive, positive inotropic, intraocular pressure lowering, adenylate cyclase stimulating, antiinflammatory, membrane transport protein binding properties of these derivatives were compared with forskolin. The majority of compounds had weaker activities compared with forskolin or were inactive. Those derivatives which are almost equipotent with forskolin or with slightly higher potency are enumerated below.

The 7-ethoxycarbonyl-7-deacetylforskolin, 7-propionyl-7-deacetylforskolin (*16, 173*), 7-glyceryl-7-deacetylforskolin, 7-dimethyl-acryloyl-7-deacetylforskolin (*43*), water soluble 6- or 7-aminoacylforskolin analogues (*41, 174, 175*) were almost equipotent with forskolin in stimulating rat brain, cardiac or platelet membrane adenylate cyclase activity.

Derivatives which had antihypertensive, hypotensive activity almost equipotent with forskolin in spontaneously hypertensive rats include 7-deacetylforskolin, 7-toluene-sulfonyl-7-deacetylforskolin, forskolin-1,9-carbonate, forskolin 1,9-sulfite, and forskolin 1,9:6,7-dicarbonate (*16*). Among the water soluble hydrochloride salts of 6- or 7-aminoacyl-7-

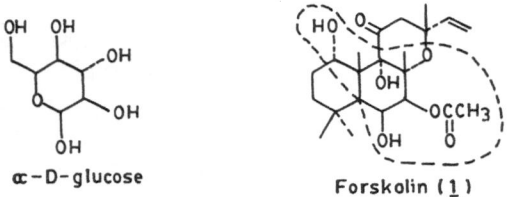

α–D–glucose   Forskolin (**1**)

Fig. 33. Structures of forskolin and α-D-glucose, the parts of A, B, C rings of forskolin that are structurally homologous to hexoses are encircled by the broken line

deacyl-forskolin derivatives containing morpholine, piperidine or N-methyl-piperazine showed potent positive inotropic activity in the guinea pig atrium model. On the basis of further *in vivo* cardiovascular hemodynamic studies and acute toxicity data 6β-piperidino-propionyl-7-deacetylforskolin and 6β-piperidino-acetylforskolin were chosen for further development (*41, 176*).

Thus the 6 and 7 positions of forskolin appear to be able to tolerate a reasonable amount of modifications with minimum effect on potency. The hydrochloride of 1-diethylaminoacetylforskolin also retained potent intraocular pressure lowering property (*177*).

The inactivity of 1,9-dideoxyforskolin and 1-deoxyforskolin and the moderate potency of 9-deoxyforskolin to produce antihypertensive, adenylate cyclase stimulating, positive inotropic, platelet aggregation inhibitory or intraocular pressure lowering properties indicated the critical nature of the 1- and 9-hydroxy groups for these activities. 1,9-Dideoxyforskolin and analogues, however, exhibited considerable amount of antiinflammatory activity.

The carbonyl group at C_{11} is not very critical for adenylate cyclase stimulation and related activities. 11-β-Hydroxy-7-deacetylforskolin is effective, though with much lower potency than (*1*), in activating adenylate cyclase and producing a hypotensive effect in spontaneously hypertensive rats (*7, 16*), whereas 11-α-hydroxy-7-deacetyl analogue is relatively ineffective. 14,15-Dihydroforskolin retains a considerable degree of antihypertensive and adenylate cyclase stimulating properties whereas forskolin 14,15-epoxide is completely inactive (*7*).

Comparable activities of prominent derivatives of (*1*) are listed in Table 4.

VIII. Concluding Remarks

The promising biological activities and unique structural features of forskolin have attracted the attention of large number of scientists.

Applications of a great variety of reagents can be observed in the description of the chemistry and synthesis of forskolin and congeners. The predominant methodology used for the synthesis of the required highly functionalized decalin system was an intramolecular Diels-Alder reaction which culminated in three total syntheses of the molecule. Many intermediates obtained by other methods also show potential for conversion to forskolin and other related polyhydroxylated natural products.

Forskolin is now established as one of the primary tools for demonstrating an involvement of cyclic AMP in physiological functions. For-

Table 4. Relative Potencies of Prominent Derivatives/Congeners of Forskolin. effects are Indicated by the Number of + Signs, – No Effect, ND Effect is not well Documented in the Literature. nAc (acetyl choline receptor) (9)

Compound	Adenylate cyclase Activation	Blood pressure lowering	Positive inotropic	Intraocular pressure lowering	Antiinflammatory	Glucose transporter	Inhibition of channels	
							nAc	GABA$_A$
Forskolin (1)	+ + +	+ + +	+ + + +	+ + +	+ +	+ + +	+ + + +	+ +
1,9-Diedoxyforskolin (3)	–	–	–	–	+ + +	+	+ + + +	+ + +
IAPS-Forskolin (68)	+	ND	ND	ND	ND	+ + + + +	ND	ND
7-(N-Methylpiperazinobutyryl)-7-deacetyl-forskolin (59);	+ +	+ +	+ +	+ +	ND	+	–	ND
6-Piperidinopropionyl-7-deacetyl-forskolin (60)e	+ +	+ +	+ +	+ +	ND	+	ND	ND
Glucose	–		–	–	–	+	ND	ND

skolin and analogues have the potential to be novel therapeutic agents which are described in this report. The ability of forskolin and analogues to affect membrane transport proteins through cyclic AMP dependent as well as independent mechanisms has opened another broad area of research on the nature and therapeutic applications of forskolin binding to membrane channel proteins.

Acknowledgements

The author wishes to thank Messrs Kailash Sharma, P. Sri Nageshkumar, H.B. Shreeshailkumar & R. Padmakumar for updating literature references on the subject, Dr. S.R. Desai for the photograph of plant, Mr. Avdhoot Parab for efficient typing and Mr. Pradeep Sawant for drawings in the present report.

References

1. BHAT, S.V., B.S. BAJWA, H. DORNAUER and N.J. DESOUZA: Structure and Stereochemistry of New Labdane Diterpenoids from *Coleus forskohlii* Briq. Tetrahedron Lett., 1669–72. (1977).
2. SEAMON, K.B. and J.W. DALY: Forskolin: A Unique Diterpene Activator of Cyclic AMP Generating System. J. Cyclic Nucleotide Res., 7, 201–24 (1981).
3. SEAMON, K.B. and J.W. DALY: Forskolin:; Cyclic AMP and Cellular Physiology. Trends Pharmacol Sci. 4, 120–23 (1983).
4. DESOUZA, N.J., A.N. DOHADWALLA and J. REDEN: Forskolin: A Labdane Diterpenoid with Antihypertensive, Positive Inotropic, Platelet Aggregation Inhibitory and Adenylate Cyclase Activating Properties. Med. Res. Rev. 3, 201–19 (1983).
5. GEORGIEVA, Z. and P. UZUNOV: Diterpene Forskolin – a Valuable Agent for Demonstration of the Role of Cyclic 3′-5′-Adenosine Monophosphate in Physiological Processes. Eksp. Med. Morfol., 27, 56–61 (1988) (Bulg.) Chem. Abstc. 109, 66182m.
6. a) SEAMON, K.B.: Forskolin and Adenylate Cyclase, New Opportunities in Drug Design. Ann. Rep. Med. Chem., 19, 293–302 (1984).
   b) DALY J.W.: Forskolin, Adenylate Cyclase and Cell Physiology, An Overview. Adv. Cyclic Nucleotide Res. 17, 81–89 (1984).
7. SEAMON, K.B. and J.W. DALY: Forskolin: Its Biological and Chemical Properties. Adv. Cyclic Nucleotide Res. 20, 1–38 (1986).
8. DOHADWALLA, A.N.: Natural Product Pharmacology, Strategies in Search of Leads for New Drug Design Trends in Pharm. Sci. 6, 49–53 (1985).
9. LAURENZA, A., E.M. SUTKOWSKI and K.B. SEAMON: Forskolin: A Specific Stimulator of Adenylyl Cyclase or a Diterpene with Multiple Sites of Action. Trends in Pharmacol. Sci., 10, 442–47 (1989).
10. DESOUZA, N.J., A.N. DOHADWALLA and R.H. RUPP (eds.): Forskolin: Its Chemical, Biological and Medical Potential. Hoechst India Limited (1986).
11. COLOMBO, M.I., J. ZINCZUK and E.A. RÚVEDA: Synthetic Routes to Forskolin, Tetrahedron Report No 306, 48, 963–1037 (1992).

12. WOLFGANG, J.: Total Synthesis of Forskolin: Synform, **1990**, 149–168; Synthesis of Intermediates for Forskolin, Synform, **1990**, 169–206.

13. References to the medicinal use of "Pashanbhedi" are found in Shaligram Nighantu Bhushman Khemraj Srikrishnadas Shresti, Bombay, India. Parts 7–8, p. 194 (1904); Chunekar Krishnachandra Bhavaprakasa Nighantu, Hindi translation, Chaukhamba Sanskrit Sansthan, Varanasi, India, p. 104 (1969).

14. BHAKUNI, D.S., M.L. DHAR, M.M. DHAR, B.N. DHAWAN, B. GUPTA and R.C. SRIMAL: Screening of Indian Plants for Biological Activity. Indian J. Exp. Biol., **9**, 91–102 (1971).

15. BHAT, S.V., B.S. BAJWA, H. DORNAUER and N.J. DESOUZA: Reactions of Forskolin, a Biologically Active Diterpenoid from *Coleus forskohlii*. J. Chem. Soc., Perkin Trans. 1, **1982**, 767–71

16. BHAT, S.V., A.N. DOHADWALLA, B.S. BAJWA, N.K. DADKAR, H. DORNAUER and N.J. DESOUZA: The Antihypertensive and Positive Inotropic Diterpene Forskolin. Effects of Structural Modifications on its Activity. J. Med. Chem. **26**, 486–92 (1983).

17. TANDON, J.S., M.M. DHAR, S. RAMAKUMAR and K. VENKATESAN: Structure of Coleonol, a Biologically Active Diterpene from *Coleus forskohlii*. Indian J. Chem., **15 B**, 880–83 (1977).

18. a) TANDON, J.S., P.K. JAUHARI, R.S. SINGH and M.M. DHAR: Structures of Three New Diterpenes, Coleonol B, Coleonol C and Deoxycoleonol Isolated from *Coleus forskohlii*. Indian J. Chem., **16B**, 341–45 (1978).

b) KATTI, S.B., P.K. JAUHARI and J.S. TANDON: New Diterpenes from *Coleus forskohlii*, Structure of the Diterpenes, Coleonol D, Coleol and Coleonone. Indian J. Chem., **17B**, 321 (1979).

c) PAINULY, C.P., S.B. KATTI and J.S. TANDON: Diterpenes from *Coleus forskohlii*: Structures of Coleonol E and Coleonol F. Indian J. Chem. **18B**, 214–16 (1979).

19. TANDON, J.S., P. PAINULY, S.B. KATTI and S. SINGH: Chemistry and Antihypertensive Activity of Coleonol Derivatives. Indian J. Chem., **23B**, 67–69 (1984).

20. SINGH, S. and J.S. TANDON: Coleonol and Forskolin from *Coleus forskohlii*. Planta Med., **45**, 62–63 (1982).

21. SAKSENA, A.K., M.J. GREEN, H.J. SHUE, J.K. WONG and A.T. MCPHAIL: Identity of Coleonol with Forskolin, Structure Revision of a Base Catalysed Rearrangement Product. Tetrahedron Lett., **26**, 551–54 (1985).

22. a) PRAKASH, O., R. ROY and M.M. DHAR: A Nuclear Magnetic Resonance Study of Coleonol. J. Chem. Soc., Perkin Trans. II, **1986**, 1779–83.

b) PRAKASH, O., R. ROY, J.S. TANDON and M.M. DHAR: Carbon-13 and Proton Two Dimentional NMR Study of Diterpenoids of *Coleus forskohlii*. Magn. Reson. Chem. **26**, 117 (1988).

23. VISWANATHAN, N. and D.H. GAWAD: Identity of Forskolin with Coleonol. Indian J. Chem., **24**, 583 (1985).

24. GABETTA, B., G. ZINI and B. DANIELI: Minor Diterpenoids of *Coleus forskohlii*. Phytochemistry, **28**, 859–62 (1989).

25. JAUHARI, P.K., S.B. KATTI, J.S. TANDON and M.M. DHAR: Coleosol, a New Diterpene from *Coleus forskohlii*. Indian J. Chem. **16B**, 1055–57 (1978).

26. AKHILA, A., K. RANI and R.S. THAKUR: Biogenetic Relationship of Polyoxygenated Diterpenes in *Coleus forskohlii*. Phytochemistry, **29**, 821–24 (1990).

27. TANDON, J.S., R. ROY, S. BALACHANDRAN and R.A. VISHWAKARMA: 13-Epi-9-deoxy-forskolin. Bioorg. Med. Chem. Lett., **2**, 249–54 (1992).

28. PAULUS, E.F.: Molecular and Crystal Structure of 1-Benzyl-7-deacetyl-7-bromoisobutyryl forskolin. Z. Kristallogr., **153**, 43–49 (1980).

29. PAULUS, E.F.: Molecular and Crystal Structure of Forskolin, Z. Kristallogr., **152**, 239–45 (1980).
30. VELDAS III L.J. and M. KOREEDA: Synthesis of the 6-benzoyl Derivative of 1-Deoxy-1-oxo-7-deacetylforskolin and an Unambiguous Assignment of the Absolute Stereochemistry of Forskolin, J. Org. Chem., **56**, 844–46 (1991).
31. KOGLER, H., and H.W. FEHLHABER: NMR investigations of Forskolin, Complete Assignment of Proton and ^{13}C-NMR Spectra and Conformational Analysis. Magn. Res. Chem. **29**, 993–8 (1991).
32. VISHWAKARMA, R.K., B.R. TYAGI, B. AHMED and A. HUSAIN: Variations in Forskolin Content. Planta Med., **54**, 471–72 (1988).
33. MERSINGER, R., H. DORNAUER and E. REINHARD: Formation of Forskolin by Suspension Cultures of Coleus forskohlii. Planta Med., **54**, 200–04 (1988).
34. SHARMA, N., K.P.S. CHONDEL and V.K. SRIVASTAVA: In Vitro Propagation of Coleus forskohlii Briq, a Threatened Medicinal Plant. Plant Cell Rep., **10**, 67–70 (1991)
35. INAMDAR, P.K., H. DORNAUER and N.J. DESOUZA: GLC method for Assay of Forskolin, a Novel Positive Inotropic and Blood Pressure Lowering Agent. J. Pharmaceut. Sci., **69**, 1449 (1980).
36. INAMDAR, P.K., P.V. KANITKAR, J. REDEN and N.J. DESOUZA: Quantitative determination of Forskolin by TLC and HPLC. Planta Med., **43**, 30–34 (1984).
37. SHAH, V., S.V. BHAT, B.S. BAJWA, H. DORNAUER and N.J. DESOUZA: The Occurrence of Forskolin in Labiatae. Planta Med., **39**, 183–85 (1980).
38. BASLAS-PRADEEPKUMAR, R.K.: Phytochemical Studies of Plants of Coleus Genera, Herba Hungarica, **20**, 213–21 (1981).
39. RÜEDI, P. and C.H. EUGSTER: Diterpenoide Drüsenfarbstoffe: Coleon L, ein neues Diosphenol aus Coleus somaliensis S. Moore, Revision der Strukturen von Coleon H,I,I',K. Helv. Chim. Acta. **60**, 1233–38 (1977) and references cited therein.
40. WANG, A.H.-J., I.C. PAUL, R. ZELNIK, D. LAVIE and E.C. LEVY: Structure and Stereochemistry of Cyclobutasin, a Diterpenoid containing a Four-membered Ring. J. Am. Chem. Soc., **96**, 580–81 (1974) and references cited therein.
41. KHANDELWAL, Y., K. RAJESHWARI, R. RAJGOPALAN, LAKSHMI SWAMY, A.N. DOHADWALLA, N.J. DESOUZA and R.H. RUPP: Cardiovascular Effects of New Water Soluble Derivatives of Forskolin. J. Med. Chem., **31**, 1872–79 (1988).
42. KOSLEY, R.W. JR. and R.J. CHERILL: Regioselective Acylations of 7-Deacetylforskolin. J. Org. Chem., **54**, 2972–75 (1989).
43. SEQUIN, E., N. FERRY, J. HANOUNE and M. KOCH: Effect of Novel 7β-Derivatives of Forskolin upon Human Platelet Adenylate cyclase System. Planta Med., **54**, 4–6 (1988).
44. O'MALLEY, G.J., B. SPAHL, R.J. CHERILL and R.W. KOSLEY, JR.: Regiocontrolled Reactions of 7-deacetylforskolin, Synthesis of 6- and 7-carbamate. J. Org. Chem. **55**, 1102–1105 (1990).
45. WADZINSKI, B.E., M.F. SHANAHAN and A.E. RUOHO: Derivatization of the Human Erythrocyte Glucose Transporter Using a Novel Forskolin Photoaffinity Label. J. Biol. Chem. **262**, 17683–89 (1987).
46. WATT, D.S., K. KAWADA, E. LEYVA and M.S. PLATZ: Exploratory Photochemistry of Iodinated Aromatic Azides. Tetrahedron Lett., **30**, 899–902 (1989).
47. MORRIS, D.I., L.A. SPEICHER, A.E. RUOHO, K.D. TEW and K.B. SEAMON: Interaction of Forskolin with the P-Glycoprotein Multidrug Transporter. Biochemistry, **30**, 8371–79 (1991) and references cited there in.
48. PFEUFFER, E. and T. PFEUFFER: Affinity Labelling of Forskolin Binding Proteins:

Comparison Between Glucose Carrier and Adenylate Cyclase. FEBS Letters, **248**, 13–17 (1989).

49. SHANAHAN, M.F., D.P. MORRIS and B.M. EDWARDS: [3][H]-Forskolin, Direct Photo-affinity Labelling of the Erythrocyte D-Glucose Transporter. J. Biol. Chem., **262**, 5978–84 (1987).

50. SEAMON, K.B., R. VAILLANCOURT, M. EDWARDS and J.W. DALY: Binding of [3H]-Forskolin to Rat Brain Membranes. Proc. Nat. Acad. Sci. USA, **81**, 5081–85 (1984).

51. SALFE, S. and D.R. STORM: Deacetylation of Forskolin Catalyzed by Bovine Brain Membranes. Biochem. Biophys. Res. Comm. **133**, 52–59 (1985).

52. PFEUFFER, T. and H. METZGER: 7-O-Hemisuccinyl-7-deacetyl-forskolin-sepharose a Novel Affinity Support for Purification of Adenylate Cyclase. FEBS Letters, **146**, 369–75 (1982).

53. KHANDELWAL, Y., G. MORAES, N.J. DESOUZA, H.-W. FEHLHABER and E.F. PAULUS: Oxidation-Reduction Studies with Forskolin. Tetrahedron Lett., 1986, 6249–52 (1986).

54. DELPECH, B. and R. LETT: Synthesis of 14,15-Dehydro-forskolin Via Dimethyl Diazomethylphosphonate Anion Reaction with an Aldehyde. Tetrahedron Lett., **30**, 1521–24 (1989).

55. HIRB, N.J.: Synthesis of 12-Oxygenated Forskolin, J. Chem. Soc., Chem. Comm., 1987, 1338–40.

56. HIRB, N.J.: A Synthesis of Forskolin, Hydroxylation of 9-deoxyforskolin. Tetrahedron Lett., **28**, 19–22 (1987).

57. SCHERKENBECK, J., D. BOTTGER and P. WELZEL: 1,9-Dideoxyforskolin, Enolate Formation, Oxidation, Tetrahedron, **43**, 3797–3802 (1987).

58. SHUTSKE, G.M.: Synthesis of 12-Bromo-, 12-Chloro- and 12-Fluoroforskolins. J. Chem. Soc., Perkin Trans. 1, **1989**, 1544–46.

59. NADKARNI, S.R., P.M. AKUT, B.N. GANGULI, Y. KHANDELWAL, N.J. DESOUZA and R.H. RUPP: Microbial Transformation of 1,9-Dideoxyforskolin to Forskolin. Tetrahedron Lett., **27**, 5265–68 (1986).

60. INAMDAR, P.K., Y. KHANDELWAL, M. GARKHEDKAR, R. RUPP and N.J. DESOUZA: Identification of Microbial Transformation Product of 1,9-Dideoxyforskolin and 7-Deacetyl-1,9-dideoxy-forskolin. Planta Med. **55**, 386–7 (1989).

61. KHANDELWAL, Y., P.K. INAMDAR, N.J. DESOUZA, R.H. RUPP, S. CHATTERJEE and B.N. GANGULI: Novel 1,9-Dideoxyforskolin Analogues through Microbial Trans-formations. Tetrahedron, **44**, 1661–66 (1988).

62. ARETZ, W., D. BOETTGER and K. SAUBER: Production of Forskolin Derivatives by Microbial Hydroxylation or Oxidation with *Neurospora crassa* for Pharmaceutical Use. Ger. Offen. DE 3,527,336 (Cl. C07D311/94) 05, Feb. 1987; Chem Abstr. **106**, 212559w.

63. ARETZ, W., D. BOETTGER and K. SAUBER: Production of Forskolin Derivatives by Microbial Hydroxylation or Oxidation with Strains of *Mortierella* for Pharmaceut-ical Use. Ger. Often. DE 3,527,335, (Cl. CO7D311/94) 05, Feb. 1987; Chem. Abstr. **106**, 212558v.

64. a) KHANDELWAL, Y., N.J. DESOUZA, S. CHATTERJEE, B.N. GANGULI and R.H. RUPP: Synthesis of Metabolites of Forskolin. Tetrahedron Letters, **28**, 4089–92 (1987).
b) LAL, B.R., H. GIDWANI and R.H. RUPP: Aluminium Chloride as a Powerful Catalyst for the Preparation of O-Isopropylidene and O-Benzylidene Derivatives of Labdanes. Synthesis, 1989, 711–13.

65. GARCIA-GRANADOS, A., A. MARTINEZ, M.B. JIMENEZ, M.E. ONORATO, F. RIVAS and J.M. ARIAS: Microbial Transformation of an *ent*-13-epi-Manoyl Oxide by *Curvularia*

lunata, a Possible Route to the Synthesis of *ent*-Forskolin Analogs. J. Chem. Res. Synop., **1990**, 94–5.

66. VISHWAKARMA, R.A. and J.S. TANDON: Stereoisomers of Coleonol (Forskolin) and Related Diterpenoids. Tetrahedron Lett., **31**, 7493–94 (1990).

67. VISHWAKARMA, R.A.: Spiroforskolin; Acid Catalyzed Rearrangement Product of Forskolin. Tetrahedron Lett., **30**, 131–32 (1989).

68. SAKSENA, A.K., M.J. GREEN, H.J. SHUE and J.K. WONG: Forskolin: A Convenient Degradation to 14-15-Dinor-8-13-epoxy-1α-6β-7β-9α-tetrahydroxylabd-12-en-11-one-7-acetate-1,9-carbonate, via β-Elimination of an Aldoxime. J. Chem. Soc., Chem. Comm. 1985, 1748.

69. DELPECH, B. and R. LETT: Retrosynthetic Studies with Forskolin. Tetrahedron Lett., **281**, 4061–64 (1987).

70. HASHIMOTO, S., M. SONEGAWA, S. SAKATA and S. IKEGAMI: A Stereocontrolled Synthesis of (±)-1,6,7-Trideoxyforskolin, J. Chem. Soc., Chem. Comm., 1987 24–25.

71. ZIEGLER, F.E. and B.H. JAYNES: Reconstruction of Forskolin from a Ring C Dihydro-pyran-4-one Degradation Product Thereof, Tetrahedron Letters, **28**, 2339–42 (1987).

72. ZIEGLER, F.E. and B.H. JAYNES: Formation of (±) Forskolin via a Cuprate Addition of Synthetic Dihydropyran-4-one, Tetrahedron Lett., **29**, 2031–32 (1988).

73. SCHERKENBECK, J., W. DIETRICH, D. MÜLLER, D. BÖTTGER and P. WELZEL: Forskolin Studies. Tetrahedron, **42**, 5949–59 (1986).

74. JENKINS, P.R., K.A. MENEAR, P. BARRACLAUGH and M.S. NOBBS: An Intramolecular Diels-Alder Approach to Forskolin. J. Chem. Soc., Chem. Comm., 1984, 1423–24.

75. MAGNUS, P., C. WALKER, P.R. JENKINS and K.A. MENEAR: Mechanistic Realization of an Apparently Non-stereospecific Intramolecular Diels-Alder Reaction. Tetrahedron Lett., **27**, 651–54 (1986).

76. NICOLAOU, K.C. and W.S. LI: An Intramolecular Diels-Alder Strategy to Forskolin. J. Chem. Soc., Chem. Commun. 1985, 421.

77. LIU, Z.Y., X.-R. ZHOU and Z.M. WU: An Efficient Approach to the AB Ring System of Forskolin, J. Chem. Soc., Chem. Comm., 1987, 1868–69.

78. KANEMATSU, K. and S. NAGASHIMA, An Efficient Synthesis of a Key Intermediate of Forskolin. J. Chem. Soc., Chem. Comm., 1989, 1028–29.

79. NAGASHIMA, S. and K. KANEMATSU: A Synthesis of an Optically Active Forskolin Intermediate via Allenyl Ether Intramolecular Cycloaddition Strategy. Tetrahedron Asymmetry, **1**, 743–49 (1990).

80. TROST, B.M. and R.C. HOLCOMB: An Unusual Stereochemical Directing Effect of a propargylic Oxygen Substituent on an Intramolecular Diels-Alder Reaction. Tetrahedron Lett. **30**, 7157–60 (1989).

81. TSANG, R. and B. FRASER-REID: Pyranose α-Enones Provide Ready Access to Functionalised *trans*-Decalins via *bis*-Annulated Pyranosides Obtained by Intramolecular Diels-Alder Reaction. A Key Intermediate of Forskolin. J. Org. Chem. **57**, 1065–67 (1992).

82. KULKARNI, Y.S. and B.B. SNIDER: A Synthetic Approach to the AB Ring system of Forskolin. Org. Prep. Proced. Int., **18**, 7–15 (1986).

83. BOLD, G., S. CHAO, R. BHIDE, S.-H. WU, D.V. PATEL, C.J. SIH and C. CHIDESTER: A Chiral Bicyclic Intermediate for the Synthesis of Forskolin. Tetrahedron Lett., **28**, 1973–76 (1987).

84. MUKHOPADHYA, A., S.M. ALI, M. HUSAIN, S.N. SURYAWANSHI and D.S. BHAKUNI: Diels-Alder Reaction of *in situ* Generated 2-Methoxy-carbonyl-*p*-quinone with D-Glucose Based Dienes; A New Approach to Forskolin. Tetrahedron Lett., **30**, 1853–56 (1989).

85. BHAKUNI, D.S.: Synthetic Studies Towards Forskolin. Pure Appl. Chem., **62**, 1389–92 (1990).

86. BERALDI, P.G., A. BARCO, S. BENETTI, G.P. POLLINI, E. POLO and D. SIMONI: The Intramolecular Nitrile Oxide Cycloaddition Route to Forskolin J. Chem. Soc., Chem. Commun. 1986, 757–58.

87. BERALDI, P.G., A. BARCO, S. BENETTI, V. FERRETTI, G.P. POLLINI, E. POLO and V. ZANIRATO: Synthetic Studies Towards Forskolin. Tetrahedron, **45**, 1517–32 (1989).

88. KOZIKOWSKI, A.P., S.H. JUNG and J.P. SPRINGER: A [(4 + 2) + (3 + 2)] Approach to a Forskolin Intermediate: A Further Understanding of π – Facial Selection in Diels-Alder Reactions. J. Chem. Soc., Chem. Comm., **1988**, 167–69.

89. HUTCHINSON, J.H., G. PATTENDEN and P.L. MYERS. Tandem Radical Cyclization – Intramolecular Mukaiyama Aldolisation Approach to Forskolin. Tetrahedron Lett., **28**, 1313–16 (1987).

90. BEGLEY, M.J., H. BHANDAL, J.H. HUTCHINSON and G. PATTENDEN: Dichotomous Reactivity in Stannane and Cobalt Mediated Radical Cyclization. Tetrahedron Lett., **28**, 1317–20 (1987).

91. BEGLEY, M.J., D.R. CHESHIRE, T. HARRISON, J.H. HUTCHINSON, P.L. MYERS and G. PATTENDEN: A New Synthetic Route to (±) Forskolin. Tetrahedron, **45**, 5215–46 (1989).

92. KOFT, E.R., A.S. KOTNIS and T.A. BROADBENT: Synthesis of a Potential Forskolin A-B Ring Precursor by Tandem Michael-Aldol Reactions. Tetrahedron Lett., **28**, 2799–2800 (1987).

93. LI, T.T. and Y.-L. WU: An Approach to Forskolin by an Efficient Synthesis of a Tricyclic Lactone Intermediate. Tetrahedron Lett., **29**, 4039–40 (1988).

94. COLOMBO, M.I., J. ZINCZUK, J.A. BACIGALUPPO, C. SOMOZA and E.A. RÚVEDA: Synthesis of Ziegler Key Intermediate and Related Precursors for the Synthesis of Forskolin and Erigerol. J. Org. Chem., **58**, 5631–39 (1990).

95. SOMOZA, C., J. DARIAS and E.A. RÚVEDA: Intramolecular Michael-Aldol Condensation Approach to the Construction of Advanced Intermediates in the Synthesis of Forskolin. J. Org. Chem., **54**, 1539–43 (1989).

96. VENKATARAMAN, H. and J.K. CHA: A Formal Synthesis of Forskolin: An Electrocyclization Approach. J. Org. Chem., **54**, 2505–6 (1989).

97. LECLAIRE, M. and J.Y. LALLEMAND: Simple Access to a Forskolin Precursor. Tetrahedron Lett., **30**, 6331–34 (1989).

98. NICOLAOU, K.C., S. KUBOTA and W.S. LI: A Synthetic Route to Forskolin. J. Chem. Soc., Chem. Comm., **1989**, 512–14.

99. MC MURRY, J.E. and M.D. ERION: Stereoselective Total Synthesis of the Complement Inhibitor K-76. J. Am. Chem. Soc., **107**, 2712–20 (1985).

100. a) NEUNERT, D., H. KLEIN and P. WELZEL: Forskolin, Some Studies on Ring B Forming Reactions. Tetrahedron, **45**, 661–72 (1989).
b) LIU, Z. and J. YANG: Synthesis of Methyl Δ^7-(1β, $4a\beta$, $8a\beta$)-Octahydro-5,5,8a-trimethyl-2-oxonaphthalene-carboxylate by Allylic Cation Promoted Cyclization: Synth. Commun., **17**, 1617–28 (1987).

101. OPLINGER, J.A. and L.A. PAQUETTE: Synthesis of Forskolin Skeleton via Anionic Oxy-Cope rearrangement. Tetrahedron Lett., **28**, 5441–44 (1987).

102. PAQUETTE, L.A. and J.A. OPLINGER: Limitations in the Application of Anionic Oxy-Cope Sigmatropy to Elaboration of the Forskolin Nucleus. Tetrahedron, **45**, 107–24 (1989).

103. SCHERKENBECK, J., M. BARTH, U. THIEL, K.-H. MITTEN, F. HEINEMANN and P. WELZEL: Model Studies Directed Towards Forskolin: Synthesis of a Tricyclic Model Compound from Farnesol. Tetrahedron, **44**, 6325–36 (1988).

104. ZIEGLER, F.E., B.H. JAYNES and M.T. SAINDANE: A C_6, C_7,-Oxygen Functionalized Intermediate for the Synthesis of Forskolin: Stereochemical Control in an Intramolecular Diels-Alder Reaction. Tetrahedron Lett., 26, 3307–10 (1985).

105. ZIEGLER, F.E., B.H. JAYNES and M.T. SAINDANE: A Synthetic Route to Forskolin. J. Am. Chem. Soc., 109, 8115–16 (1987).

106. HASHIMOTO, S., S. SAKATA, M. SONEGAWA and S. IKEGAMI: A Total Synthesis of (±)-Forskolin. J. Am. Chem. Soc., 110, 3670–72 (1988).

107. COREY, E.J., P. DA SILVA JARDINE and J.C. ROHLOFF: Total Synthesis of (±)-Forskolin. J. Am. Chem. Soc., 110, 3672–73 (1988).

108. COREY, E.J. and P. DA SILVA JARDINE: A Short and Efficient Enantioselective Route to a Key Intermediate for the Total Synthesis of Forskolin. Tetrahedron Lett., 30, 7297–7300 (1989).

109. COREY, E.J., R.K. BAKSHI and S. SHIBATA: High Enantioselective Borane Reaction of Ketones Catalyzed by Chiral Oxazaborolidene Mechanism and Synthetic Applications. J. Am. Chem. Soc., 109, 5551–53 (1987).

110. a) COREY, E.J., P. DA SILVA JARDINE and T. MOHRI: Enantioselective Route to a Key intermediate in the Total Synthesis of Forskolin,. Tetrahedron Lett., 29, 6409–12 (1988).
b) HAYASHI, Y. and K. NARASAKA: Asymmetric Reduction of Ketones and Total Synthesis of Forskolin. Kagaku (Kyoto), 43, 700–1, (1988).

111. METZGER, H. and E. LINDNER: Forskolin, a Novel Adenylate Cyclase Activator. IRCS Med. Sci. Biochem., 9, 99 (1981).

112. SEAMON, K.B., W. PADGETT and J.W. DALY: Forskolin – Unique Diterpene Activator of Adenylate Cyclase in Membranes and in Intact Cells. Proc. Natl. Acad. Sci. USA, 78, 3363–67 (1981).

113. KHANUM, A. and M.L. DUFAU: Inhibitory Action of Forskolin on Adenylate Cyclase Activity and Cyclic AMP Generation. J. Biol. Chem., 261, 11456–9 (1986).

114. SEAMON, K.B.: Forskolin and Adenylate Cyclase. 1st Atlas Sci. Pharmacol., 1, 250–3 (1987), Chem. Abstr. 108 : 87421h.

115. LINDNER, E., A.N. DOHADWALLA and B.K. BATTACHARYA: Positive Inotropic and Blood Pressure Lowering Activity of a Diterpene Derivative Isolated from *Coleus forskohlii*, Forskolin. Arzneim. Forsch., 28, 284–89 (1978).

116. DUBEY, M.P., R.C. SHRIMAL, S. NITYANAND and B.N. DHAWAN: Pharmacological Studies on Coleonol, a Hypertensive Diterpene form *Coleus forskohlii*. J. Ethnopharmacol., 3, 1–13 (1981).

117. LINDNER, E. and H. METZGER: The Action of Forskolin on Muscle Cells is Modified by Hormones, Calcium ions and Calcium Antagonists. Arzneim. Forsch., 33, 1436–41 (1983).

118. ADNOT, S., M. DESMIER, N. FERRY, J. HANOUNE and T. SEVENET: Forskolin, a Powerful Inhibitor of Human Platelet Aggregation. Biochem. Pharmacol., 31, 4071–74 (1982).

119. AGARWAL, K.C. and R.E. PARKS: Synergistic Inhibition of Platelet Aggregation by Forskolin plus PGE_1, or 2-Fluoroadenosine: Effects of 2',5'-Dideoxyadenosine and 5'-Methylthioadenosine. Biochem. Pharmacol., 31, 3713–16 (1982).

120. SIEGL, A.M., J.W. DALY and J.B. SMITH: Inhibition of Aggregation and Stimulation of Cyclic AMP Generation in Intact Human Platelets by the Diterpene Forskolin. Mol. Pharmacol., 21, 680–87 (1982).

121. SIEGL, A.M. and J.W. DALY: Receptor (Norepinephrine), P-Site (2',5'-Dideoxyadenosine) and Calcium Mediated Inhibition of Prostaglandin and Forskolin-Activated Cyclic AMP Systems in Human Platelets. J. Cyc. Nucl. Proc. Phos. Res. 10, 229–46 (1985).

122. METZGER, H. and E. LINDNER: The Positive Inotropic Activating Forskolin, a Potent Adenylate Cyclase Activator. Arzneim. Forsch., **31**, 1248–50 (1981).
123. CAPRIOLI, J. and M. SEARS: Forskolin Lowers Intraocular Pressure in Rabbits, Monkeys, and Man. Lancet, **1**, 958–60 (1983).
124. CAPRIOLI, J. and M. SEARS: Combined Effect of Forskolin and Acetazolamide on Intraocular Pressure and Aqueous Flow in Rabbit Eyes. Exp. Eye Res., **39**, 47–50 (1984).
125. CAPRIOLI, J. and M. SEARS: The Adenylate Cyclase Receptor Complex and Aqueous Humor Formation. Yale J. Biol. Med., **57**, 283–300 (1984).
126. CAPRIOLI, J., M. SEARS, L. BAUSHER, D. GREGORY and A. MEAD: Forskolin Lowers Intraocular Pressure by Reducing Aqueous Inflow. Invest. Ophthalmol. Vis. Sci., **25**, 268–77 (1984).
127. BARTELS, S.P., S.R. LEE and A.H. NEUFELD: Forskolin Stimulates Cyclic AMP Synthesis, Lowers Intraocular Pressure and Increases Outflow Facility in Rabbits. Curr. Eye Res., **2**, 673–81 (1983).
128. LEE, P.Y., S.M. PODOS, T. MITTANG and C. SEVERIN: Effect of Topically Applied Forskolin on Aqueous Humor Dynamics in *Cynomolgus* Monkey. Invest. Ophthalmol. Vis. Sci., **25**, 1206 (1984).
129. POTTER, D.E., J.A. BURKE and J.R. TEMPLE: Forskolin Suppreses Sympathetic Neuron Function and Causes Ocular Hypotension. Currr. Eye Res., **4**, 87–96 (1985).
130. SMITH, B.R., R.N. GASTER, I.H. LEOPOLD and L.D. ZELEZNICK: Forskolin, a Potent Adenylate Cyclase Activator, Lowers Rabbit Intraocular Pressure. Arch. Ophthalmol., **102**, 146–48 (1984).
131. BURKA, J.F.: Inhibition of Antigen and Calcium Ionophore A 23187 Induced Contractions of Guinea Pig Airways by Isoprenaline and Forskolin. Can. J. Physiol Pharmacol., **61**, 581–89 (1983).
132. BURKA, J.F.: Effects of Selected Bronchodilators and Antigen Induced and A 23187-Induced Contraction of Guinea Pig Trachea. J. Pharmacol. Exp. Ther., **255**, 427–35 (1983).
133. LICHEY, J., T. FRIEDRICH, M. PRIESNITZ, G. BIAMINO, P. USINGER and H. HUCKAUF: Effect of Forskolin on Methacholine-Induced Bronchoconstriction in Extrinsic Asthmatics. Lancet, **2**, 167 (1984).
134. KREUTNER, W., R.W. CHAPMAN, A. GULBENKIAN and S. TOZZI: Bronchodilator and Antiallergic Activity of Forskolin. Eur. J. Pharmacol., **111**, 1–8 (1985).
135. CHANG, J., J.M. HAND, S. SCHWALM, A. DERVINIS and A.J. LEWIS: Bronchodilating Activity of Forskolin *in vitro* and *in vivo*. Eur. J. Pharmacol., **101**, 271–74 (1984).
136. JOOST, H.G., A.D. HABBERFIELD, I.A. SIMPSON, A. LAURENZA and K.B. SEAMON: Activation of Adenylate Cyclase and Inhibition of Glucose Transport in Rat Adipocytes by Forskolin Analogs: Structural Determinants for Distinct Sites of Action. Mol. Pharmacol., **33**, 449–53 (1988).
137. KASHIWAGI, A., T.P. HUECKSTEADT and J.E. FOLEY: The Regulation of Glucose Transport by cAMP Stimulators via Three Different Mechanisms in Rat and Human Adipocytes. J. Biol. Chem., **258**, 13685–92 (1983).
138. SERGEANT, S. and H.D. KIM: Inhibition of 3-O-Methylglucose Transport in Human Erythrocytes by Forskolin. J. Biol. Chem., **260**, 14677–82 (1985).
139. KIM, H.D., S. SEARGEANT and S.D. SHUKLA: Glucose Transport in Human Platelets and Its Inhibition by Forskolin. J. Pharmacol. Exp. Ther., **236**, 585–9 (1986).
140. JOOST, H.G. and H.J. STEINFELDER: Forskolin Inhibits Insulin Stimulated Glucose Transport in Rat Adipocyte Cells by a Direct Interaction with the Glucose Transporter. Mol. Pharmacol., **31**, 279–83 (1987).

141. KLIP, A., T. RAMLAL, A.G. DOUEN, P.J. BILAN and K.L. SKORECKI: Inhibition by Forskolin of Insulin Stimulated Glucose Transport in L_6 Muscle Cells. J. Bio. Chem., **265**, 1023–9 (1988).

142. WAGONER, P.K. and B.S. PALLOTTA: Modulation of Acetylcholine Receptor Desensitization by Forskolin is Independent of cAMP. Science, **240**, 1655–57 (1988).

143. WHITE, M.M.: Forskolin Alters Acetylcholine Receptor Gating by Mechanism Independent of Adenylate Cyclase. Mol. Pharmacol., **34**, 427–30 (1988).

144. HAEGGBLAD, J., H. ERIKSSON, B. HEDLUND and E. HEILBRONN: Forskolin Blocks Carbachol-Mediated Ion-Permeability of Chick Myotube Nicotinic Receptors and Inhibits Binding of ^{3}H-Phencyclidene to Terpedo Microsac Nicotinic Receptors. Naunyn-Schmiedeberg's Arch. Pharmacol., **336**, 381–6 (1987).

145. MCHUGH, E.M. and R.M. MCGEE Jr.: Direct Anesthetic Like Effect of Forskolin on the Nicotinic Acetylcholine Receptors of PC 12 Cells. J. Biol. Chem., **261**, 3103–6 (1986).

146. MIDDETON, P., F. JARAMILLO and S.M. SCHUETZE: Forskolin Increases the Rate of Acetylcholine Receptor Desensitization at Rat Soleus Endplates. Proc. Natl. Acad. Sci. USA, **83**, 4967–71 (1986).

147. ALLGAIER, C., B.K. CHOI and G. HERTTING: Forskolin Modulates Acetylcholine Release in the Hippocampus Independently of Adenylate Cyclase Activation. Eur. J. Pharmacol., **181**, 279–82 (1990).

148. ALBUQUERQUE, E.X., S.S. DESHPANDE, Y. ARACAVA, M. ALKONDON and J.W. DALY: A Possible Involvement of Cyclic AMP in the Expression of Desensitization of the Acetyl Choline Receptor, A Study with Forskolin and Its Analogues. FEBS Letters, **199**, 113–20 (1986).

149. HOSHI, T., S.S. GARBER and R.W. ALDRICH: Effect of Forskolin on Voltage-Gated Potassium Channels is Independent of Adenylate Cyclase Activation. Science, **240**, 1652–55 (1988).

150. GALIETTA, L.J.V., A. RASOLA, V. BAVONE, D.C. GRUENERT and G. ROMEO: Forskolin and Verapamil Sensitive Potassium Current in Human Tracheal Cells, Biochem. Biophys. Res. Commun. **179**, 1155–60 (1991).

151. HARRIS-WARRICK, R.M.: Forskolin Reduces a Transient Potassium Current in Lobster Neurons by a cAMP Independent Mechanism. Brain Res. **489**, 59–66 (1989).

152. GARBER, S.S., T.A. HOSHI and R.W. ALDRICH: Interaction of Forskolin with Voltage Gated Potassium Channels in PC 12 Cells. J. Neurosci., **10**, 3361–8 (1990).

153. KRAUSE, D., S.C. LEE and C. DEUTSCH: Forskolin Effects on Voltage Gated Potassium Conductance of Human T Cells. Pflüegers Arch., **412**, 133–40 (1988).

154. COOMBS, J. and S. THOMPSON: Forskolin's Effect on Transient Potassium Current in Nudibrach Neurons is not Reproduced by cAMP. J. Neurosci., **7**, 443–52 (1987).

155. ZÜNKLER, B.J., G. TRUBE and T. OHNO-SHOSAKU: Forskolin Induced Block of Delayed Rectifying Potassium Channels in Pancreatic β Cells is not Mediated by AMP. Pflüegers Arch., **411**, 613–19 (1988).

156. WATANABE, K. and M. GOLA: Forskolin Interaction with Voltage Dependent Potassium Channels in Helix is not Mediated by Cyclic Nucleotides. Neurosci. Letters, **78**, 211–16 (1987).

157. MARGIOTTA, M., C. ARDIZZONE and C. LIPPE: Effects of Forskolin on Active Sodium Transport and Permeability of Isolated Bufo bufo Skin. Boll. Soc. Ital. Biol. Sper., **62**, 729–35 (1986), Chem. Abstr. **105**: 127435n.

158. ARDIZZONE, C., M. MARGIOTTA and C. LIPPE: Action of Forskolin on Non-Electrolyte Permeability Across the Frog Skin as Compared to that of Vasopressin and Isoprenaline. Arch. Int. Physiol. Biochem., **95**, 105–12 (1987).

159. BOUTJDIR, M., P.F. MERY, R. HANF, A. SHRIER and R. FISCHMEISTER: High Affinity Forskolin Inhibition of L Type Calcium Current in Cardiac Cells. Mol. Pharmacol., **38**, 758–65 (1990).

160. DREUX, C. and V. IMHOFF: Forskolin, a Tool for Rat Parotid Secretion Studies. Calcium-45-efflux is not Related to cAMP. Am. J. Physiol., **251**, C754–C762 (1986).

161. YANAGIBASHI, K., V. PAPADOPOULOS, E. MASAKI, T. IWAKI, M. KAWAMURA and P.E. HALL: Forskolin Activates Voltage Dependent Calcium Channels in Bovine but not in Rat Fasciculata Cells. Endocrinology (Baltimore), **124**, 2383–91 (1989).

162. FILIPPOV, A.K. and V.L. POROTIKOV: Effect of Forskolin on Action Potential Slow Inward Current and Tension of Frog Artial Fibers. J. Physiol. (Paris), **80**, 163–7 (1985).

163. MORITA, K., T. DOHI, S. KITAYAMA: Y. KOYAMA and A. TSUJIMOTO: Stimulation-evoked Calcium Fluxes in Cultured Bovine Adrenal Chromaffin Cells are Enhanced by Forskolin. J. Neurochem., **48**, 248–52 (1987).

164. HARTZELL, H.C. and R. FISCHMEISTER: Effect of Forskolin and Acetylcholine on Calcium Current in Single Isolated Cardiac Myocytes. Mol. Pharmacol., **32**, 639–45 (1987).

165. CLEMENT, E., H. SCHEER, D. ZACCHETTI, C. FOSOLATE, T. POZZAN and J. MELDOLESI: Receptor Activated Calcium Influx, Two Independently Regulated Mechanisms of Influx Stimulation Coexist in Neurosecretary PC 12 Cells. J. Biol. Chem., **267**, 2164–72 (1992).

166. HEUSCHNEIDER, G. and R.D. SCHWARTZ: cAMP and Forskolin Decrease γ-Aminobutyric Acid Gated Chloride Flux in Rat Brain Synaptoneurosomes. Proc. Natl. Acad. Sci. USA, **86**, 2938–42 (1989).

167 a) DREHER, R.M. and H.E. KNOELL: Stimulation of Epithelial Cell Proliferation by Forskolin. Ger. Offen. D.E., 3,716,907. (Cl. C12 N5/02) 01 Dec 1988, Chem. Abstr., **111**, P170652s.

b) ARTUC, M., C. REINHOLD and H. KAPPUS: Effect of Forskolin on Growth of Human Epidermal Keratinocytes in Culture, Med. Sci. Res., **16**, 1027–8 (1988).

168. SZABO, M., N.E. STAIB, B.J. COLLINS, and L. CUTTLER: Biphasic Action of Forskolin on Growth Hormone and Prolactin Release by Rat Anterior Pituitary Cells in vitro. Endocrinology (Baltimore), **127**, 1811–17 (1990).

169. SCHEUERMANN, S.E.: Forskolin as Growth Promoter for Live-stock, Ger. Offen D.E. 3,815,718 (Cl. A 23 K 1/16), 16 Nov. 1989, Chem. Abstr. **112**: 197062g.

170. THULESIUS, O. and J. CHRISTENSON: Forskolin Treated Antithrombogenic Surgical Implants. Eur. Pat. Appl. Ep. 308,802 (Cl A 61 L 33/00), 29 March 1989. Chem. Abstr., **112**, P. 62684r.

171. SUGIYAMA, K.T., F.A. KOJI and M. EGAWA: Forskolin Containing Composition for Hair Graying Suppression. Eur. Pat. Appl. Ep. 293, 837 (Cl. A 62 K7/06) Dec. 1988, Chem. Abstr. **109**: 102499x.

172. LAL, B., J. BLUMBACH, A.N. DOHADWALLA and N.J. DESOUZA: Pharmaceutical Compositions Comprising Labdane Diterpenoid Derivatives and Pyrimido-(6, 1a)-isoquinoline-4-one derivatives and their Use. Eur. Pat. Appl. Ep. 370, 379 (Cl A 61 K7/06) 30 May 1990, Chem. Abstr., **114**, 234867n.

173. SEAMON, K.B., J.W. DALY, H. METZGER, N.J. DESOUZA and J. REDEN: Structure Activity Relationships for Activation of Adenylate Cyclase by the Diterpene Forskolin and Its Derivatives J. Med. Chem., **26**, 436–39 (1983).

174. LAURENZA, A., Y. KHANDELWAL, N.J. DESOUZA: R.H. RUPP, H. METZGER and K.B. SEAMON: Stimulation of Adenylate Cyclase by Water Soluble Analogues of Forskolin. Mol. Pharmacol., **32**, 133–9 (1987).

175 a) TATEE, T., T. TAKAHIRA, K. YAMASHITA, M. SAKURAI, A. SHIOZAWA, K. NARITA and
H. UCHIDA: Preparation of Novel Forskolin Derivatives as Cardiotonic and Hypotensive Agents and as Adenylate Cyclase Stimulants. Eur. Pat. Appl. EP 222, 413 (Cl
CO7D 311/92) 20 May 1987, Chem. Abstr., **108**: 131427r.
b) TACHIE, T., T. TAKAHIRA, A. FUJITA and F. ISHIKAWA: Forskolin Derivatives as
Antihypertensive and Adenylate Cyclase Activitors. Jpn Kokai Tokyo Koho, Jp 01
09, 986 (89 09, 986) (Cl C07D 311/92) 13 Jan 1989, Chem. Abstr. **111**: 134567j.

176. HOECHST, A.-G.: Process for the Preparation of 7-Acyloxy-6-(aminoacyloxy)-
polyoxylabdenones and their Use as Antihypertensives. Austrian AT 389,114 (Cl.
CO7D311/92) 25, Oct, 1989., Chem. Abstr. **112**: 235649p.

177. KOSLEY, R.W. Jr. and R.J. CHERILL: Preparation of 1-(Aminoacyloxy)-8,13-epoxy-
labd-14en-11-ones for Treatment of Glaucoma. U.S. US 4,639,446 (CL 514-222; A 61
K31/35), 27 Jan 1987, Chem Abstr. **111**: 134568k.

178. DOHADWALLA, A.N., S.S. MANDREKAR, N.K. DADKAR, Y. KHANDELWAL, R.H. RUPP
and N.J. DESOUZA: Oxygenated Labdane derivatives for Treatment of Antiinflammatory Diseases. U.S. US 4,724,238 (Cl. 514-475; A61k31/35), 09 Feb. 1988, Chem.
Abstr. **110**, 147859k.

Additional References

The additional references are listed in this section according to tissue, cell type and
function used in the investigations. Studies carried out to elucidate the mechanism of action
of forskolin and analogues are listed under adenylate cyclase and protein kinase.

Adenylate Cyclase and Protein Kinase

1. JONES. S.B. and D.B. BYLUND: Characterisation and Possible Mechanisms of α_2-
Adrenergic receptor-mediated Sensitization of Forskolin Stimulated Cyclic AMP
Production in HT29 Cells, J. Biol. Chem. **263**, 14236–44 (1988).

2. JONES. S.B. and D.B. BYLUND: Effects of α_2-Adrenergic Agonist Preincubation on
Subsequent Forskolin-stimulated Adenylate Cyclase Activity and [^{3}H]-Forskolin
Binding in the membranes from HT 29 Cells. Biochem. Pharmacol., **40**, 871–7 (1990).

3. KNOPP. J., Z. STRAKOVA and J. BRTKO: Effect of Forskolin and Triiodothyronine on
cAMP Production in Rat Thymocytes Endocrinol. Exp., **22**, 29–34 (1988).

4. ANDERSON. R.J., R. BRECKON and D. COLSTON: Regulation by Forskolin of Cyclic
AMP Phosphodiesterase and Protein Kinase C Activity in LLC-PK$_1$ Cells. Biochem.
J., **279**, 23–7 (1991).

5. CHIJIWA. T., A. MISHIMA. M. HAGIWARA. M. SANO; K. HAYASHI. T. INOUE. K. NAITO. T.
TOSHIOKA and H. HIDAKA: Inhibition of Forskolin Induced Neurite Outgrowth and
Protein Phosphorylation by a newly Synthesised Selective Inhibitor of Cyclic AMP-
Dependent Protein Kinase, N-[2(-p-Bromocinnamylamino) ethyl]-5-isoquinoline
Sulfonamide (H-89) of PC 12D Pheochromocytoma Cells. J. Biol. Chem., **265**,
5267–72 (1990).

6. HO. A.K., C.L. CHIK and D.C. KLEIN: Forskolin Stimulates Pinealocyte cGMP
Accumulation, Dramatic Potentiation by an α-Adrenergic-[Ca^{2+}]-Mechanism Involving Protein Kinase C. FEBS Lett., **249**, 207–12 (1989).

7. HO. R.J., Q.H. SHI and J. RUIZ: Conditional Inhibition of Forskolin-Activated

Adenylate Cyclase by Guanosine Diphosphate and Its Analog. Arch. Biochem. Biophys., **251**, 148–55 (1986).

8. GOLE, J.W.D., G.L. ORR and R.G.H. DOWNER: Forskolin Insensitive Adenylate Cyclase in the Cultured Cells of *Choristoneura fumiferana* (Insecta). Biochem Biophys. Res. Commun., **145**, 1192–7 (1987).

9. JACKMAN G.P. and A. BOBIK: Forskolin-Mediated Activation of Cardiac, Liver and Lung Adenylate Cyclase in the Rat. Relation to [^{3}H]-Forskolin Binding Sites, Biochem. Pharmacol., **35**, 2247–51 (1986).

10. WRIGHT, M.S., B.J. COOK and G.M. HOLMAN: Regulation of Adenylate Cyclase from *Leucophea maderae* by Calcium, Calmodulin and Forskolin, Comp. Biochem. Physiol. C. Comp. Pharmacol. Toxicol. 85C, 357–62 (1986).

11. SHI Q.H., J.A. RUIZ and R.J. HO: Forms of Adenylate Cyclase, Activation and/or Potentiation by Forskolin, Arch. Biochem. Biophys., **251**, 156–65 (1986).

12. SMIGEL M.D.: Purification of the Adenylate Cyclase. J. Biol. Chem., **261**, 1976–82 (1986).

13. DE FORESTA, B., M. ROGARD and J. GALLEY: Adenylate Cyclase of Bovine Adrenal Cortex Plasma Membranes, Divergence Between Corticotropin and Fluoride, Combined Effects with Forskolin FEBS Letters, **216**, 107–12 (1987).

14. NELSON, C.A. and K.B. SEAMON: Binding of [^{3}H]-Forskolin to Solubilised Preparations of Adenylate Cyclase. Life Sci., **1988**, 1375–83.

Blood Cells

15. DE CHAFFOY DE COURCELLES, D., P. ROEVENS and H. VAN BELLE: Prostagladin E$_1$, and Forskolin Antagonise C-kinase Activation in the Human Platelet. Biochem, J., **244**, 93–9 (1987).

16. SANNOMIYA, Y., N. TATSUMI and K. OKUDA: Effects of Forskolin and Cilostazol on Colt Retraction. Biochem. Int., **17**, 1059–70 (1988).

17. SIEGL, A.M. and G. MOROFF: Effects of Forskolin on the Maintenance of Platelet Properties During Storage. J. Lab. Clin. Med., **108**, 354–9 (1986).

18. AGARWAL, K.C., B.A. ZICLINSKI and R.S. MAITRA: Significance of Plasma Adenosine in the Antiplatelet Activity of Forskolin: Potentiation by Dipyridamole and Dilazep. Thromb. Haemost., **61**, 106–10 (1989).

19. WATANABE, Y. and K.H. JAKOBS: Forskolin Sensitizes Human Platelet Adenylate Cyclase to Modulation of Substrate (Magnesium ATP) Affinity by Hormones. Biochem. J., **237**, 273–6 (1986).

20. MOKHTARI, A., K.C. DO and S. HARBON: Forskolin Alters Sensitivity of the cAMP Generating System to Stimulating as well as Inhibitory Agonists. A Study with Intact, Human Platelets and Guinea Pig Myometrium. Eur. J. Biochem., **176**, 131–7 (1988).

21. APITZ, C.R. and J.E. AJOENE: The Antiplatelet Principle of Garlic Synergistically Potentiates The Antiaggregatory Action of Prostacyclin. Forskolin, Indomethacin and Dipyridamole on Human Platelet. Thromb. Res. **42**, 303–11 (1986).

22. GRAY, S.J. and S. HEPTINSTALL: Interactions between Prostaglandin E$_2$ and Inhibitors of Platelet Aggregation Which Act Through Cyclic AMP. Eur. J. Pharmacol., **194**, 63–70 (1991).

23. TENG, C.M., M.L. HUNG, T.F. HUANG and C. OUYANG: Triwaglerin, a Potent Platelet Aggregation Inducer Purified from *Trimeresurus wagleri* Snake Venom. Biochem. Biophys. Acta, **992**, 258–64 (1989).

24. LAURENZA, A., D. MORRIS and K.B. SEAMON: Irreversible Loss of [^{3}H]-Forskolin Binding Sites in Human Platelets by 2-Haloacyl Analogs of Forskolin, Mol. Pharmacol., **37**, 69–74 (1990).

25. FURUI, H., K. SUZUKI, K. TAKAGI and T. SATAKE: Effect of Colforsin (Forskolin) on Human Neutrophil Superoxide Production and Intracellular Calcium Mobilization. Clin. Exp. Pharmacol. Physiol., 16, 199–209 (1989).

26. KHAN, M.M., A.C. TRAN and K.M. KEANEY: Forskolin and Prostaglandin E_2 Regulate the Generation of Human Cytolytic T Lymphocytes. Immunopharmacology, 19, 151–61 (1990).

27. GRIESE, M., S. GRIESE and D. REINHARDT: Inhibitory Effects of Portussis Toxin on the cAMP Generating System in Human Mononuclear Leukocytes. Eur. J. Clin. Invest., 20, 317–22 (1990).

28. MENETSKI, J.P. and M. GELLERT: Recombination Activity in Lymphoid Cell Lines is Increased by Agents that Elevate cAMP. Proc. Natl. Acad. Sci., USA, 87, 9324–8 (1990).

29. HALLEK, M., T. KAMP, E. HAEN U. GOEHLY, B. EMMERICH and J. REMIEN: Reduced Responsiveness of Adenylate Cyclase to Forskolin in Human Lymphoma Cells. Biochem. Pharmacol, 42, 1329–34 (1991).

30. WEBSTER, H.K., W.P. WIESMANN, G.S. WARD, B. PEMPANICH and C.S. PAVIA: Reversible Defect in cAMP Metabolism in Lymphocytes in Malaria Infection. Immunopharmacology, 19, 169–75 (1990).

31. EBSTEIN, R.P., J. MINTZER, Y.LIPSCHITZ, Z.SHEMESH and J. STESSMAN: Hormone and Forskolin Stimulated Cyclic AMP Accumulation in Human Lymphocytes, Reliability of Longitudinal Time Measurement. Experientia, 42, 838–41 (1986).

Adipocytes and Lipolysis

32. GONZALEZ-NICOLAS, J., J. JIMENEZ, A. PAGE-PANUECLAS, M.T. ZABALA and F.J. MORENO: Regulation of Lipid Metabolism by Dipyridamole and Adenosine Antagonists in Rat Adipocytes. Int. J. Biochem. 21, 883–8 (1989).

33. DE PERGOLA, G., A. HOLMANG, J. SVEDBERG, R. GIORGINO and P. BJORNTORP: Testosterone Treatment of Ovariectomized Rats: Effects on Lipolysis Regulation in Adipocytes. Acta Endocrinol, 123, 61–66 (1990).

34. ALLEN, D.O., B. AHMED and K. NASEER: Relationships between Cyclic AMP Levels and Lipolysis in Fat Cells after Isoproterenol and Forskolin. J. Pharmacol, Exp. Ther. 238, 659–64 (1986).

35. ALLEN, D.O. and J.T. QUESENBERRY: Quantitative Differences in the Cyclic AMP Lipolysis Relationships for Isoproterenol and Forskolin. J. Pharmacol. Exp. Ther., 244, 852–8 (1988).

36. LEHMANN, M. and J. KOOLMAN: The Influence of Forskolin on the Metabolism of Ecdysone and 20-Hydroxyecdysone in Isolated Fat Body of the Blowfly, *Calliphora vicina*. Biol. Chem. Hoppe-Seyler, 367, 387–93 (1986).

37. FATEMI, S.H.: Evaluation of the Effects of Forskolin and the Antilipolytic Agents Insulin and Nicotinic Acid on Cyclic AMP Levels in Rat Epididymal Adipocytes. Biomed. Biochem. Acta, 45, 539–47 (1986).

38. FATEMI, S.H.: Interaction between Prostaglandin E_1 and Forskolin in Modulation of Cyclic AMP Levels in Rat Epididymal Adipocytes. Prostaglandins Leukotrienes Med., 18, 151–61 (1985).

Heart

39. ENGLAND, P.J. and M. SHAHID: Effects of Forskolin on Contractile Responses and Protein Phosphorylation in the Isolated Perfused Rat Heart. Biochem. J., 246, 687–95 (1987).

40. HISAJIMA, H., T. HAMA, K. KURAHASHI, H. USUI and M. FUJIWARA: Vasodilation Produced by Forskolin Compared with that Produced by Adenosine in Rabbit Coronary Artery. J Cardiovasc. Pharmacol, **8**, 1262–7 (1986).

41. ANTONOV, A.S., M.E. LUKASHEV, Y.A. ROMANOV, V.A. TKACHUK, V.S. REPIN and V.N. SMIRNOV: Morphological Alterations of Endothelial Cells from Human Aorta and Umbilical Vein Induced by Forskolin and Phorbol-12-myristate-13-acetate: A Synergistic Action of Adenylate Cyclase and Protein Kinase C Activators. Proc. Natl. Acad. Sci. USA; **83**, 9704–8 (1986).

42. HAI, C.M. and R. PHAIR: Forskolin and Caffeine Induce Ca^{2+} Release from Intracellular Stores in Rabbit Aorta. Am. J. Physiol., **257**, C_{413}–C_{418} (1989).

43. MCMAHON, E.G. and R.J. PAUL: Effects of Forskolin and Cyclic Nucleotides on Isometric Force in Rat Aorta. Am. J. Physiol., **250** C_{468}–C_{473} (1986).

44. BUSCHMANS, E., D.J. HEARSE and A.S. MANNING: Forskolin Effects on Cyclic AMP and Contractile Function in the Isolated Rat and Guinea Pig Heart. Can. J. Cardiol., **1**, 385–94 (1985).

45. JACKMAN, G.P. and A. BOBIK: Differential Forskolin Activation of Rat Heart and Lung Adenylate Cyclase, Dependence on Membrane Protein Interactions. Biochem. Pharmacol., **38**, 1091–5 (1989).

46. ALAM, S.Q., Y.F. REN, and B.S. ALAM: [^{3}H]-Forskolin and [^{3}H]-Dihydroaloprenolol Binding Sites and Adenylate Cyclase Activity in Heart of Rats Containing Different Oils. Lipids, **23**, 207–13 (1988).

47. VADEN, S.L. and H.R. ADAMS: Ionotropic, Chronotropic and Coronary Vasodialator Potency of Forskolin. Eur. J. Pharmacol.,**118**, 131–7 (1985).

48. HEARSE, D.J. and R. ZUCCHI: Forskolin and Myocardial Function in the Normal Ischemic and Perfused Rat Heart. Can. J. Cardiol., **2**, 303–12 (1986).

49. EGAN, T.M., D. NOBLE, S.J. NOBLE, T. POWELL, V.M. TWIST and K. YAMAOKA: On the Mechanism of Isoprenaline and Forskolin Induced Depolarisation of Single Guinea Pig Ventricular Myocytes. J. Physiol., **405**, 785 (1988).

50. EGAN, T.M., D. NOBLE, S.J. NOBLE, T. POWELL, V.M. TWIST and K. YAMAOKA: On the Mechanism of Isoprenaline and Forskolin Induced Depolarisation of Single Guinea Pig Ventricular Myocytes. J. Physiol., **400**, 299–300 (1988).

51. WEST, G.A., G. ISENBERG and L. BELARDINELLI: Antagonism of Forskolin Effects by Adenosine in Isolated Heart and Ventricular Myocytes. Am. J. Physiol., **250**, H769–77 (1986).

52. ZALUPS, R.K. and S.S. SHEU: Effects of Forskolin on Intracellular Sodium Activity in Resting and Stimulated Cardiac Purkinge Fibre from Sheep. J. Mol. Cell. Cardiol., **19**, 887–90 (1987).

53. BOWLLING, N., V.L. WYSS, P.J. GENGO, B. UTTERBACK, R.F. KAUFFMAN and J.S. HAYES: Cardiac Inotropic Responses to Calcium and Forskolin are not Altered by Prolonged Isoproterenol Infusion. Eur.J. Pharmacol., **127**,155–64 (1990).

54. SHUDDY, R.E., C. MAK and M.R. BRISTOW: Comparative *in vitro* Myocardial Inotropic Effect and *in vivo* Hemodynamic Effect of Forskolin and Isoproterenol in Young Lambs. Pediatr. Res. **25**, 580–4 (1989).

55. SCARPACE, P.J.: Forskolin Activation of Adenylate Cyclase in Rat Myocardium with Age, Effects of Guanine Nucleotide Analogues. Mech. Ageing Dev., **52**, 169–78 (1990).

56. MACLID, K.M.: The Influence of Neat Nitroprusside on Positive Inotropic Responses of Rabbit Popillary Muscle to Forskolin. Proc. West. Pharmacol. Soc., **29**, 81–3 (1986).

57. BROWN, L., C. SERNIA, R. NEWLIRG and P. KLETCHEV: Comparison of Inotropic and Chronotropic Responses in Rat Isolated Atria and Ventricles. Clin. Exp. Pharmacol. Physiol. **18**, 753–60 (1991).

58. IBAYASHI. S., A.C. NGAI, J.R.MENO and H.R. WINN: Effects of Tropical Adenosine Analogs and Forskolin on Rat Pial Arterioles *in vivo*. J. Cerb., Blood flow Metab., 11, 72–6 (1991).

59. LINDGREN. S. and K.E. ANDERSSON: Comparison of the Effects of Milrinone and OPC 3911 with those of Isoprenaline, Forskolin and Dibutyryl-cAMP in Rat Aorta. Gen. Pharmacol, 22, 617–24 (1991).

60. WILLIAMS. J.L. Jr. and K.V. MALIK: Forskolin stimulates prostaglandin Synthesis in Rabbit Heart by a Mechanism that Requires Calcium and is Independent of Cyclic AMP, Circ. Res., 67, 1247–56 (1990).

61. GAUTHIER. C. and H. SOUSTRE: Forskolin Effects on Slow Inward Current and Phasic Tension of Frog Atrial Fiber Modulation of Adenosine and Phosphodiesterase Inhibitors, Eur. J. Pharmacol, Mol. Pharmacol. Sect., 225, 129–35 (1992).

62. SONOKI. H.V. and Y.MASUO: Effects of Forskolin on Canine Congestive Heart Failure. Nippon Yakurigaku Zasshi., 88, 389–94 (1986).

63. LEMMER. B., H. BISSENGER and P.H. LARY: Effects of Forskolin on cAMP Levels in Rat Heart at Different Times of Day. IRCS Med. Sci., 14, 1103 (1986).

64. VEGESNA. R.V.K. and J. DIAMOND: Effects of Prostaglandin E_1, Isoproterenol and Forskolin on cAMP Levels and Tension in Rabbit Aortic Rings. Life Sci., 39, 303–11 (1986).

65. VEGESNA. R.V.K. and J. DIAMOND: Activation of cAMP Dependent Protein Kinase in Rabbit Aortic Rings by Prostaglandin E_1, and Forskolin is Accompanied by Contraction and Relaxation, Respectively. Proc. West. Pharmacol. Soc., 29, 39–43 (1986).

66. HATJIS. C.G.: Forskolin, Unique Diterpene Activator of Adenylate Cyclase in Pregnant and Nonpregnant Guinea Pig Myometrial Membranes. Am. J. Obstet. Gynecol., 155, 1202–8 (1986).

67. HATJIS. C.G.: Forskolin-stimulated Adenylate Cyclase Activity in Fetal and Adult Rabbit Myocordial Membranes. Am. J. Obstet. Gynecol., 155, 1326–31 (1986).

68. HON. M., Y. KORETSUME. T. KAGIYA. Y. WATNABE. K. IWAKURA, K. IWAI. A. KITABATAKE. H. YOSHIDE. M. INOUE and T. KAMADA: An Increase in Myocordial β-Adrenoceptors to Compensate for Postischemic Dysfunction Following Coronary Microembolization in Dogs, Cardiovasc. Res., 23, 424–31 (1989).

Stomach and Parotid Gland

69. CHOQUET. A., R. MAGOUS. J.C. GALLEYRAND and J.P. BALI: Is Forskolin a True Stimulant of Gastric acid Secretion. C.R. Seances Soc., Biol. Sec, Fil., 182, 335–43 (1988).

70. CORUZZI. G., M. ADAMI and G. BERTACCINI: Effect of Forskolin on Gastric Acid Secretion '*in vitro*' Interaction with Different Secretagogues, Gen. Pharmacol., 19, 767–70 (1988).

71. FONG. J.C., L.T. HO and F.F. WANG: Forskolin Stimulation of Pepsinogen Secretion from Frog Esophageal Mucosa is Partly Mediated by Intrinsic Cholinergic Neurons. Biochem. Int., 19, 1165–72 (1989).

72. REN. Y.F., S.N. AHMED. B.S. ALAM and S.Q. ALAM: [^{3}H]-Forskolin Binding Sites in Rat Submandibular Salivary glands. Arch. Oral. Biol., 33, 779–82 (1988).

73. MURATA. K and Y. OGURA: Effect of Isoproterenol and Forskolin on Amylase Release from Parotid Tissue after Chronic Pilocarpine Administration in Rats Following Ligation removal. Jpn. J. Pharmacol., 45, 545–9 (1987).

74. HERLING. A.W. and M. BICKEL: The Stimulatory Effect of Forskolin on Gastric Acid Secretions in Rats. Eur. J. Pharmacol., 125, 233–9 (1986).

75. POAT. J.A., H.E. CRIPPS. R. COWBURN and L.L. IVERSON: Synergistic Interactions

Between Forskolin, Isoprenaline and Substance P as Secretagogues in Rat Parotid Glands. Eur. J. Pharmacol., **144**, 317–26 (1987).

76. WILSON. G.A. and I.H.M. MAIN: Stimulatory Effect of Forskolin Gastric Acid Secretion in the Rat. *in vitro* and *in vivo*, Eur. J Pharmacol.,**123**, 371–7 (1986).

77. TAKUMA. T.: Propranolol Inhibits Cyclic AMP Accumulation and Amylase Secretion in Parotid Acinar Cells Stimulated by Isobutylmethylxanthine and Forskolin. Biochem. Biophys. Acta, **1052**, 461–6 (1990)

78. ISHIKAWA. T., H. OMOTANI. Y. NEZANI and T. ETO: Effects of Cimetidine and Ranitidine in Forskolin Stimulated Acid Secretion in Isolated Parietal Cells. Ther. Res., **11**, (Suppl 1), 120–3 (1990).

79. WATSON. E.L. and K.L. JACOBSON: Forskolin Activation of Adenylate Cyclase in Mouse Parotid Membranes. Life Sci., **39**, 693–7 (1986).

Eye

80. SHIBATA. T. and H. MISHIMO: Ocular Pigmentation and Intraocular Pressure Response to Forskolin. Curr. Eye Res., **7**, 667–74 (1988).

81. ZARBIN. M.A., J. BARABAN and P. WORLEY: Autoradiographic Distribution of Forskolin and Phorbol Ester Binding Sites in the Retina. Brain Res., **497**, 334–43 (1989).

82. SAETTONE. M.F., S. BURGALASSI and B.G. IANNACCINI: Preparation and Evaluation in Rabbits of Topical Solution containing Forskolin. J. Ocul. Pharmacol., **5**, 111–18 (1989).

83. BRUBAKER, R.F., K.H. CARLSON. H. KEITH, J. KULLERSTRAND and J.W. MACLAREN: Topical Forskolin (Colforsin) and Aqueous Flow in Humans. Arch. Ophthalmol, **105**, 637–41 (1987).

84. CHU. J.C. and A. CANDIO: Effects of Forskolin, Prostaglandin $F_{2\alpha}$ and Barium (2^+) Short Circuit Current of the Isolated Rabbit Iris Ciliary Body. Curr. Eye Res., **5**, 511–16 (1986).

85. GOLDMAN. H.E., P. MELLORGA. D.J. PETTIBONE and M.F. SUGRUE: Characterization of [^{3}H]-Forskolin Binding Sites in the Iris Ciliary Body of the Albino Rabbit. Life Sci., **42**, 1307–14 (1988).

86. JARKMAN. S.: Effects of Low Doses of Forskolin on the C-Wave of the Direct Current Electroretinogram and on the Standing Potential of Eye. Doc. Ophthalmol., **67**, 305–14 (1987).

87. EGUCHI. S., S. MATSUMOTO. S. KOYANO and M. TAKASE: Effect of Topical Forskolin on Intraocular Pressure of Normal Volunteers. Atarashii Ganka, **3**, 537–9 (1986) (Japan), Chem. Abstr. **105**: 127424h.

88. HOSOKAWA. T., T. YAMASHITA. Y. KASUYA. S. YANURA. S. MATSUMOTO. M. ARAIE and M. TAKASE: Effects of Forskolin and Timolol on Cyclic AMP in Aqueous Humor and Intraocular Pressure. Atarashii, Ganka, **4**, 543–6 (1987) (Japan) Chem. Abstr. 107: 89867m.

89. SHIBATA. T.: Experimental Studies on Changes of Intraocular Pressure by the Application of Forskolin II. Intraocular Pressure Changes in Albino and Pigmented Rabbits after Treatment with 1% Forskolin. Hiroshima Daigaku Igaku Zasshi, **35**, 171–7 (1987) (Japan) Chem. Abstr 107 168772e.

90. SHIBATA. T.: Experimental Studies on Changes of Intraocular Pressure by the Application of Forskolin III. Binding to Melanin and Ocular Toxicity of Forskolin, Hiroshima Daigaku Igaku Zasshi, **35**, 179–87 (1987) (Japan). Chem. Abstr. **107**: 168773f.

91. MEYER. B.H., A.A. STULTING, F.O. MUELLER, H.G. LUUS and M. BADIAN: The Effects of Forskolin Eye Drops on Intraocular Pressure. SAMJ, **71**, 570–1 (1987).

92. SETO, C., S. EGUCHI, M. ARAIE, S. MOTASUMOTO and M. TAKASE: Acute Effects of Topical Forskolin on Aqueous Humor Dynamics in Man. Jpn. J. Ophthalmol., **30**, 238–44 (1986).

Kidney

93. TAMAKI, T., K. HASUI, T. SHOJI, Y. AKI, H. KIYOMOTTO, H. IWAO and Y. ABE: Forskolin Preferentially Dilates the Afferent Arteriole in the Canine Kidney. Jpn. J. Pharmacol. **55**, 161–4 (1991).
94. NAKAMURA, K.T., B.M. ALDEN, G.P. METHERNE, P.A. JOSE and J.E. ROBELLARD: Ontogeny of Renal Hemodynamic Response to Terbutaline and Forskolin in Sheep. J. Pharmacol. Exp. Ther., **247**, 453–9 (1988).

Liver

95. HAMLIN, S., K. RAHMAN, M. CARRELLA and R. COLEMAN: Modulation of Biliary Lipid Secretion by Forskolin and Cyclic AMP Analogues. Biochem. J. **265**, 879–85 (1990).
96. AL-TURK, W.A., O. SHAHEEN and S. OTHMAN: Effect of Forskolin on cAMP Accumulation and Ketogenesis in Isolated Hepatocytes. Gen. Pharmacol., **17**, 577–80 (1986).
97. TZANAKAKIS, G.N., K.C. AGARWAL and M.P. VEZERIDIS: Inhibition of Hepatic Metastasis from a Human Pancreatic Adenocarcinoma (RWP-2) in the Nude Mouse by Prostacyclin, Forskolin and Ketoconazole. Cancer (Philadelphia), **65**, 446–51 (1990).
98. SAITOH, R., Z. MINAMI, S. KAWATA, S.MIYOSHI K. TAJIMA, K. MASHITA, K. MORIWAKI and S. TARUI: The Inhibitory Effect of Forskolin on Antibody Dependent Cell Mediated Cytotoxicity Using Chang Liver Cells as Target Cells. Life Sci., **42**, 239–45 (1988).

Muscle

99. ABE, A. and H. KARAKI; Inhibitory Effects of Forskolin on Vascular Smooth Muscle of Rabbit Aorta. Jpn. J. Pharmacol., **46**, 293–301 (1988).
100. ENEINA, J.L. and F. HARTUNG: Analysis of the Hyperpolarising Effects of Forskolin in Guinea Pig Atrial Heart Muscle, Naunyn-Schmiedeberg's Arch. Pharmacol., **337**, 435–8 (1988).
101. MORITA, T., S. KONDO, S. DOHKITA and S. TSUCHIDA: The Time course of Changes in Force and Cyclic AMP Levels Produced by Isoproterenol and Forskolin in Isolated Rabbit Detrusor Muscle. Tohoku J. Exp. Med., **151**, 201–4 (1987).
102. TSUKAWAKI, M., K. SUZUKI, R. SUZUKI, K. TAKAGI and T. SATAKE: Relaxant Effects of Forskolin on Guinea Pig Tracheal Smooth Muscle. Lung. **165**, 225–37 (1987).
103. MORITA, T. and M.A. WHEELER: Relaxant Effect of Forskolin in Rabbit Detrusor Smooth Muscle; Role of cAMP. J. Urol. (Baltimore), **135**, 1293–5 (1986).
104. MORITA, T., M.A. WHEELER, I. MIYAGAWA, S. KONDO and R.M. WEISS: Effects of Forskolin on Contractibility and Cyclic AMP Levels in Rabbit Detrusor Muscle, Tohoku J. Exp. Med., **149**, 283–5 (1986).
105. KEMPASKI, O., B. WROBLEWSKA and M. SPATZ; Effects of Forskolin on Growth and Morphology of Cultured Glial and Cerebrovascular Endothelial and Smooth Muscle Cells. Int. J. Dev. Neurosci, **5**, 435–45 (1987).
106. POAT, J.A., H. CRIPPS and L.L. IVERSON: Interactions of Dopamine and Forskolin in Rat Streatal Tissue. Symp. Neurosci, **5**, 161–9 (1988).
107. XIAO, R.P. and W.C. DeMELLO: Intracellular Resistance in Rat Papillary Muscle:

Interaction Between Cyclic AMP and Calcium. J. Cardiovasc. Pharmacol., **17**, 754–60 (1991).

108. ABE, A. and H. KARAKI: Effect of Forskolin on Cystolic Calcium Level and Contraction in Vascular Smooth Muscle, J. Pharmacol. Exp. Ther. **249**, 895–900 (1989).

109. OZAKI, H., S.C. KWON, M. TAJIMI and H. KARAKI: Changes in Cystolic Calcium and Contraction Induced by various Stimulants and Relaxants in Canine Tracheal Smooth Muscle. Pflüegers Arch., **416**, 351–9 (1990).

Nervous Tissue

110. WOMBLE, M.D. and W.O. WICKELGREN: Activation of Adenylate Cyclase by Forskolin Prolongs Calcium Action Potential Duration in Lamprey Sensory Neurons. Brain Res., **485**, 89–94 (1989).

111. NAKAGAWA-YAGI, Y., Y. SAITO, Y. TAKADA and M. TAKAYAMA: Carbachol Enhances Forskolin-stimulated Cyclic AMP Accumulation via Activation of Calmodulin System in Human Neuroblastoma SH-SY5Y Cells. Biochem. Biophys. Res. Commun. **178**, 116–23 (1991).

112. LANDO, M., E. ABEMAYOR, M.A. VERIETY and N. SIDELL: Modulation of Intracellular Cyclic Adenosine Monophosphate Levels and the Differentiation Response of Human Neuroblastoma Cells. Cancer Res. **50**, 722–7 (1990).

113. MURPHY, M.G. and Z. BYCZKO; Effects of Adenosine Analogues on Basal, Prostagland in E_2 and Forskolin-stimulated Cyclic AMP Formation in Intact Neuroblastoma Cells. Biochem. Pharmacol., **38**, 3289–95 (1989).

114. WEISS, S.: Forskolin Alternates the Evoked Release of [^{3}H]-GABA From Striated Neurons in Primary Culture. Brain Res., **463**, 182–6 (1988).

115. UHLEN, S. and J.E.S. WIKBERG: α-Adrenoceptor Mediate Inhibition of cAMP Production in the Spinal Cord after Stimulation of cAMP with Forskolin but not after Stimulation with Capscicin or Vasoactive Intestinal Peptide. J. Neurochem., **52**, 761–7 (1989).

116. UHLEN, S and J.E.S. WIKBERG: Relationship between Forskolin and Calcium Calmodulin Stimulation of Rat Cerebral Cortex Adenylate Cyclase: Enzyme Activation Modulates Substrate (Magnesium-ATP) Affinity. Pharmacol. Toxicol (Copenhagen) **63**, 90–5 (1988).

117. UHLEN, S. and J.E.S. WIKBERG: Inhibition of Cyclic AMP Production by α_2-Adrenceptor Stimulation in the Guinea Pig Spinal Cord Slices. Pharmacol. Toxicol. (Copenhagen), **63**, 178–82 (1988).

118. WAKADE, A.R., S.V. BHAVE, R.K. MALHOTRA and T.D. WAKADE: Forskolin Mediates the Survival of Nerve Growth-factor-dependent Sympathetic Neurons of Chick Embryo by a Cyclic AMP Independent Mechanism, J. Neurochem. **54**, 1281–7 (1990).

119. TISCHLER, A.S., L.A. RUZICKA and R.L. PERLMAN: Mimicry and Inhibition of Nerve Growth Factor Effects: Interactions of Staurosporine, Forskolin and K252a in PC 12 Cells and Normal Rat Chromaffin Cells *in vitro*. J. Neurochem., **55**, 1159–65 (1990)

120. GEORGIEVA, Z.: Effects of the Diterpene Sclareol Glycol on Convulsive Seizures. Methods Find. Exp. Clin. Pharmacol., **11**, 335–40 (1989).

121. MARRIOTT, D., M. ADAMS and M.R. BOARDER: Effect of Forskolin and Prostaglandin E_1 on Stimulus Secretion Coupling in Cultured Bovine Adrenal Chromaffin Cells. J. Neurochem., **50**, 616–23 (1988).

122. MOREK, A. and A. GEISLER: Calmodulin-dependent Adenylate Cyclase Activity in Rat Cerebral Cortex. Effects of Divalent Cations, Forskolin and Isoprenaline. Arch. Int. Physiol. Biochem., **97**, 259–71 (1989).

123. DONALDSON, J., D.A. KENDALL and S.J. HILL: Discriminatory Effects of Forskolin and EGTA on the Indirect Cyclic AMP Responses to Histamine, Noradrenaline, 5-Hydroxytryptamine and Glutamate in Guinea Pig Cerebral Cortical Slices. J. Neurochem., **54**, 1484–91 (1990).

124. MOBBS, C.V., J.M. ROTHFELD, R. SALUJA and D.W. PFAFF: Phorbol Esters and Forskolin Infused into Midbrain Central Gray Facilitate Lordosis. Pharmacol. Biochem. Behav., **34**, 665–7 (1989).

125. DANURA, T., T. KUROKAWA, A. YAMASHITA and S. ISHIBASHI: Effective Inhibition by Pentobarbital of Forskolin Stimulated Adenylate Cyclase Activity in Rat Brain. Chem. Pharm. Bull. **37**, 3142–4 (1989).

126. GUILLEN, A., A. HARO and A.M. MUNICO: Regulation by Forskolin of Octopamine Stimulated Adenylate Cyclase from Brain of the Dipterous *Ceratitis capitata*. Arch. Biochem. Biophys. **254**, 234–40 (1987).

127. SRIVASTAVA, A.M. and A.K. SRIVASTAVA: Modulation of Adenylate Cyclase Activity by Calcium Phospholipid Dependent Protein Kinase in Rat Brain. Mol. Cell. Biochem., **92**, 91–8 (1990).

128. STÜBNER, D. and R.A. JOHNSON: Forskolin Decreases Sensitivity of Brain Adenylate Cyclase to Inhibition of 2',5'-Dideoxyadenosine. FEBS Letters, **248**, 155–61 (1989).

129. WYSHAM, D.G., A.F. BROTHERTON and D.D HEISTAD: Effects of Forskolin on Cerebral Blood Flow; Implications for a Role of Adenylate Cyclase. Stroke (Dallas) **17**, 1299–303 (1986).

130. POAT, J.A., H.E. CRIPPS and L.L. IVERSEN: Differences between High-affinity Forskolin Binding Sites in Dopamine-rich and Other Regions of Rat Brain. Proc. Natl. Acad. Sci. U.S.A., **85**, 3216–20 (1988).

Pancreas

131. IWATSUKI, K., A. HORIUCHI, F. YAMAGISHI and S. CHIBA: Effects of Forskolin on Pancreatic Exocrine Secretion and Cyclic Nucleotide Concentration of the Dog Pancreas. Arch. Int. Pharmacodyn. Ther., **286**, 320–8 (1987).

132. ANAZODO, M.I., A.B. MUELLER, H. SAFAYHI and H.P.T. AMMON: Potentiation of Forskolin Induced Increase of cAMP by Diamide and N-Ethylmaleimide in Rat Pancreatic Islets. Horm. Metab. Res. **22**, 61–4 (1990).

Bone

133. CONAWAY, H.H., R.L. ABRAHAM and G.L. WADKINS: Effects of Forskolin on Bone Resorption in the Absence and Presence of Parathyroid hormone and Calcitonin. Calcif. Tissue Int., **40**, 276–81 (1987).

134. KSIEGER, N.S. and P. STERN: Effects of Forskolin on Bone in Organ Culture. Am. J. Physiol., **252**, E 44–48 (1987).

135. LORENZO, J.A., S. SOUZA and J. QUINTON: Forskolin has Both Stimulatory and Inhibitory Effects on Bone Resoption in Fetal Rat Long Bone Cultures. J. Bone Miner. Res., **1**, 313–17 (1986).

136. MALEMUD, C.J., T.M. MILLS, R. SHUCKETT and R.S. PAPAY: Stimulation of Sulfated Proteoglycan Synthesis by Forskolin in Monolayer Cultures of Rabbit Articular Chondrocytes. J. Cell. Physiol. **129**, 51–9 (1986).

137. HU, L.M., S.F. KEMP, M.J. ELDERS and W.G. SMITH: Effect of Forskolin on Synthesis of Xyloside-initiated Glucosaminoglycans in Embryonic Chick Chondrocytes. Biochem. Biophys. Acta, **1051**, 112–14 (1990).

138. HU, L.M., S.F., KEMP, M.J. ELDERS and W.G. SMITH: Metabolic Effects of Forskolin in Chick Chondrocytes. Biochem. Biophys. Acta, **1013**, 294–9 (1989).
139. SECHENSKA, M. and A. TIMANOVA: Forskolin as an Activator of Adenylate Cyclase Complex Differentiating Erythroid Bone-marrow Cells. Acta Physiol. Pharmacol. Bulg., **16**, 50–6 (1990).
140. HAKEDA, Y., Y. NAKATANI, T. YOSHINO, N. KURIHORA, K. FUJITA, N. MEEDA and M. KUMEGAWA: Effect of Forskolin on Collagen Production in Clonal Osteoblastic MC_3T_3-EI Cells. J. Biochem., **101**, 1463–9 (1987).

Respiratory Tissue

141. FULLER, R.W., G.O.MALLEY, A.J. BAKER and J. MACDERMOT: Human Alveolar Macrophage Activation Inhibition by Forskolin but not by β-Adrenoreceptor Stimulation or Phosphodiesterase Inhibition. Pulm. Pharmacol., **1**, 101–6 (1988).
142. HEASLIP, R.J., F.R. GIESA, T.J. RIMELE and D. GRIMES: Sensitivity of the $PGF_{2\alpha}$ versus Carbachol-contracted Trachea to Relaxation by Salbutamol, Forskolin and Prenalterol. Eur. J. Pharmacol., **128**, 73–9 (1986).
143. GIEMBYCZ, M.A. and J. DIAMOND: Partial Characterization of cAMP Dependent Protein Kinase in Guinea Pig Lung Employing the Synthetic Heptapeptide Substrate, Kemptide. *In vitro* Sensitivity of the Soluble Enzyme to Isoprenaline, Forskolin, Methacholine and Leukotriene D_4. Biochem. Biopharmacol., **39**, 1297–312 (1990).
144. UNDEM, B.J. and C.K. BUCKNER: Effects of Forskolin Alone and in Combination with Isoproterenol on Antigen Induced Histamine Release from Guinea Pig Minced Lung. Arch Int. Pharmacodyn. Ther., **281**, 110–19 (1986).

Reproductive Tissue

145. NISHIKAWA, T., A. SATO, T. KANAI, T. KASAJIMA, Y. NAKAJIMA, M. ANDO and A. TAKAO: The Teratogenic Effect of Forskolin on Cardiovascular Development in the Chick Embryo. Reprod. Toxicol, **3**, 139–42 (1989).
146. FUCHS, V. and H.H. RIEDAL: The Effect of Forskolin on Sperm Motility. Andrologia, **21**, 293–6 (1989).
147. RUI, H., P. KENNETH, and J.O. GORDDADZE: Sperm Adenylyl cyclase in Young and Middle-aged men. Andrologia, **21**, 131–5 (1989).
148. RANKIN, J.H.G., M. LANDAUER, Q. TIAN and T.M. PHERNETTON: Cardiovascular Responses to Forskolin in the Ovine Fetus, J. Dev. Physiol, **11**, 7–10 (1989).
149. CORUZZI, G., E. POLI, C. MONTANARI and G. BERTACCINI: Pharmacological Charaterization of Mare Uterus Motility with Special Reference to Calcium Antagonists and Beta-2-adrenergic Stimulants. Gen. Pharmacol, **20**, 513–18 (1989).
150. REID, D.L., M.C. HOLLISTER, T.M. PHERNETTON and J.H.G. RARKIN: Effects of Forskolin on Placental Vascular Resistance in Rabbits. Proc. Soc. Exp Biol. Med., **188**, 451–4 (1988).
151. BAUM, M.G.S. and E.B. AHREN: Effects of Forskolin, Luteinizing Hormone and Prostaglandin F_2 on Isolated Rat Corpora Lutea. Acta Endocrinol. (Copenhagen), **112**, 571–8 (1986).
152. REID, D.L., T.M. PHERNETTON and J.H.G. RANKIN: Ovine Fetal Coronary and Cerebral Vascular Responses to Forskolin. J. Dev. Physiol., **12**, 63–5 (1989).
153. DEKLAC, L., A. MOKHTARI and S. HARBON: A Re-evaluated Role for cAMP in uterine Relaxation. Differential Effect of Isoproterenol and Forskolin. J. Pharmacol. Exp. Ther., **239**, 236–42 (1986).

Urinary Bladder

154. CASAVOLA, V., G. CALAMITA, G. VALENTI and M. SVELTO: Some Characteristics of Forskolin Actions on Osmotic Water Flow in Frog Urinary Bladder. Med. Sci. Res., **16**, 373–4 (1988).

155. LIPPE, C. and C. ARDIZZONE: Action of Forskolin on Non-electrolyte Permeability Across the Urinary Bladder of Bufo bufo as Compared to that of Various Hormones. Gen. Pharmacol., **19**, 513–14 (1988).

Hormone

156. GUILD, S., Y. ITOH, J.W. KEBABIAN, A. LUIM and J. REISINE: Forskolin Enhances Basal and K^+ Evoked Hormone Release from Normal and Malignant Pituitary Tissues: The role of Ca^{2+}. Endocrinology (Baltimore), **118**, 268–79 (1986).

157. RAY, K.P., J.J. GOMM, G.J. LAW, C. SIGOURNAY and M. WALLIS; Dopamine and Somatostatin Inhibit Forskolin Stimulated Prolactin and Growth Hormone Secretion but not Stimulated Cyclic AMP Levels in Sheep Anterior Pituitary cell culture. Mol. Cell. Endocrinol., **45**, 175–82 (1986).

158. YUN, K., S. YAMASHITA, K. IZUMI, N. YONEMITSU and H. SUGIHARA: Effects of Forskolin on the Morphology and Function of the Rat Thyroid Cell Strain FRTL-5: Comparison with the Effects of Thyrotropin. J. Endocrinol, **111**, 397–405 (1986).

159. GUILLON, G., M.N. BALESTRE, C. LOMBARD, F. RESSENDREN and C.J. KIRK; Influence of Bacterial Toxins and Forskolin upon Vasopressin–induced Inositol Phosphate Accumulation in WRK1 Cells. Biochem. J., **260**, 665–72 (1989).

160. AHMAD, F., M.M. KHAN, A.K. RASTOGI and J.R. KIDWAI: Insulin and Glucagon Releasing Activity of Coleonol (Forskolin) and Its Effect on Blood Glucose Level in Normal and Alloxan Diabetic Bats. Acta Diabetal Lat., **28**, 71–77 (1991).

161. VALENSI, P., B. LILIEVRE, D. SANDRE-BANON and J.R. ATTALI: Stimulating Effect of Forskolin on Thyroxine Secretion, Influence of Calcium. Pathol. Biol., **39**, 205–10 (1991).

162. MENENDEZ-PELAEZ, A., G.R. BUZZELL, K.O. NONAKA and R.J. REITER: *In vivo* Administration of Isoproterenol or Forskolin during the Light Phase Induces increases in the Melatonin Content of the Syrian Hamster Pineal Gland without a Rise in N-Acetyl Transferase Activity. Neurosci. Letters, **110**, 314–18 (1990).

163. O'TOOLE, L.B., K.J. ARMOUR, C. DECOURT, N. HAZON, B. LALILOU and I.W. HENDERSON: Secretory Pattern of 1α-Hydroxycarticosterone in the Isolated Perfused Internal Gland of Dog Fish. *Scyliorhinus canicula*, J. Mol. Endocrinol., **5**, 55–60 (1990).

164. ADASHI, E.Y. and C.E. RESNICK; 3',5'-cyclic Adenosine Monophosphate as an Intracellular Second Messenger of luteinizing Hormone; Application of the Forskolin Criteria. J. Cell. Biochem., **31**, 217–18 (1986).

165. CLARK, K.L., G.M. DREW and A.HILDITCH: Potentiation of the Effects of Dopamine in the Rabbit Isolated Splenic Artery by 3-Isobutyryl-1-methylxanthine or Forskolin. Nauny-Schmiedeberg's, Arch. Pharmacol., **340**, 533–40 (1989).

166. WHALEN, M.M. and A.D. BANKHURST: Effect of β-Adrenergic Receptor Activation, Cholera Toxin and Forskolin on Human Natural Killer Cell Function, Biochem. J., **272**, 327–31 (1990).

167. SANTANA, C., J.M. GUERRERO, R.J. REITER, A. GONZALEZ-BRITO and A. MENENDEZ-PELAEZ: Forskolin, an Activator of Adenylate Cyclase Activity, Promotes Large Increase in N-Acetyl Transferase Activity and Melatonin Production in the Syrian

Hamster Pineal Gland only during the Late Dark period. Biochem. Biophys. Res. Commun **115**, 200–15 (1988).

168. FRANKYLN, J.A., M. WILSON and J.R. DATRES: Demonstration of Thyrotropin B Subunit Messenger RNA in Rat Pituitary Cells in Primary Culture. Evidence for Regulation by Thyrotropin Releasing Hormone and Forskolin. J. Endocrinol., **111**, R_1-R_2 (1986).

169. NIKULA, H. and I. HUHTANIEMI: Effect of Protein Kinase C Activation on Cyclic AMP and Testosterone Production of Rat Leyding Cells *in vitro*. Acta Endocrinol., **121**, 327–33 (1989).

Ion Channels

170. NISHZAWA, Y., K.B. SEAMON, J.W. DALY and R.S. ARONSTAM: Effects of Forskolin and Analogues on Nicotinic-receptor Mediated Sodium Flux, Voltage Dependent Calcium Flux and Voltage Dependent Rabidium Efflux in Pheochromocytoma PC 12 Cells. Cell. Mol Neurobiol., **10**, 351–68 (1990).

171. SEN, R.P., E.G. DELICADO and M.T. MIRAS-PORTUGAL: Effect of Forskolin and Cyclic AMP Analogues on Adenosine Transport in Cultured Chromaffin Cells. Neurochem. Int., **17**, 523–8 (1990).

Transcription-Translation

172. SHUPNIK, M.A., B.A. ROSENZWEIG and M.O. SHOWERS: Interactions of Thyrotropin Releasing Hormone, Phorbol Ester and Forskolin Sensitive Regions of the Rat Thyrotropin-β Gene. Mol Endocrinol, **4**, 829–36 (1990).

173. MUEHL, H., T. GEIGER, W. PIGNAT, F. MAERKI, H. V. BOSCH, K. VASBECK and J. PFEILSCHIFTER: PDGF Supresses the Activation of Group II Phospholipase A_2 Gene Expression by interleukin 1 and Forskolin in Mesangial Cells. FEBS Lett., **291**, 249–52 (1991).

174. BERTRAND, S., D. KARIM. D. BRIGITTE, J. CLAUDE and C. CLAUDE: Cyclic AMP Regulation of Gs Protein. Thyrotropin and Forskolin Increase the Quantity of Stimulatory Guanine Nucleotide Binding Proteins in Cultured Thyroid Follicles. J. Biol. Chem., **265**, 19942–6 (1990).

175. AIZAWA, T. and A. NOZAWA: Phorbol Ester Regulates the Abundance of Enkephalin Precursor mRNA but not of Amyloid β-Protein Precursor mRNA in Rat Testicular Peritubular Cells. Biochem. Biophys. Res. Comm., **16**, 568–75 (1989).

176. MOENS, U., A. SUNDSFJORD, T. FLAEGSTAD and T. TRAAVIK: BK virus early RNA Transcripts in Stably Transformed Cells: Enhanced Levels Induced by Dibutyryl Cyclic AMP, Forskolin and 12-O-Tetradecanoyl-phorbol-13-acetate Treatment. J. Gen Virol. **71**, 1461–71 (1990).

177. REUSE, S., I. PRISON and J.E. DUMONT: Differential Regulation of Protooncogenes C-jun and jun-D Expressions by Protein Kinase C and Cyclic AMP, Mitogenic Pathways in Dog Primary Thyrocytes: TS11 and cyclic AMP Induce Proliferation but Downgrade C-jun Expression. Exp. Cell. Res., **196**, 210–15 (1991).

178. LOEFFLER, J.P., N. KLEY, C.W. PITTIUS and V. HOELLT: Corticotropin Releasing Factor and Forskolin Increase Proopiomelanocortin Messenger RNA Levels in Rat Anterior and Intermediate Cells *in vitro*. Neurosci., Letters, **62**, 383–7 (1986).

179. KIIL, B.H., E. RUUD. S. FUNDERUD and T. GODAL: Distinct Effect of Forskolin and Interferon-y on Cell Proliferation and Regulation of Histocompatibility Antigen Expression in Hematopoietic Cells. Biochem. Biophys. Acta, **887**, 150–6 (1986).

180. SCHALKWIJK, C. J. PFEILSCHIFTER, F. MAERKI and H. VAN DEN BOSCH: Interleukin-1β, Tumor Necrosis Factor and Forskolin Stimulate the Synthesis and Secretion of Group II Phospholipase A_2 in Rat Masangial Cells. Biochem. Biophys. Res., Commun., **174**, 268–75 (1991).

181. HARSH, G.R., W.M. KAVANAUGH and N.F. STARKSEN: Cyclic AMP Blocks Expression of the C-sis Gene in Tumor Cells. Oncg. Res., **4**, 65–73 (1989).

182. CHOI, H.S., B. LI, Z. LIN, L.E. HUANG and A.Y.C. LIU: cAMP and cAMP Dependent Protein Kinase Regulate the Human Heat Shock Protein 70 Gene Promoter Activity. J. Biol. Chem., **266**, 11858–65 (1991).

Skin

183. MATSUO, S. and H. IIZUKA: Cholera Toxin and Forskolin Induced Cyclic AMP Accumulations of Pig Skin (Epidermis). Modulation by Chemicals which Reveal the beta-Adrenergic Augmentation Effect. J. Dermatol. Sci., **1**, 7–13 (1990).

184. BONIC, A., K. KOURIS, N. RAJACIC, M. NAZAL and O. THULESIUS: Increased Skin Flap Viability after Treatment with Forskolin or with Ridogrel, a Thromboxane Synthesis Inhibitor and Receptor Blocker. Res. Exp Med., **190**, 223–7 (1990).

185. IIZUKA, H., S. MATSUO, T. TAMURA and N. OHKUMA: Increased Cholera Toxin and Forskolin Induced cAMP Accumulations in Psoriatic Involved versus Uninvolved or Normal Human Epidermis, J. Invest. Dermatol., **91**, 154–7 (1988).

Miscellaneous

186. BROOKER, G. and C. PEDONE: Maintenance of Whole Cell Isoproterenol and Forskolin Responsiveness in Adenylate Cyclase of Permeabilised Cells. J. Cyclic Nucleotide protein phosphorylation Res., **11**, 113–21 (1986).

187. HO, L.J. and R.J. HO: Production and Assay of Antibodies to an Activator of Adenylate Cyclase Forskolin. J. Cyclic Nucleotide Protein Phosphorylation Res., **11**, 421–32 (1987).

188. KITAJIMA, S., M. SANO, K. KATO, A. SETO-OHSHIMA and A. MIZUTANI: Changes of Calmodulin and S-100 Protein in C_6 Glioma Cells as a Result of Cellular Differentiation Induced by Forskolin. Acta Histochem. Cytochem., **19**, 365–9 (1986).

189. McCULLOCH, A.J., T.A. THOMSON, R. DEACON and C.R. GARDNER: Hypoxic Amnesia and its Reversal with Forskolin. Biochem. Soc. Trans. **17**, 212–13 (1989).

190. CORNFIELD, I.J., D.L. NELSON, P.J. MONROE, E.W. TAYLOR and S.S. NIKAM: Use of Forskolin Stimulated Adenylate Cyclase in Rat Hippocampus as a Screen for Compounds that Act Through 5-HTR_{1A} Receptors, Proc. West. Pharmacol. Soc., **31**, 265–7 (1988).

191. LIPPINCOTT-SCHWARTZ, J., J. GLICKMAN, J.G. DONALDSON, J. ROBBINS T.E. KREIS, K.B. SEAMON, M.P. SHEETZ and R.D. KLAUSNER: Forskolin Inhibits and Reverses the Effects of Brefeldin A on Golgi Morpholgy by cAMP Independent Mechanism. J. Cell. Biol., **112**, 567–77 (1991).

192. TERTRIN-CLARY, C., M. ROY, L. DE and P. LLOSA: Action of Manganese (2^+) and Vanadium Compounds on Hormone and Forskolin Induced Stimulation of Juncile Rat Ovarian Adenylate Cyclase. Biochem. Int., **13**, 1019–35 (1986).

193. TAKEDA, O., K. YAMASA, K. KOHDA, A. YAMASHITA, T. KUROKAWA and S. ISHIBASHI: The Inhibitory Effect of Methanol on Forskolin Activated Adenylate Cyclase in Rat Erythrocyte Membrane Dependent on the State of the Guanine Nucleotide Binding Stimulatory and Regulatory Protein. J. Pharmacobio. Dyn., **11**, 377–80 (1988).

194. Marone, G., M. Columbo, M. Triggiani, S. Vigorita and S. Formisano: Forskolin Inhibits the Release of Histamine from Human Basophil and Mast Cells. Agents Actions, **18**, 96–9 (1986).
195. Szabo, G., P.L. Hoffman and B. Tabakoff: Forskolin Promotes the Development of Ethanol Tolerance in 6-Hydroxydopamine Treated Mice. Life Sci., **42**, 615–21 (1988).

(Received June 22, 1992)

Steroidal Oligoglycosides and Polyhydroxysteroids from Echinoderms

L. MINALE, R. RICCIO, and F. ZOLLO

Dipartimento di Chimica delle Sostanze Naturali Università di Napoli
"Federico II", Napoli, Italy

Contents

1. Introduction

The phylum Echinodermata, which comprises about 6,000 living species distributed in all seas from the tropics to the antarctic zone, is divided into five classes: Crinoidea (sea lilies and feather stars), Holothuroidea (sea cucumbers or holothurians), Echinoidea (sea urchins), Asteroidea (sea stars or starfishes) and Ophiuroidea (brittle stars). Among the echinoderms, sea cucumbers and starfishes invariably contain saponins which are generally responsible for their toxicity. Saponins, complex water soluble compounds composed of a carbohydrate moiety attached to a steroid or triterpenoid aglycone, have been isolated from a great number of terrestrial plants, but are uncommon animal constituents. In the animal kingdom they are almost ubiquitous in sea cucumbers and starfishes (*1*), while found only rarely in alcyonarians (*2*), gorgonians (*3–5*), sponges (*6–10*) and as potent shark-repelling compounds in fishes (*11*). Chemically, saponins from sea cucumbers are triterpenoid glycosides whereas those from starfishes are steroidal glycosides.

The presence of oligoglycosides in both Holothuroidea and Asteroidea supports the opinion that sea cucumbers and starfishes are phylogenetically closely related. In starfishes and sea cucumbers, Δ^7-sterols, which are probably a consequence of the presence of haemolytic saponins, are also predominant, whereas the other three classes of Echinoderms contain the usual Δ^5-sterols (*12*). Haemolysis is a consequence of the abstraction of membrane cholesterol by the saponins. Saponins show a much lower affinity for Δ^7-sterols and this helps to explain the apparent immunity of starfishes and sea cucumbers to their own saponins (*1*). It also makes it highly reasonable to ascribe the presence of Δ^7-sterols in these organisms to biochemical convergence, *i.e.* an adaptation to the action of inherent cytotoxins (*13*). During our investigation of the chemistry of Echinoderms, we examined some species of ophiuroids, which have received only moderate attention from chemists in comparison with the above classes, and found two steroidal glycosides along with a series of sulphated polyhydroxysteroids (*14*). These findings provide biochemical support for the opinion that the phylogenetic relationship of starfishes and ophiuroids is closer than that between starfishes and holothuroids. The distribution of secondary metabolites in the different classes of echinoderms, in connection with the phylogenetic relation among them, appears complex and offers a contradictory picture. A detailed presentation of the structures of some secondary metabolites of the phylum Echinodermata and their significance as chemotaxonomic markers can be found in an article of STONIK and ELYAKOV (*13*).

In the past few years a large number of metabolites with cytotoxic, antifungal and antineoplastic activity have been isolated from echinoderms, mainly steroidal glycosides and polyhydroxysteroids from starfishes, some polyhydroxysteroids and two steroid glycosides from brittle stars and some triterpene glycosides from sea cucumbers.

The present review deals with the steroidal oligoglycosides and polyhydroxysteroids found in starfishes and brittle stars. We have attempted to make it comprehensive because it is the first to be devoted entirely to this subject. Nevertheless, various aspects of the subject are found in a number of monographs. Among these are articles by HASHIMOTO (1979) (*15*), BURNELL and APSIMON (1983) (*1*), KREBS (1986) (*16*), MINALE, RICCIO, PIZZA and ZOLLO (1986) (*17*), QUINN (1988) (*18*), STONIK and ELYAKOV (1988) (*13*) and HABERMEHL and KREBS (1990) (*19*). All compounds are presented in a standardised form in order to allow easy reference.

2. Steroidal Oligoglycosides from Asteroidea (Starfishes): General Characteristics

The toxic properties of starfishes have been known for many years (*15*), but it was only in 1960 that HASHIMOTO and YASUMOTO recognized that the toxicity is associated with compounds similar to plant saponins. They extracted dried *Asterina pectinifera* by a method developed for plant saponins; the extract, which appeared to be a mixture of saponins, proved to be ichthyotoxic and haemolytic (*20*). Toxicity has also been noted toward anellids, mollusks, arthropods and verterbrates (*1, 15, 21, 22*). Because of their general toxicity, it is probable that saponins act primarily as chemcial defense agents against infectious aquatic fungi, protists, parasites and predators. The "avoidance reactions" and the "escape responses" exhibited by many organisms, such as some sea anemones, brittle stars, sea urchins and especially many species of mollusks, when in the presence of, or contacted by, starfishes, is a well-known and fascinating biological phenomenon. MAKIE *et al.* found that at least in some species the substance eliciting the escape response is a saponin (23). In addition, these oligoglycosides, at least in some species of starfish, participate in reproduction processes. Ikegami et al. identified saponins as the spawning inhibitors in the ovaries of Asterias amurensis (24). More recently Fujimoto et al. observed that three steroidal saponins designated Co-ARIS I (6), II (40) and III (38), isolated from the egg jelly of the starfish Asterias amurensis, are essential for inducing the acrosome reaction (25). Starfish extracts and purified saponin fractions have shown a variety of pharmacological activities: haemolytic activity (1, 17, 23), in vitro cytotoxicity toward tumor cells (26, 27, 28), antiviral activity (29), blockage of neuromuscolar transmission in mammals (30, 31) and antiinflammatory, analgesic and hypotensive activities (32). Starfish saponin extracts are also known to inhibit development of fertilized sea urchin eggs (33). Asteroid saponins have produced the same cytotoxic effects on sea urchin eggs as have holothurins and the cytotoxic action takes the form of an interruption of cell multiplication, cytolysis and animalization of larvae (differentiation effect). These effects can be attributed to a disturbance in the biosynthesis of proteins, DNA and RNA. A recent investigation of the effects of seventeen individual starfish saponins on fertilized sea urchin and starfish eggs has shown that all compounds inhibited sea urchin embryos from further development in the morula stage, that the saponins with an ergostane type side chain were more active than those with no methyl group at C-24 and that pentaglycosides are somewhat more active than hexaglycosides (34). With respect to the inhibitory activity on cell division in fertilized echinoderm eggs, Fusetani et al. (34) have emphasized that the cytologic modifications observed are similar to those produced by cytochalasin, and therefore concluded that it is likely that asterosaponins inhibit active polymerization during embryogenesis. The inhibition values for each saponin are reported below. Fusetani et al. also showed that higher doses of saponins were required to inhibit or kill starfish embryos. In an analogous study extended to the cyclic glycosides, glycosides of polyhydroxysteroids (mono- and diglycosides) and polyhydroxysteroids from starfishes, we have shown that the sulphated penta- and hexaglycosides (asterosaponins) are more active than the other groups of related steroids (35). Among active saponins, asteroside C (45), a pentaglycoside with a methyl group at C-24 in the aglycone side-chain, was the most active, while the tetrasaccharide myxodermoside A (20) was least active, thus parallelling the results of Fusetani et al. (34). A more recent study of the biological activity of representative isolated saponins and related steroids from starfishes confirmed a high incidence of cytotoxicity and inhibition of Gram-positive bacteria but only weak antiviral activity and no inhibition of Gram-negative bacteria (36). Dubois et al. (37) have recently reported that the asterosaponins pectiniosides A (12), C (51) and E (16) were active at *ca.* 10 μg/ml on L 1210 and KB cells, whereas pectiniosides B (48), F (17), as well as pectinioside D, which corresponds to the asterone analog of pectinioside A, did not show any activity against either type of tumor cell below 20 μg/ml. A review of current knowledge about the pharmacologically active substances of echinoderms is to appear shortly (*38*).

Thus, most of the chemical work on starfish saponins has been initially prompted by their toxic and, more generally, biological properties. In recent years the structural studies of these molecules has increased rapidly, largely exceeding biological studies dealing with individual compounds. Almost two hundred fifty steroidal constituents, which include steroidal glycosides and polyhydroxysteroids, have been isolated from *ca.* fifty different starfish species, belonging to fifteen families, representative of the three major orders (Paxillosida, Valvatida and Forcipulata) of the class Asteroidea (Table 1). On the basis of their chemical structures, the steroidal glycosides have been subdivided into three main groups (*39*): the asterosaponins, which are sulphated glycosides, usually penta- and hexaglycosides, based on $\Delta^{9(11)},3\beta,6\alpha$-dioxysteroidal aglycones with a sulphate at C-3 and the oligosaccharide moiety at C-6; the cyclic glycosides, so far only found in two species of the genus *Echinaster*, and the glycosides of polyhydroxysteroids, which are as widespread as the asterosaponins among starfishes. These substances which usually occur only in minute amounts consist of a polyhydroxysteroidal aglycone linked to one or two sugar units and can be found in both sulphated and non-sulphated forms. Analysis of the polar extractives of the starfish *Tremaster novaecaledoniae* has now led to the discovery of a new class of glycosides, in which the polyhydroxysteroids are conjugated with a phosphate residue to which the sugars are glycosidically attached (*40*). Beside steroidal glycosides, starfishes also proved to be a rich source of non-glycosidated highly oxygenated steroids which are also found in sulphated and non-sulphated forms. Sulphation is a typical process in the biosynthesis of secondary metabolites in many marine invertebrates, especially in echinoderms.

2.1 Isolation Procedures

The isolation of water soluble products from natural sources, particularly from the marine environment with the presence of sodium chloride and other salts in all the aqueous extracts, poses numerous problems. The recent advent of an array of modern separation techniques has now allowed the more complex mixtures to be tackled (*18, 41*);

Table 1. *Distribution of Steroidal Constituents in Starfishes by Species*

Order and Family	Place of collection	Asterosaponins	Glycosides of polyhydroxy steroids	Polyhydroxy steroids
ORDER PAXILLOSIDA				
ASTROPECTINIDAE				
Astropecten indicus	Clifton Coast of Karachi, Pakistan		85, 107, 121	
Astropecten scoparius	Okkairi Bay, Jawate Prefecture, Japan	14	83, 84, 90, 135	180, 182, 183, 188, 228
Astropecten latespinosus	Hakata Gulf, Fukuoka, Japan	7, 14		
LUIDIIDAE				
Luidia maculata	New Caledonia; Genkai-Sea, Fukuoka, Japan	4, 8, 15, 27, 28, 29, 30		194, 195, 196
ORDER VALVATIDA				
ARCHASTERIDAE				
Archaster typicus	New Caledonia			216–224
GONIASTERIDAE				
Sphaerodiscus placenta	Mediterranean Sea, Bay of Policastro, Italy		93, 94, 103, 114, 125, 126	193, 199, 202
Rosaster sp.	New Caledonia			173, 174, 175, 194, 200, 201
OREASTERIDAE				
Choriaster granulatus	New Caledonia		138, 139, 142	
Culcita novaeguineae	Zampa, Okinawa, Japan		92, 93, 101, 102, 112, 117, 118, 119, 120, 125, 126, 166, 167	187, 199, 200, 227, 229

Halityle regularis	New Caledonia	4, 11, 47	93, 102, 111, 125, 126, 127, 128, 166	180, 181
Oreaster reticulatus	Bay of Cartagena, Colombia		64, 65	
Pentaceraster alveolatus	New Caledonia	4, 37	76, 146	173, 174, 175, 176
Protoreaster nodosus	New Caledonia	4, 37	146	173, 174, 175, 176, 177, 178
Poraster superbus	New Caledonia		77, 78, 162	173, 174, 175, 179
ASTEROPSEIDAE				
Dermasterias imbricata	Gulf of California		105, 106	180, 184, 185, 186, 189, 190, 227, 230
OPHIDIASTERIDAE				
Fromia monilis	New Caledonia		59, 60, 129, 130, 131, 132, 168, 169	
Gomophia watsoni	New Caledonia	31, 32	122, 166, 167	227, 229
Hacelia attenuata	Bay of Napoli, Italy		74, 82, 99, 100, 157–159, 102	192, 203, 204, 205
Linckia guildingi			146	
Linckia laevigata	New Caledonia	4, 8, 9, 10, 14		
Nardoa gomophia	New Caledonia	4, 14	93, 102, 111, 125, 126, 127, 128, 166	180, 227
Nardoa novaecaledoniae	New Caledonia		102, 125, 126	
Nardoa tuberculata	Zampa, Okinawa, Japan,			
Ophidiaster ophidianus	Bay of Napoli, Italy	4, 9, 31, 32	93, 102, 125, 126, 166	227, 229
ASTERINIDAE				
Asterina pectinifera	Fukuoka and Sendai, Japan	7, 12, 16, 17, 18, 48, 51		173, 174, 175, 182, 183, 225
Patiria pectinifera	Posyet Bay, Sea of Japan		64, 79, 81	173, 174, 175
Patiria miniata	Gulf of California	7, 18, 19, 48, 49	64, 65, 66, 80	173, 174, 175, 176, 183, 188, 191
Tremaster novaecaledoniae	New Caledonia		170, 171, 172 (phosphated)	197, 198, 233, 234, 235, 236, 238

Table 1. (Contd.)

Order and Family	Place of collection	Asterosaponins	Glycosides of polyhydroxy steroids	Polyhydroxy steroids
ECHINASTERIDAE				
Echinaster sepositus	Bay of Napoli, Italy	53, 54, 55, 56*	147, 148, 152, 153	
Echinaster luzonicus	New Caledonia	57*		
Echinaster brasiliensis	Grand Bahama Island	14		
Henricia laeviuscola	Gulf of California	50	58, 140, 141, 143, 144, 149, 150, 151, 152, 153, 154	
ACANTHASTERIDAE				
Acanthaster planci	Okinawa, Japan; Ehine Prefecture, Japan and New Caledonia	4, 7, 14, 24, 25, 26	138, 145, 146	
MITHRODIIDAE				
Thromidia catalai	New Caledonia	4, 9, 11	142, 163	
SOLASTERIDAE				
Solaster borealis	Mutsu Bay, Aomori Prefecture, Japan	39	69, 70, 71, 75	174, 175, 176, 183, 188, 194, 206, 207, 208, 209
Crossaster papposus	West Coast of Sweden		67, 68, 72, 73, 164, 165	199
MIXASTERIDAE				
Myxoderma platyacanthum	Gulf of California	20		194, 210, 211, 212, 213, 214, 215

PTERASTERIDAE				
Euretaster insignis	New Caledonia			239–244
PORANIIDAE				
Porania pulvillus	West Coast of Sweden	94		
ORDER FORCIPULATA				
ASTERIIDAE				
Aphelasterias japonica	Mutsu Bay, Aomori Prefecture, Japan	137, 185		
Asterias amurensis	Pacific coast of Hokkaido; Gulf of Shimabora, Nagasaki	4, 5, 6, 13, 35, 36, 38, 40, 42, 43, 44, 45	61, 62, 63, 98 140, 143, 160, 161	226, 237
Asterias forbesi	Bay of Funday, Canada	5, 6, 13, 21, 22, 23, 52, 6		
Asterias vulgaris				
Coscinasterias tenuispina	Bay of Napoli, Italy	4, 11, 27, 29, 33, 34, 41, 43	108, 109, 110, 113, 115, 134	181, 231, 232
Distolasterias nippon	Sendai, Japan	95, 96, 97		
Pisaster brevispinus	Gulf of California	4, 13, 14	116, 124	
Pisaster ochraceus	Gulf of California	4, 13	116, 136	
Pisaster giganteus	Gulf of California	4, 13	123, 133, 156	
Marthasterias glacialis	Bay of Napoli, Italy	14, 15, 27, 29	89, 104	
Pycnopodia helianthoides	Gulf of California	4	86, 87, 88, 134	174, 175, 180, 182

*53–57: Cyclic steroidal glycosides

this has been crucial for the advance in our knowledge of steroid glycosides from starfishes.

The "asterosaponins" usually occur, in minute amounts, as complex mixtures of very similar molecules together with other closely related polar steroids, such as "glycosides of polyhydroxysteroids" and non-glycosidated polyhydroxysteroids. An illustrative example of the complexity of polar steroid constituent mixtures in starfishes is represented by *Coscinasterias tenuispina* (42), from which a mixture of nineteen components was separated into ten asterosaponins, six glycosides of polyhydroxysteroids and three polyhydroxysteroids (Table 1). The complexity of saponin mixtures in starfishes is also well illustrated by extracts of *Asterias amurensis* (43), reinvestigation of which has led to the isolation of eleven asterosaponins and four xylopyranosides of polyhydroxysteroids (Table 1), and by re-analysis of the aqueous extracts from *Culcita novaeguineae* (44), from which eleven glycosides of polyhydroxysteroids and five polyhydroxysteroids were isolated (Table 1). We note that in this latter case all compounds were obtained in very small amounts ranging from 9.2 to 2.0 mg from 3.8 kg of fresh organisms.

The problem of isolating individual saponins and related polar steroid constituents from starfishes has been satisfactorily solved in our laboratory by using a combination of chromatographic techniques summarized in Scheme 1, as applied in the separation of nineteen constituents from *Coscinasterias tenuispina* (42). Saponins and other polar steroids are recovered from the aqueous extract by passing the solution through a column of Amberlite XAD-2 resin, washing out salts with distilled water and eluting adsorbed material with a few beds of methanol. The residue from this eluate is subjected to an initial gel-filtration on a Sephadex LH-60 column, eluting with a methanol-water (2:1) mixture. This step allows a good separation of the asterosaponin mixture from the lower molecular weight glycosides and polyhydroxy-steroids. The crude asterosaponin mixture is further fractionated by droplet counter current chromatography (DCCC) in descending mode, using a two phases solvent system made up by n-butanol-acetone-water in a 45:15:75 ratio. Owing to the relatively higher viscosity of this solvent system, the separation is made by using larger glass columns. In a typical separation, such as in the case of *C. tenuispina* asterosaponin mixture, this procedure afforded two major fractions, the first one containing the more polar asterosaponins, *i.e.* tenuispinoside C (41) together with the asterone containing saponins and a second one consisting of the remaining saponins. Although the application of DCCC is usually not sufficient to obtain complete separation of individual asterosaponins, nevertheless it represents, in our experience, a necessary step for a successful sub-

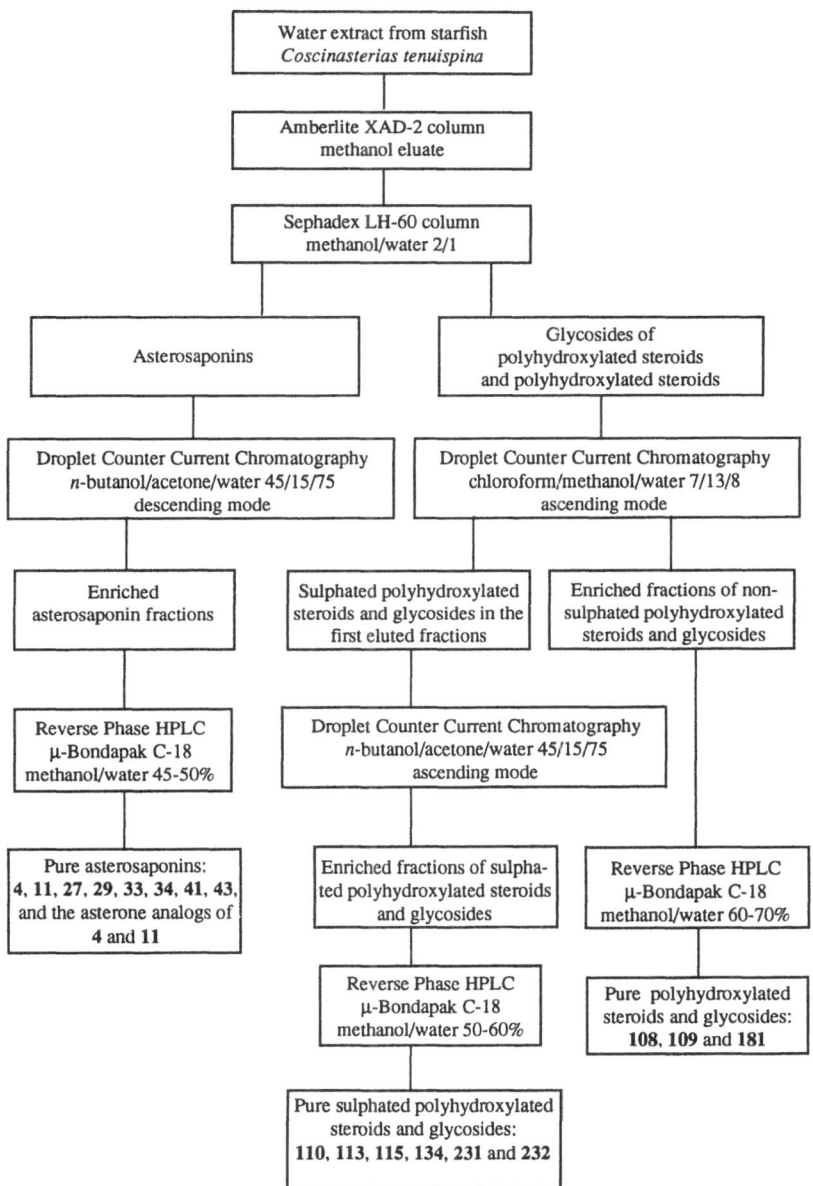

Scheme 1. General chromatographic steps used for the isolation of asterosaponins and related polar steroids from starfishes as applied to the extracts from *Coscinasterias tenuispina*

sequent final separation by reverse phase HPLC (μ-Bondapack C-18 column, methanol-water 45:55 to 50:50).

The strategy for separation of the fraction of crude monoglycosides and polyhydroxysteroids obtained from the Sephadex LH-60 column involves again a preliminary DCCC step, performed in ascending mode with a solvent system made up of chloroform-methanol-water in a 7:13:8 ratio. The more polar sulphated compounds are recovered, still as a mixture, from the first eluted fractions, while the non-sulphated steroidal glycosides and polyhydroxysteroids are gradually eluted in order of decreasing polarity and combined in fractions that are then submitted to a final purification by reverse phase HPLC. The fractionation of sulphated compounds is achieved by a further two-step procedure: again a DCCC, in ascending mode with n-butanol-acetone-water (45:15:75), to give partially resolved fractions which are then submitted to final separation by HPLC.

For the separation of the asterosaponins from ovaries of *Asterias amurensis*, OKANO and IKEGAMI (*45*) used preliminary partitioning steps of a chloroform-methanol-water extract in chloroform/methanol/water and n-butanol/water mixtures, followed by successive chromatography on a DEAE-Sephadex A-25 column, on a silanized silica gel column and finally on reverse phase kieselgel P or RP-8 columns.

In another study, KOMORI and colleagues separated 170 g of the crude glycoside mixture from *Asterina pectinifera* recovered from the aqueous extract [Amberlite XAD-2, washing with water and elution with methanol], by sequential chromatographic steps performed on silica gel with a discontinuous gradient of chloroform/methanol/water (from 7/3/0.5 to 5/5/0.7), on a reverse phase Lobar RP-8 column with 55% methanol and on Sephadex LH-20 (50% methanol) as last refining step. This procedure afforded the major pectiniosides A (**12**, 1.68 g) and B (**48**, 374 mg) together with the minor pectiniosides C (**51**), D (the asterone analog of **12**), E (**16**), F (**17**) and acanthaglycoside C (**7**), all obtained in *ca.* 50 mg amounts. Purification of the major components often required two successive chromatographic steps on silica gel and/or on Lobar RP-8 (*46, 47*).

FINDLAY and colleagues described separation of the crude glycoside mixture from *Asterias forbesi* by initial chromatography of the methanolic eluates from Amberlite XAD-2 on silica gel, with chloroform-/methanol/water 66:44:5, followed by ptlc using n-butanol/ethyl acetate/water 5:1:1. Final purification required successive chromatography on a Merck RP-18 reverse phase column, using methanol/water 7:3 as eluent, followed by a reverse phase ptlc using the same solvent system (*48*).

References, pp. 297–308

3. Asterosaponins

This group of saponins, for which the term asterosaponin was originally coined, is present in the majority of species examined (see Table 1). The main exception is represented by two *Echinaster* species, which contain cyclic steroidal glycosides. The starfish *Euretaster insignis* is also apparently devoid of asterosaponins and contains instead a group of disulphated 3,21-dihydroxysteroids along with the more usual sulphated 3β-hydroxysterols (49). In contrast with the preponderance of Δ^7-sterols in starfishes, the free sterol mixture of *E. insignis* includes only a relatively small amount of cholest-7-en-3β-ol (3% of the total sterol mixture) and large amounts of 5α-steroidal alcohols. This finding could be related to the apparent absence of asterosaponins in this species.

The asterosaponins are quite fragile molecules and usually occur as complex mixtures, fairly difficult to separate into the individual components. For this reason most of the initial studies were concerned with the analysis of aglycones obtained by acid hydrolysis of partially purified saponin mixtures. This has resulted in production of several artifacts. Asterone (3β, 6α-dihydroxy-5α-pregn-9(11)-en-20-one, **1** (50–52), which is an artifact obtained by retro-aldol cleavage of the genuine thornasterol A (**2**) sulphate aglycone (53), has been the most widely reported steroid obtained by acid hydrolysis of asterosaponins. $\Delta^{20(22)}$- and $\Delta^{17(20)}$-Steroids, along with a rearranged $\Delta^{13(14)}$-17-methyl-18-norsteroid, certainly artifacts generated during acid hydrolysis, have also been reported (54, 55). An excellent review of previous work on the aglycones can be found in the chapter of Burnell and ApSimon (1).

1

2, R=R'=H

3, R=SO$_3^-$Na$^+$, R'= Fuc $\xrightarrow{1-4}$ Fuc $\xrightarrow{1-4}$ Qui $\xrightarrow{1-4}$ Qui

To continue with the toxic saponins from starfishes, Yasumoto and Hashimoto in 1965 isolated the two major saponins from *Asterias amurensis*, which they named asterosaponins A and B, (*56*). IKEGAMI's group continued the work on asterosaponin A and in 1972 suggested structure **3**, which includes a linear tetrasaccharide chain, made up by fucose and quinovose in a 2:2 ratio, attached at C-6 of thornasterol A (*57, 58*). This remained the sole example of a linear tetrasaccharide among the *ca* fifty subsequently reported structures of asterosaponins, all including branched carbohydrate chains with five to six sugar units. About ten years later, after a ^{13}C-NMR spectrum exhibiting five anomeric carbon signals showed that the proposed structure **3** for asterosaponin A was incorrect (*45*), the same group re-examined the purification and the structures of saponins from ovaries of *A. amurensis*, and successfully isolated five pure asterosaponins. Asterosaponin A was shown to be a mixture mainly composed of ovarian asterosaponin-1 (OA-1) for which IKEGAMI et al. (*59*) successfully determined structure **6** containing the rare D-6-deoxy-*xylo*-hexos-4-ulopyranosyl unit. The latter, because of the ease with which it undergoes degradation during acid hydrolysis, appeared to be the cause of the initial incorrect structure determination. The story illustrates the importance of proper isolation and purification techniques and of the value of ^{13}C-NMR spectroscopy applied to the investigation of intact molecules.

The first complete structure of an asterosaponin appeared in the literature in 1978, when KITAGAWA and KOBAYASHI determined the structure of the major saponin from *Achantaster planci*, thornasteroside A (**4**) (*60*). Thornasteroside A turned out to be the most widely distributed asterosaponin, having been isolated from fifteen species representative of the three major orders of Asteroidea. Next followed the structure of glycoside B$_2$ (**5**), first isolated from the ovaries of *Asterias amurensis* and differing from thornasteroside A only in that the terminal fucose unit is replaced by quinovose (*61*). A little later the IKEGAMI group described the structure of ovarian asterosaponin-1 (OA-1, **6**) (*59*), again from the ovaries of *A. amurensis*.

After discovery of thornasteroside A (**4**), glycoside B$_2$ (**5**) and asterosaponin OA-1 (**6**), the literature, starting in 1983, contains a series of papers describing the structures of about fifty asterosaponins. These have revealed that all asterosaponins exhibit several common structural features, which include a $\Delta^{9(11)}$-3β-6α-dioxysteroidal aglycone, often with a 23-oxo function in the side-chain and bearing a sulphate at C-3, and an oligosaccharide moiety, made up by five to six sugar units, at C-6. A close resemblance is also evident in the saccharide portions of these asterosaponins: sugars are in their pyranose form with the β-anomeric configuration (α for arabinose) and linked with a constant pattern of interglycosidic linkages. A branching point is always located on the second

Table 2. Selected ¹H nmr data in CD_3OD from steroidal aglycones in asterosaponins

H' at C	3	11	18	19	21	22	23	24	25	26,27	28
Asterosaponin (aglycone)											
A	4.22[a]	5.37[b]	0.81	1.02	1.37 s	2.62 ABq (15)		2.42 d (7.5)		0.93,0.94 d's (6.5)	
B[c]		5.30 m	0.79	0.95	1.34 s	2.45 d (16)		6.03 s		1.92,2.16 s's	
C	4.22[a]	5.37[b]	0.70	1.04	0.94 d			6.20 s		1.94,2.11 s's	
D	4.22[a]	5.37[b]	0.70	1.04	0.94d					0.94 d (7.0)	
E	4.22[a]	5.37[b]	0.82	1.02	1.37 s	2.70 br s			2.73 m	1.08,1.10 d's (6.8)	
F	4.22[a]	5.37[b]	0.82	1.02	1.30 s	2.76 d (2.5)	2.94 dt (2.5,6.0)			1.01 d (6H's) (7.0)	
G	4.22[a]	5.37[b]	0.83	1.03	1.30 s	2.83 d (2.2)	2.76 dd (2.2,7.5)			0.98,1.04 d's (6.2)	
H	4.22[a]	5.38[b]	0.85	1.02	1.25 s			5.28 t (6.5)		1.67,1.74 s's	
I	4.22[a]	5.37[b]	0.82	1.03	1.31 s			5.13 t (6.5)		1.65,1.70 s's	
J	4.20[a]	5.40 br	0.84	1.00	1.70 s			6.16 s		1.90,2.11 s's	
K	4.22[a]	5.57[c]	0.78	1.05	1.39 s	2.55,2.75 d's (15)		2.43 d (7.0)		0.94,0.95 d's (7.0)	
L	4.22[a]	5.37[b]	0.83	1.03	1.31 s					1.07 d (6H's) (7.0)	4.72 br s
M	4.22[a]	5.37[b]	0.82	1.02	1.37 s	2.62,2.76 d's (15)		2.52 m (7.5)		0.88,0.97 d's (6.5) 1.00 d (7.5)	
N[c]			0.78	1.02	1.34 s	2.60 s				0.85,0.92 d's (6.6) 0.99 d (5.5)	
O	4.22[a]	5.38[b]	0.83	1.03	1.29 s	2.75 d (2.4)	2.78 dd (2.4,7.5)			0.93,0.97 d's (6.8) 1.02 d (6.6)	
P[d]			1.00	1.10	1.67 s	2.83 s				0.84,0.93 d's (6.8) 0.88 t (29-H_3)	

[a] m $W_{1/2}$ = 22 Hz; [b] br d, J = 5.5 Hz; [c] free aglycone synthetic model in $CDCl_3$; [d] free aglycone synthetic model in d_5-Pyr; [e] br d (J = 5 Hz), 12β-H δ 3.94 d (J = 5 Hz)

Table 3. Carbon chemical shifts[a] of steroidal aglycones of selected asterosaponins, aglycone skeleton and asterosaponin from which ^{13}C nmr data are derived

C	[A] 4 (67)	[B] 24 (86)	[C] 27 (82,83)	[D] 29 (82,83)	[E] 32 (67)	[F] 33 (42)	[G] 36 (43)	[H][b] 37 (68)	[I] 39 (93)	[K][c] 41 (42)	[L] 42 (43)	[M] 45 (43)	[N] 46 (80)	[O] 47 (69)	[P] 51 (47)
1	36.0	36.2	35.7	35.8	36.2	36.1	35.9	35.9	35.5	36.7	35.9	36.1	36.0	36.2	36.0
2	29.5	29.7	29.0	29.1	29.5	29.4	29.2	29.2	28.8	29.6	29.3	29.4	29.4	29.5	29.4
3	78.2	78.8	78.2	78.3	78.0	77.9	77.5	78.3	77.2	77.4	78.0	78.0	77.7	77.9	77.7
4	30.6	30.9	30.5	30.6	30.9	30.8	30.6	30.6	30.1	30.6	30.6	30.8	30.8	31.0	30.8
5	49.4	49.6	49.0	49.0	49.5	49.5	49.2	49.2	48.6	49.2	49.2	49.5	49.3	49.5	49.3
6	80.4	80.6	80.8	80.4	80.6	80.4	80.0	80.0	79.5	80.7	80.0	80.5	80.4	80.9	79.9
7	41.6	41.7	41.1	41.0	41.6	41.6	41.7	41.2	40.8	41.4	41.5	41.8	41.6	41.6	41.7
8	35.3	35.5	35.4	35.4	35.6	35.5	35.2	35.4	35.8	37.0	35.2	35.5	35.3	35.5	35.4
9	145.5	145.4	145.5	145.5	145.7	145.9	145.6	145.6	145.0	–	145.5	145.8	145.4	146.0	145.5
10	38.3	38.5	38.1	38.1	38.5	38.4	38.2	38.2	37.8	–	38.2	38.4	38.3	38.5	38.4
11	116.7	116.7	116.3	116.4	116.7	116.7	116.4	116.6	116.2	119.7	116.5	116.7	116.7	116.7	116.7
12	42.4	42.7	41.0	41.6	42.7	42.5	42.2	42.6	42.0	80.9	42.6	42.6	42.5	42.5	42.5
13	41.6	41.8	41.0	41.0	41.8	42.0	41.2	41.6	41.3	–	41.2	41.4	41.6	42.0	41.6
14	54.0	54.1	53.6	53.6	54.2	53.9	53.7	54.0	53.5	49.6	53.9	54.0	54.1	54.0	54.1

C	4	24	27	29	32	33	36	37	39	41	42	45	36	47	51
15	23.3	26.6	25.1	25.1	23.4	23.2	22.9	22.6	23.0	23.6	23.6	23.4	23.2	23.2	23.3
16	25.1	25.3	28.4	28.4	25.5	25.4	25.1	25.3	24.6	26.1	25.1	25.2	25.2	25.3	25.2
17	59.6	59.8	56.3	56.0	59.8	59.8	59.7	55.1	58.2	56.5	58.9	59.7	59.1	59.8	58.8
18	13.5	13.8	11.4	11.4	13.6	13.5	13.3	13.4	13.1	13.2	13.4	13.6	13.6	13.5	13.8
19	19.2	19.5	19.0	19.1	19.4	19.3	19.1	19.2	18.8	19.8	19.1	19.4	19.2	19.3	19.5
20	74.2	74.3	32.9	32.3	74.0	71.6	71.4	76.4	73.3	—	74.1	74.0	73.8	71.4	73.6
21	27.0	27.6	19.4	19.4	27.2	23.5	23.2	21.5	25.1	25.4	25.9	27.2	27.1	23.7	27.2
22	54.9	56.1	51.3	50.0	52.1	65.7	64.5	77.8	43.9	53.5	43.2	53.6	53.8	64.3	55.4
23	211.7	201.7	200.6	210.2	216.0	53.5	59.5	29.6	22.5	—	29.7	215.6	216.0	57.6	215.8
24	54.0	126.0	124.9	52.2	—	41.3	—	124.2	125.1	53.5	158.3	54.2	54.1	42.0	61.3
25	24.3	154.9	154.4	24.4	42.5	26.8	30.2	131.8	130.0	25.4	34.1	29.9	29.8	31.8	29.3
26	22.5	27.5	27.0	22.3	17.9	22.7	19.1	25.6	16.9	22.8	21.9	18.6	21.4	19.4	21.2
27	22.6	20.9	20.3	22.5	18.2	23.0	19.0	17.4	25.7	22.9	21.9	21.4	18.5	20.4	19.5
28											106.3	12.1	12.5	13.0	21.2
29															12.1

a Data (in ppm) were mostly obtained at 62.9 MHz from solution in d_5-pyridine and are referred to the central line of the C-2 solvent signal (135.5 ppm); b d_5-pyridine with a few additional drops of D_2O; c signals due to quaternary carbons not detected. Thornasteroside A (4), acanthaglycoside A (24), marthasteroside B (27), marthasteroside C (29), ophidianoside C (32), tenuispinoside A (33), asteroside B (36), protoreasteroside (37), solasteroside A (39), tenuispinoside C (41), asteroside D (42), asteroside C (45), acanthaglycoside F (36), regularoside A (47), pectinioside C (51).

Table 4. Carbon Chemical Shifts in $[^2H_5]$-Pyridine of Saccharide Chain of Selected Asterosaponins

Sugar carbon	Forbeside H 23 (48)[a]	Thornasteroside A 4 (65)[b]	Myxodermoside A 20 (85)	Glycoside B₂ 5 (43)	Acanthaglycoside C 7 (79,80)[b]	Maculatoside (luidiaglycoside B) 8 (81)[b]	Ophidianoside F 9 (67)
	Qui I	Qui I	Qui I	Qui I	Qui I	Qui I	Qui I
1	103.5	104.4	105.0	104.3	104.7	104.4	105.1
2	73.5	74.2	74.2	73.9	74.1	74.2	74.3
3	88.9	40.2	90.2	89.7	89.6	90.0	90.5
4	73.8	74.0	74.4	74.1	74.3	74.5	74.6
5	71.5	72.0	72.2	71.7	71.8	72.0	72.6
6	17.7	18.2	17.9	17.7	18.0	17.9	17.9
	Xyl	Xyl	Xyl	Xyl	Xyl	Xyl	Xyl I
1	103.6	105.1	104.2	104.0	104.4	105.0	104.5
2	83.6	82.6 (84.4)	82.6	82.5	82.8 (84.9)	82.0 (85.0)	82.5
3	75.1	75.7	75.6	75.9	75.5	75.4	75.3
4	69.5	79.2	77.9	78.5	77.6	78.6	78.2
5	66.1	64.7	64.5	64.2	64.2	64.4	64.4
6							

	Qui II	Qui II	Qui II	Qui II	Qui II	Qui II	Qui II
1	105.2	105.1 (107.3)	105.1	105.2	105.3 (107.0)	105.0 (107.1)	105.1
2	75.4	75.7	75.3	75.4	75.5	75.5	75.6
3	76.4	76.9	76.8	76.5	76.8	76.8	77.0
4	76.3	76.3	76.2	76.2	76.1	76.0	76.3
5	73.4	73.8	73.8	73.5	73.7	73.8	73.7
6	18.1	18.4	18.5	18.1	18.4	18.5	18.5

		Gal	Gal	Gal	Glc	Qui III	Xyl II
1		102.5	104.2	101.9	101.6	101.3	102.1
2		83.5	72.0	82.7	84.5	84.7	84.0
3		75.7	74.8	74.5	77.6	77.4	77.4
4		69.5	70.1	69.2	70.9	76.0	70.5
5		76.9	77.1	76.8	78.5	73.8	66.9
6		62.2	62.3	61.7	62.0	18.2	

		Fuc		Qui III	Fuc	Fuc	Fuc
1		107.2		106.0	106.8	106.9	106.6
2		72.0		75.2	71.8	72.0	72.1
3		74.7		77.3	74.7	74.9	75.0
4		72.7		76.1	72.4	73.0	73.7
5		72.0		73.0	71.8	72.5	72.1
6		17.5		18.1	17.1	17.2	17.2

[a] spectrum run in d_5-pyridine/D_2O (5:1); assignments by comparison with forbeside A (= versicoside A, 13), assisted by 500 MHz HETCOR data; [b] in parentheses the shifts observed in the spectrum of the desulphated saponin; [c] spectrum run in d_4-methanol; [d] Z = 6-deoxy-β-D-hexos-xylo-4-ulopyranosyl unit in the hydrate form; [e] spectrum obtained after equilibration with d_5-pyridine/D_2O (5:1); otherwise both signals at 201.1 ($>C=O$) for the keto form and at 92.7 ppm for the gem diol form are observed and all carbon signals due to the saccharide unit appear splitted. Assignments assisted by HETCOR data (78); [f] spectrum run in d_5-pyridine/D_2O (5:1); assignments by comparison with forbeside C (6), assisted by 500 MHz HETCOR data

Table 4. (continued)

	Regularoside B 11 (69)	OA-4 38 (43)	Henricioside A 50 (94)[c]	Versicoside A 13 (65)	Marthasteroside A$_1$ 14 (83)[b]	Marthasteroside A$_2$ 15 (83)[b]	Pectinioside C 51 (47)	Pectinioside F 17 (37)
	Qui I	Qui I	Qui I	Qui I	Qui I	Qui I	Qui I	Qui I
	105.1	104.6	105.3	104.2	105.0	105.1	105.1	105.3
	74.8	74.0	74.0	74.3	74.4	74.4	74.2	74.1
	89.9	89.7	88.9	89.9	90.5	90.5	90.3	89.6
	74.4	74.3	74.4	74.4	74.5	74.6	74.6	74.2
	72.4	71.8	72.6	72.0	72.6	72.6	72.0	71.8
	17.9	18.1	17.6	18.2	18.0	18.0	17.9	18.1
	Xyl	Xyl	Xyl	Xyl	Xyl	Xyl	Xyl	Xyl
	104.1	105.1	104.1	104.9	104.4	104.5	104.5	104.6
	82.1	82.7	80.6	82.5	82.4 (84.8)	82.3 (85.0)	82.7	83.2
	75.2	75.5	75.3	75.7	75.5	75.4	75.6	75.4
	78.4	79.1	77.8	79.0	79.2	78.7	77.9	78.0
	64.3	64.2	63.9	64.6	64.3	64.0	64.4	64.2
	Qui II	Qui II	Qui II	Qui II	Qui II	Qui II	Qui II	Qui II
	104.8	104.3	105.0	105.3	105.3 (107.4)	105.1 (107.4)	105.3	105.7
	75.6	75.5	75.7	75.7	75.7	75.7	75.6	75.5
	77.0	76.6	77.5	77.0	77.0	77.1	76.6	76.4
	76.1	76.3	76.3	76.3	76.3	76.3	76.2	76.1
	73.7	73.3	73.6	73.8	73.8	73.8	73.5	73.7
	18.3	18.3	19.0	18.7	18.5	18.2	18.5	18.5

References, pp. 297–308

Residue	1	2	3	4	5	6
Fuc I	101.7	82.4	74.2	71.5	71.8	16.8
Fuc II	106.5	71.8	74.8	73.5	71.9	17.0
Glc	101.6	84.6	76.6	70.9	78.1	62.2
Qui III	106.1	76.1	77.4	76.9	73.7	18.3
Gal	101.4	82.6	74.7	69.3	76.7	61.6
Ara	103.4	71.6	73.4	68.2	65.8	
Gal I	102.2	83.3	75.1	69.4	77.0	62.0
Fuc	106.5	71.6	84.2	71.7	72.7	17.2
Gal II	106.5	73.1	75.1	70.2	77.1	62.4
Gal	102.4	82.8	74.7	69.8	76.7	62.3
Fuc I	105.9	71.8	84.4	71.9	71.9	17.1
Fuc II	106.1	72.2	75.0	72.8	72.3	17.1
Qui III	101.4	84.3	76.1	77.6	73.2	18.5
Fuc I	106.0	71.7	84.3	71.8	71.9	17.1
Fuc II	106.2	72.2	75.1	72.9	72.4	17.0
Glc	101.8	84.3	77.8	71.0	78.4	62.2
Fuc I	106.5	71.8	85.1	72.7	72.1	17.2
Fuc II	106.7	71.8	75.1	72.6	72.1	17.3
Glc	101.4	83.0	75.6	78.2	75.8	61.9
T Fuc (1→2)	107.1	72.0	74.8	73.7	71.3	17.2
T Gal (1→4)	104.3	71.8	74.4	69.8	75.8	61.8

Sugar carbon: 1, 2, 3, 4, 5, 6

Table 4. (continued)

Sugar carbon	Patiriioside A 49 (64)	OA-1 = Co ARIS I = forbeside C 6 (77)[e]	Forbeside F 21 (48)[f]	Forbeside G 22 (48)[f]	Laevigatoside 10 (62)	Pectinioside A 12 (46)	Pectinioside E 16 (37)
	Qui I	Z^d	Z^d	Qui I	Qui I	Qui I	
1	104.5	104.5	104.5	103.8	105.2	104.1	104.8
2	74.1	73.1	73.1	73.0	73.9	74.0	74.3
3	89.4	90.4	90.4	90.0	91.1	90.2	90.6
4	74.2	92.7	92.7	92.2	74.6	74.4	74.5
5	72.1	73.1	72.1	72.6	72.5	72.2	72.2
6	17.8	13.4	13.4	12.9	18.3	18.0	18.0
	Xyl	Qui I	Qui I	Qui I	Qui II	Qui II	Qui II
1	103.9	103.5	104.4	103.2	103.9	103.3	103.5
2	82.9	83.4	83.4	83.9	82.3	83.1	82.8
3	75.1	74.7	74.9	74.8	75.2	74.0	73.9
4	78.0	84.8	84.5	73.4	84.8	85.1	85.5
5	64.0	73.1	73.5	72.2	73.7	71.7	71.8
6		18.1	18.1	17.5	17.9	18.2	18.2
	Qui II	Qui II	Qui II	Qui II	Qui III	Qui III	Qui III
1	105.3	105.8	105.7	105.2	104.9	105.5	105.2
2	75.5	76.0	75.9	75.3	75.4	75.5	75.4
3	76.9	76.5	76.5	75.9	76.7	76.6	76.7
4	76.3	75.5	75.5	74.9	76.2	75.8	75.9
5	73.7	73.9	74.0	73.4	73.6	74.0	73.8
6	18.2	18.0	17.6	17.2	18.5	18.4	18.5

	Glc	Glc	Ara	Fuc I	Fuc I	Glc
1	102.5	102.2	102.7	103.4	102.3	101.3
2	82.4	83.5	82.0	73.7	81.6	83.1
3	76.4	76.6	73.4	74.4	74.5	76.0
4	79.2	70.9	68.2	71.8	71.6	78.9
5	76.1	78.8	66.1	71.5	71.6	76.0
6	61.1	61.7	–	16.7	16.6	61.1

	Glc — T Xyl (1→2)	Glc — Fuc	Ara — Fuc	Fuc I — T Ara (1→2)	Fuc I — Fuc II	Glc — T Fuc (1→4)	T Fuc (1→4)
1	104.8	105.9	106.8	104.9	106.4	106.3	105.3
2	75.2	72.2	71.8	71.5	73.3	71.7	72.3
3	77.8	74.9	75.1	74.0	74.5	74.5	74.8
4	71.9	73.2	73.4	69.2	72.2	71.8	73.6
5	67.5	71.7	71.9	67.2	71.7	73.6	71.5
6		16.8	17.1		16.9	16.9	17.1

Table 4. (*continued*)

Pectinioside G	Patirioside B	Marthasteroside B	Luidiaglycoside C
18 (63)	**19** (64)	**27** (83)[b]	**28** (90)[b]
Qui I	Qui I	Glc	Glc
104.4	104.7	105.1	103.6
74.1	73.7	75.2	73.8
90.4	90.2	91.8	91.4
74.1	74.2	70.1	69.6
72.2	72.0	77.4	77.6
17.7	17.7	62.9	62.2
Qui II	Qui II	Qui I	Qui I
103.3	103.1	104.0	105.0
83.0	82.8	83.0 (85.0)	82.8 (84.7)
74.3	74.5	75.2	75.0
85.0	84.9	85.7	85.7
72.1	71.9	73.8	71.6
17.9	17.9	18.0	18.0
Qui III	Qui III	Qui II	Qui II
105.3	105.1	105.2 (107.5)	105.0 (107.2)
75.5	75.3	75.7	75.8
76.9	76.7	77.1	77.6
76.1	75.9	76.5	76.3
73.7	73.6	73.7	73.8
18.1	18.1	18.4	18.2
Glc	Glc	Fuc I	Qui III
102.1	101.9	102.6	102.3
82.2	81.9	82.0	84.5
75.8	75.9	74.1	76.8
79.0	79.8	72.0	75.8
76.0	76.7	71.9	73.8
61.0	61.7	16.7	18.2
T Fuc (1– > 2)	T Fuc (1– > 2)	Fuc II	Fuc
106.0	105.7	106.7	107.1
71.6	71.2	71.8	71.9
74.7	74.8	75.2	75.0
73.4	73.2	72.7	73.0
71.8	71.4	72.0	72.5
17.7	16.6	17.0	17.1

References, pp. 297–308

Table 4. (*continued*)

Pectinioside G	Patirioside B
18 (63)	**19** (64)
T Ara (1– > 4)	T Gal (1– > 4)
104.9	104.2
71.4	71.6
73.8	74.4
69.2	69.6
67.1	75.9
	61.0

monosaccharide (xylose or quinovose) starting from the aglycone and a terminal quinovose is always found 2-linked to the branched sugar. The more common sugars are D-fucose, D-quinovose, D-xylose, D-galactose and D-glucose. Other less common monosaccharides are D-6-deoxy-*xylo*-hex-4-ulose and L-arabinose, the latter having been occasionally found in laevigatoside (**10**) from *Linckia laevigata* (*62*) and very recently in henricioside A (**50**) from *Henricia laeviuscola* (*94*), in pectinioside G (**18**) from *Asterina pectinifera* (*63*) and in patirioside A (**49**) from *Patiria miniata* (*64*). Thornasteroside A (**4**) is a good example of the general structure of pentaglycosides, which are generally the more common components among asterosaponins; versicoside A (**13**), first isolated from *Asterias amurensis* versicolor (*65*), is representative of the structure of the first reported group of hexaglycosides, while pectinioside E (**16**), isolated from *Asterina pectinifera* (*37*), is representative of the more recently described group of asterosaponins whose hexasaccharide chain has two branches.

Greater differences are observed in the structure of the steroidal side chain, although thornasterol A is by far the more common aglycone. The structures of the asterosaponins isolated so far are described in the following pages, arranged according to the chemical structure of the aglycone. Abbreviations used in the formulae of asterosaponins are:

Qui = D-quinovose, Fuc = D-fucose, Xyl = D-xylose, Ara = L-arabinose, Gal = D-galactose, Glc = D-glucose. All monosaccharides are in their pyranose forms and the glycosidic linkages are β (α in the arabinopyranose).

The ^{1}H- and ^{13}C-NMR data of the steroid aglycones (A–P) encountered since now in asterosaponins are in Tables 2 and 3; the ^{13}C-NMR data of the twenty six saccharide chains found since now in asterosaponins are in Table 4.

A

4 Thornasteroside A:

$$Fuc \xrightarrow{1-2} Gal \xrightarrow{1-4} Xyl \xrightarrow{1-3} Qui \xrightarrow{1-6} A$$
$$\uparrow^{1-2}$$
$$Qui$$

Occurrence: *Acanthaster planci (60), Asterias amurensis (65), Luidia maculata (66), Ophidiaster ophidianus (67), Linckia laevigata (62), Protoreaster nodosus* and *Pentaceraster alveolatus (68), Halityle regularis (69), Nardoa gomophia (70), Coscinasterias tenuispina (42), Thromidia catalai (71), Pisaster ochraceus* and *Pisaster brevispinus (72), Pisaster giganteus (73), Pycnopodia helianthoides (74).*
Activity: Inhibits cell division in fertilized sea urchin eggs [75–100% inhibition at 10^{-7} M (*35*); ED_{50} 10 μg/ml; LD_{99} 5 μg/ml (*34*)]; and showed antifungal activity (minimum inhibitory amount against *Cladosporium cucumerinum*, 20 μg) (*35*).
Physical data: m.p. 203–204°C, $[\alpha]_D = -40°$ (H$_2$O), FAB-MS (− ve ion), m/z 1243 [M_{SO_3}-]. FAB-MS (+ ve ion), m/z 1289 [M_{SO_3Na} + Na]$^+$.

Thornasteroside A (**4**) is the first asterosaponin whose structure was fully elucidated. Instead of acid hydrolysis Kitagawa and coworkers (*53*) used a glycosidase mixture to remove the sugars, followed by solvolysis of the sulphate to yield the genuine thornasterol A (**2**) and its 24-methyl derivative, thornasterol B (minor). On treatment with acid, thornasterol A afforded both asterone (**1**, minor) and the corresponding $\Delta^{20(22)}$-steroid (major). Thus, it was suggested that the pregnane type sapogenin (**1**), isolated not only from *A. planci* but also from many other species of starfishes, may be an artifact. Sulphate was shown to be attached to 3β-OH upon conversion of the sapogenol sulphate mixture to the corresponding 6-keto derivative, characterized by CD spectroscopy.

The 20S configuration was later proposed on the basis of the chemical shift of the 21-methyl protons of thornasterol itself (δ 1.34), the 23-OH derivative (δ 1.41) and its triacetate (δ 1.29) [(20S)- and (20R)-hydroxycholesterol: δ (CH$_3$-21): 1.28 and 1.13 ppm respectively] (17) and confirmed by stereoselective synthesis of (20S)-thornasterol A (75). The structure of the saccharide chain was determined by partial methanolysis of the permethylated saponin followed by complete methanolysis (60). The ^{13}C-NMR assignments of thornasteroside A, made essentially by comparison with reference substances, were reported in successive papers (65, 67).

5 Glycoside B$_2$ (= Forbeside B):

$$\text{Qu i} \xrightarrow{1\text{-}2} \text{Ga l} \xrightarrow{1\text{-}4} \text{Xy l} \xrightarrow{1\text{-}3} \text{Qu i} \xrightarrow{1\text{-}6} \textbf{A}$$

$$\uparrow 1\text{-}2$$
$$\text{Qu i}$$

OCCURRENCE: *Asterias amurensis* (43, 61), *Asterias forbesi* (76).
PHYSICAL DATA: m.p. 206–208°C; $[\alpha]_D = -2.8°$ (H$_2$O), $+4.1°$ (MeOH); FAB-MS (− ve ion), m/z 1243 [M$_{SO_3}-$].

The structure of the saccharide chain was determined by partial hydrolysis to give a tri- and mono-glycoside, followed by characterization of the shorter triglycoside by methylation, hydrolysis and identification of the derived monosaccharides as alditol acetates (61). Treatment of **5** with a glycosidase mixture of the mollusc *Charonia lampas* afforded thornasterol A sulphate. FINDLAY and coworkers in their continuing studies of the extracts of *Asterias forbesi* reisolated glycoside B2, which they named forbeside B, and assigned the complete sugar sequence and the sites of interglycosidic linkages by using two-dimensional (2D) nuclear proton correlation (COSY), relayed correlation spectroscopy (RECSY), Nuclear Overhauser enhancement (NOESY) and hydrogen-carbon correlation (HETCOR) experiments (76). The spectra were run in d$_6$-DMSO; the ^{13}C-NMR spectral data in d$_5$-pyridine were reported in (43).

6 Ovarian Asterosaponin I (= Co-Aris I; = Forbeside C):

$$\text{Fuc} \xrightarrow{1\text{-}2} \text{Fuc} \xrightarrow{1\text{-}4} \text{Qu i}$$
$$\uparrow 1\text{-}2$$
$$\text{Qu i}$$

OCCURRENCE: *Asterias amurensis* (*25, 43, 59, 77*), *Asterias forbesi* (*78*), *Asterias vulgaris* (*78*).

ACTIVITY: Induces acrosome reaction in egg jelly of *A. amurensis* (*25*) and inhibits cell division in fertilized sea urchin eggs (ED_{50} 15µg/ml; LD_{99} 5 µg/ml) (*34*).

PHYSICAL DATA: m.p. 203–204°C, $[\alpha]_D = -2.0°$ (MeOH), FAB-MS (– ve ion), m/z 1257 (hydrate form) and 1239 (keto form) $[M_{SO_3^-}]$.

The ^{13}C-NMR spectrum in media containing H_2O (d_5-pyridine-H_2O) showed a signal at 92.8 ppm, whereas for samples dried on P_2O_5, the spectrum (d_6-DMSO) contained the carbonyl signal at 200.7 ppm, in agreement with the presence of a keto sugar easily converted into the hydrate form. On alkaline treatment, **6** gave asterone 3β-sulphate and a tetrasaccharide because of the lability of glucopyranosyluloses in alkaline media, which results in the release of the substituents at C-1 and C-3, thus establishing the direct attachment of the keto sugar to the aglycone. The structure of the tetrasaccharide chain was determined by a) reduction with $NaBH_4$ followed by methanolysis affording quinovitol, methyl quinovoside and methyl fucoside in a 1:1:2 ratio, and b) methylation of the reduced tetrasaccharide chain and identification of the derived alditol acetates. Reduction with $NaBD_4$ of the shortened triglycoside obtained by enzymatic hydrolysis followed by sugar analysis led to the identification of the ketosugar as 6-deoxy-*xylo*-hex-4-ulose (*59*). More recently FINDLAY *et al.* (*78*) determined the complete structure of the oligosaccharide chain of **6**, re-isolated from *A. forbesi*, by using 2D-NMR techniques.

7 Acanthaglycoside C:

$$Fuc \xrightarrow{1-2} Glc \xrightarrow{1-4} Xyl \xrightarrow{1-3} Qui \xrightarrow{1-6} A$$
$$\uparrow {\scriptstyle 1-2}$$
$$Qui$$

OCCURRENCE: *Astropecten latespinosus* (*79*), *Acanthaster planci* (*80*), *Asterina pectinifera* (*47, 63*), *Patiria miniata* (*64*).

ACTIVITY: inhibits cell division in fertilized sea urchin eggs (ED_{50} 15 µg/ml; LD_{99} 2 µg/ml) (*34*).

PHYSICAL DATA: m.p. 225–230°C (dec); $[\alpha]_D = -6.4°$ (MeOH), FAB-MS (+ ve ion), m/z 1305 $[M_{SO_3Na} + K]^+$, 1289 $[M_{SO_3Na} + Na]^+$, FAB-MS (– ve ion), m/z 1243 $[M_{SO_3^-}]$.

Acanthaglycoside C was first isolated as a desulphated derivative after solvolysis of the crude glycoside mixture from *Astropecten latespinosus* (*79*). The structure of the saccharide chain was determined by

^{13}C-NMR spectroscopy and comparison with that of thornasteroside A (**4**) which differs from **7** only for the internal galactose here replaced by glucose. Native acanthaglycoside C was then re-isolated from *A. planci* (*80*) and later from *A. pectinifera* (*47, 63*) and *Patiria miniata* (*64*).

8 Maculatoside (= Luidiaglycoside B):

$$\text{Fuc} \xrightarrow{1-2} \text{Qu i} \xrightarrow{1-4} \text{Xy l} \xrightarrow{1-3} \text{Qu i} \xrightarrow{1-6} \textbf{A}$$
$$\uparrow^{1-2}$$
$$\text{Qu i}$$

OCCURRENCE: *Luidia maculata* (*66, 81*), *Linckia laevigata* (*62*).
ACTIVITY: Inhibits cell division in fertilized sea urchin eggs; (ED_{50} 15 μg/ml; LD_{99} 5 μg/ml) (*34*).
PHYSICAL DATA: m.p. 207–210°C (dec); $[\alpha]_D = +2.7°$ (MeOH); FAB-MS (+ ve ion), m/z 1273 $[M_{SO_3Na} + Na]^+$.

Maculatoside was isolated almost simultaneously from *L. maculata* by KOMORI et al. (*81*) and MINALE et al. (*66*). The structure was assigned on the basis of ^{13}C-NMR spectroscopy and enzymatic hydrolysis which afforded a shortened triglycoside, the prosapogenol Qui → Xyl → Qui → A, identical with a sample obtained by the same procedure from known saponins. The oligosaccharide chain closely resembles the previous ones, differing from the thornasteroside A (**4**) in that internal galactose is replaced by quinovose. It is identical with that of acanthaglycoside A (**24**), and the difference between **8** and **24** resides only in the aglycone side chain.

9 Ophidianoside F:

$$\text{Fuc} \xrightarrow{1-2} \text{Xy l} \xrightarrow{1-4} \text{Xy l} \xrightarrow{1-3} \text{Qu i} \xrightarrow{1-6} \textbf{A}$$
$$\uparrow^{1-2}$$
$$\text{Qu i}$$

OCCURRENCE: *Ophidiaster ophidianus* (*67*), *Linckia laevigata* (*62*), *Thromidia catalai* (*71*).
PHYSICAL DATA: $[\alpha]_D = +0.4°$ (MeOH); FAB-MS (+ ve ion), m/z 1259 $[M_{SO_3Na} + Na]^+$.

Ophidianoside F is a homologue of the major constituent ophidianoside C (**32**), and has the same pentasaccharide chain and the common thornasterol A 3β-sulphate as the aglycone. Thus, the structure was entirely derived by ^{1}H- and ^{13}C-NMR data and comparison with thornasteroside A (**4**) for the aglycone portion and with ophidianoside C (**32**) for the saccharide moiety (*67*).

10 Laevigatoside:

$$Fuc \xrightarrow{1-2} Ara \xrightarrow{1-4} Qui \xrightarrow{1-3} Qui \xrightarrow{1-6} A$$
$$\uparrow 1-2$$
$$Qui$$

OCCURRENCE: *Linckia laevigata* (*62*).
PHYSICAL DATA: $[\alpha]_D = +4.2°$ (MeOH); FAB-MS (+ ve ion), m/z 1273 $[M_{SO_3Na} + Na]^+$, 1267 $[M_{SO_3K} + H]^+$, 1251 $[M_{SO_3Na} + H]^+$.

Laevigatoside is the first reported asterosaponin containing an arabinopyranosyl unit; later arabinose was found in the saccharide chains of pectinioside G (**18**), patirioside A (**49**) and henricioside A (**50**). Arabinose is a common sugar among the group of glycosides of polyhydroxysteroids, but is found in the furanose form. Enzymatic hydrolysis with a glycosidase mixture from *Charonia lampas* gave a triglycoside made up only from quinovose; the interglycosidic linkages in both **10** and in the derived shortened prosapogenol were entirely derived by ^{13}C-NMR spectroscopy and comparison with reference compounds.

11 Regularoside B:

$$Fuc \xrightarrow{1-2} Fuc \xrightarrow{1-4} Xyl \xrightarrow{1-3} Qui \xrightarrow{1-6} A$$
$$\uparrow 1-2$$
$$Qui$$

OCCURRENCE: *Halityle regularis* (*69*), *Coscinasterias tenuispina* (*42*), *Thromidia catalai* (*71*).
PHYSICAL DATA: $[\alpha]_D = +4.2°$ (MeOH); FAB-MS (+ ve ion), m/z 1289 $[M_{SO_3Na} + K]^+$, 1273 $[M_{SO_3Na} + Na]^+$

After permethylation followed by methanolysis which determined that fucose and quinovose are the terminal monosaccharides, the structure of the oligosaccharide chain was then derived by ^{13}C-NMR spectroscopy, and FAB mass spectrometry.

12 Pectinioside A:

$$Fuc \xrightarrow{1-2} Glc \xrightarrow{1-4} Qui \xrightarrow{1-3} Qui \xrightarrow{1-6} A$$
$$\uparrow 1-2$$
$$Qui$$

OCCURRENCE: *Asterina pectinifera (46)*.

ACTIVITY: Cytotoxic against L1210 (IC$_{50}$ value 10 µg/ml) and KB (IC$_{50}$ value 10 µg/ml) cells (*37*).

PHYSICAL DATA: $[\alpha]_D = -6.6°$ (pyridine); FAB-MS ($-$ ve ion), m/z 1257 $[M_{SO_3^-}]$.

After laevigatoside (**10**) pectinioside A is the second reported asterosaponin which has quinovose, instead of xylose, as the branched sugar. The sequence Qui (1 → 2) Qui (1 → 3) Qui—A of the triglycoside obtained by partial hydrolysis of **12** was preferred to the alternative Qui (1 → 3) Qui (1 → 2) Qui—A by determining the spin lattice relaxation time (T1) of each Qui. The average T1 value of the 3-linked Qui was shortest, thus indicating to be attached to the aglycone. The complete saccharide sequence of **12** followed from the results of methanolysis of the permethylated sample (*46*). In a subsequent paper KOMORI and coworkers described the isolation in small amounts (*ca.* 50 mg) of the asterone analog of the major pectinioside A (1.68 g from 170 g extract), possibly an artifact, and named pectinioside D (*47*).

13 Versicoside A (= Forbeside A):

$$\text{Ga l} \xrightarrow{1\text{-}3} \text{Fuc} \xrightarrow{1\text{-}2} \text{Ga l} \xrightarrow{1\text{-}4} \text{Xy l} \xrightarrow{1\text{-}3} \text{Qu i} \xrightarrow{1\text{-}6} \text{A}$$
$$\uparrow {\scriptstyle 1\text{-}2}$$
$$\text{Qu i}$$

OCCURRENCE: *Asterias amurensis versicolor (65), Pisaster ochraceus and Pisaster brevispinus (72), Pisaster giganteus (73), Asterias forbesi (76)*.

ACTIVITY: Inhibits cell division in fertilized sea urchin eggs; (ED$_{50}$ 30 µg/ml; LD$_{99}$ 10 µg/ml) (*34*).

PHYSICAL DATA: m.p. 198.5–200.5°C (dec); $[\alpha]_D = -1.7°$ (H$_2$O), FAB-MS (+ ve ion), m/z 1451 $[M_{SO_3Na} + Na]^+$, 1445 $[M_{SO_3K} + H]^+$, 1349 $[M_{SO_3H} + Na]^+$.

Versicoside A is the first example of a hexaglycoside isolated from starfishes. The structure was determined by ^{13}C-NMR and FAB mass spectrometry applied to the native saponin itself and to the shortened glycosides obtained by partial hydrolysis with *Charonia lampas* glycosidase mixture, the pentaglycoside originating from **13** by loss of the terminal galactose and the common triglycoside Qui (1 → 2) Xyl (1 → 3) Qui (1 → 6)—A. The sites of the interglycosidic linkages were confirmed by the usual methylation analysis (*65*). A 2D-NMR analysis (COSY, RECSY and NOESY) was later applied to a sample of versicoside A, re-isolated from *Asteria forbesi* (*76*), which allowed determination of the complete sugar sequence and the sites of interglycosidic linkages, thus obviating the need for hydrolysis.

14 Marthasteroside A$_1$:

$$Fuc \xrightarrow{1-3} Fuc \xrightarrow{1-2} Gal \xrightarrow{1-4} Xyl \xrightarrow{1-3} Qui \xrightarrow{1-6} A$$
$$\uparrow 1-2$$
$$Qui$$

OCCURRENCE: *Marthasterias glacialis* (*82, 83*), *Astropecten latespinosus* (*79*), *Linckia laevigata* (*62*), *Nardoa gomophia* (*70*), *Pisaster brevispinus* (*72*), *Acanthaster planci* (*80*), *Astropecten scoparius* (*84*).

ACTIVITY: Cytotoxic against bovine turbinate cells at concentration of 10 μg/ml (*36*); antitumor activity against human limphoma cells (JUR-CAT) and mouse limphoma cells (YAC-1) at a dose of 50 μg/ml (50% inhibition) (*36*); also inhibits cell division in fertilized sea urchin eggs (ED$_{50}$ 15 μg/ml; LD$_{99}$ 5 μg/ml) (*34*).

PHYSICAL DATA: $[\alpha]_D = +3.6°$ (MeOH); FAB-MS (+ ve ion), m/z 1451 $(M_{SO_3K} + Na)^+$, 1435 $[M_{SO_3Na} + Na]^+$, 1429 $[M_{SO_3K} + H]^+$, 1413 $[M_{SO_3Na} + H]^+$.

Marthasteroside A$_1$ was first isolated as a native saponin from *Marthasterias glacialis* (*82, 83*) and as a desulphated saponin from *Astropecten latespinosus* after solvolysis of a glycoside mixture (*79*). The interglycosidic linkages in the saccharide chain were determined by methanolysis of the native saponin followed by acid methanolysis and *p*-bromobenzoylation. The permethylated sugars (permethylated methyl quinovosides and methyl fucosides) were identified by GLC, and the monobenzoates [methyl 3,4,6-tri-O-methyl-2-O-(*p*-bromobenzoyl)-α-D-galactopyranoside, methyl 2,4-di-O-methyl-3-O-(*p*-bromobenzoyl)-α-D-quinovopyranoside, methyl 2,4-di-O-methyl-3-O-(*p*-bromobenzoyl)-α-D-fucopyranoside] and the dibenzoate [methyl 3-O-methyl-2,4-di-O-(*p*-bromobenzoyl)-α-D-xylopyranoside] were identified, after hplc separation, by ^{1}H-NMR spectroscopy. The sugar sequence was derived by enzymatic hydrolysis with the glycosidase mixture from *Charonia lampas*, which gave the triglycoside Qui (1 → 2) Xyl (1 → 3) Qui (1 → 6) A along with minor amount of a tetraglycoside containing one galactose unit in addition to quinovose (× 2) and xylose (× 1) (*83*).

15 Marthasteroside A$_2$ (= luidiaglycoside A):

$$Fuc \xrightarrow{1-3} Fuc \xrightarrow{1-2} Qui \xrightarrow{1-4} Xyl \xrightarrow{1-3} Qui \xrightarrow{1-6} A$$
$$\uparrow 1-2$$
$$Qui$$

Occurrence: *Luidia maculata* (*66, 81*), *Marthasterias glacialis* (*82, 83*).
Activity: Cytotoxic against bovine turbinate cells at concentration of 10 µg/ml level (*36*); also inhibits cell division in fertilized sea urchin eggs (ED_{50} 15 µg/ml; LD_{99} 10 µg/ml) (*34*).
Physical data: $[\alpha]_D = + 6.5°$ (MeOH); FAB-MS (+ ve ion), m/z 1419 $[M_{SO_3Na} + Na]^+$.

The structure of marthasteroside A_2, which differs from marthasteroside A_1 (**14**) by replacement of galactose with quinovose was determined by the same steps (*83*). The ^{13}C-NMR assignments were made essentially by comparison with reference substances.

16 Pectinioside E:

$$\text{Fuc} \xrightarrow{1-4} \text{Glc} \xrightarrow{1-4} \text{Qui} \xrightarrow{1-3} \text{Qui} \xrightarrow{1-6} \textbf{A}$$
$$\qquad\qquad \uparrow_{1-2} \qquad \uparrow_{1-2}$$
$$\qquad\qquad \text{Xyl} \qquad \text{Qui}$$

Occurrence: *Asterina pectinifera* (*37*).
Activity: Cytotoxic against L1210 (IC_{50} 8.0 µg/ml) and KB cells (IC_{50} 11.5 µg/ml) (*37*).
Physical data: m.p. 225–230°C; $[\alpha]_D = - 13.0°$ (H_2O), FAB-MS (– ve ion), m/z 1389 $[M_{SO_3^-}]$.

Pectinioside E is the second reported asterosaponin having an hexasaccharide chain possessing two branches, pectinioside B (**48**) being the first one. The nature of the interglycosidic linkages were determined by combining the ^{13}C-NMR spectra and methylation analysis by GC-MS of the partially methylated alditol acetates. Enzymatic hydrolysis with glycosidase mixture from *Charonia lampas* gave the triglycoside Qui (1 → 2) Qui (1 → 3) Qui (1 → 6)thornasterol A sulphate, already obtained from pectinioside A (*12*), and a pentaglycoside arising from **16** by removal of the terminal fucose.

17 Pectinioside F:

$$\text{Gal} \xrightarrow{1-4} \text{Glc} \xrightarrow{1-4} \text{Xyl} \xrightarrow{1-3} \text{Qui} \xrightarrow{1-6} \textbf{A}$$
$$\qquad\qquad \uparrow_{1-2} \qquad \uparrow_{1-2}$$
$$\qquad\qquad \text{Fuc} \qquad \text{Qui}$$

Occurrence: *Asterina pectinifera* (*37*).
Activity: No antitumor activity below 20 µg/ml.
Physical data: m.p. 225–230°C; $[\alpha]_D = - 5.8°$ (H_2O); FAB-MS (– ve ion), m/z 1405 $[M_{SO_3^-}]$.

Structure analysis of the saccharide chain followed the same criteria as described earlier. Enzymatic partial hydrolysis removed the terminal galactose to give a pentaglycoside which had ^{13}C-NMR signals due to the sugar moiety fully superimposable on those of acanthaglycoside C (7). The hexasaccharide chain of pectinioside F is identical with that of pectinioside B (48), the difference between the two saponins residing only in the structure of the side chain of the aglycone. This is the third asterosaponin in which the oligosaccharide has two branches.

18 Pectinioside G:

$$
\text{A r a} \xrightarrow{1-4} \text{G l c} \xrightarrow{1-4} \text{Qu i} \xrightarrow{1-3} \text{Qu i} \xrightarrow{1-6} \text{A}
$$
$$
\uparrow^{1-2} \qquad \uparrow^{1-2}
$$
$$
\text{F u c} \qquad \text{Qu i}
$$

PHYSICAL DATA: $[\alpha]_D = +5.8°$ (MeOH); FAB-MS (− ve ion), m/z 1389 $[M_{SO_3^-}]$.

OCCURRENCE: *Asterina pectinifera (63)*, *Patiria miniata (64)*.

In addition to the major peak corresponding to the molecular anion (m/z 1389), the negative ion FAB mass spectrum exhibited fragments arising from separate loss of arabinose (m/z 1257; loss of 132 mass units) and fucose (m/z 1243; loss of 146 mass units) and a minor fragment at m/z 949, interpreted as due to the loss of arabinose, glucose and fucose. These data were assumed as the first indication of the presence of the second branching point on the glucose unit. This was confirmed by identification of the terminal arabinose, fucose and quinovose. The interglycosidic linkages were deduced from ^{13}C-NMR data. In particular in the penta-glycoside, obtained by removal of the terminal fucose on enzymatic hydrolysis, the signal of the anomeric carbon of glucose moved downfield to 104.3 ppm (102.1 ppm in **18**). This established that the terminal fucose is located at C-2 and accordingly that the arabinose is at C-4 of glucose (63).

19 Patirioside B:

$$
\text{G a l} \xrightarrow{1-4} \text{G l c} \xrightarrow{1-4} \text{Qu i} \xrightarrow{1-3} \text{Qu i} \xrightarrow{1-6} \text{A}
$$
$$
\uparrow^{1-2} \qquad \uparrow^{1-2}
$$
$$
\text{F u c} \qquad \text{Qu i}
$$

OCCURRENCE: *Patiria miniata (64)*.

ACTIVITY: Antifungal (minimum inhibitory amount against *Cladosporium cucumerinum* 40 µg) (35).

PHYSICAL DATA: $[\alpha]_D = +6.0°$ (MeOH); FAB-MS ($-$ve ion), m/z 1419 $[M_{SO_3^-}]$.

The structure of **19** differs from that of pectinioside G (**18**) only in that the terminal arabinose is here replaced by galactose, and was similarly derived.

20 Myxodermoside A:

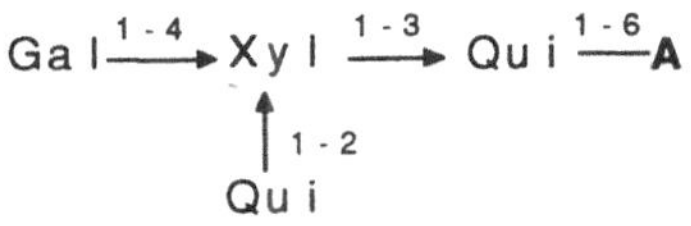

$$Ga\ l\xrightarrow{1-4}Xy\ l\xrightarrow{1-3}Qu\ i\xrightarrow{1-6}A$$
$$\uparrow 1-2$$
$$Qu\ i$$

OCCURRENCE: *Myxoderma platyacanthum* (*85*).
ACTIVITY: Inhibits cell division in fertilized sea urchin eggs (15% inhibition at 10^{-5} M).
PHYSICAL DATA: $[\alpha_D = +3.1°$ (H_2O), FAB-MS ($-$ve ion), m/z 1097 $[M_{SO_3^-}]$.

The structure was entirely derived from spectral data [FAB-MS, ^{1}H- and ^{13}C-NMR] and comparison with known asterosaponins.

Myxodermoside A is one of the very few examples of a tetrasaccharide isolated from starfishes. It is formally related to thornasteroside A (**4**) by loss of the terminal fucose; whether this shortened glycoside originates by enzymatic hydrolysis of penta- or hexasaccharides or it is a biological precursor is an open question.

21 Forbeside F:

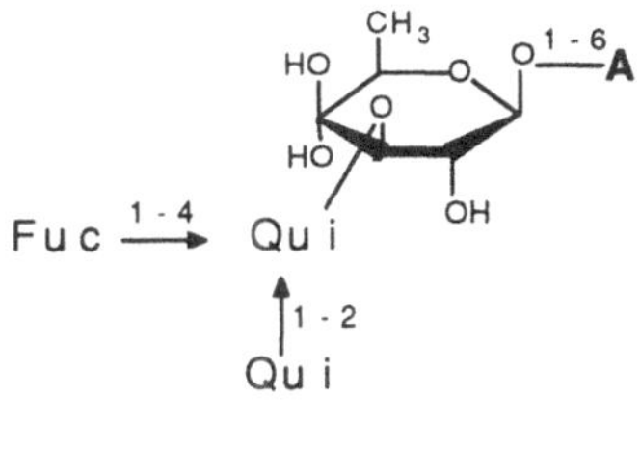

$$Fuc\xrightarrow{1-4}Qu\ i$$
$$\uparrow 1-2$$
$$Qu\ i$$

OCCURRENCE: *Asterias forbesi* (*48*).
PHYSICAL DATA: m.p. 208°C; $[\alpha]_D = +11.6°$ (H_2O); FAB-MS ($+$ve ion), m/z 1157 $[M_{SO_3Na} + Na]^+$.

The structure was derived by ^{1}H- and ^{13}C-NMR spectroscopy and comparison with forbeside C [= ovarian asterosaponin I (OA-I), **6**] to which it is structurally related by loss of the terminal fucose.

22 Forbeside G:

$$Qu\,i \xrightarrow{1\text{-}2} Qu\,i$$

Occurrence: *Asterias forbesi* (*48*).
Physical data: m.p. 194°C; $[\alpha]_D = -4.7°$ (H_2O), FAB-MS (− ve ion), m/z 965 [$M_{SO_3^-}$].

The structure was derived by the close correlation of its ^1H- and ^{13}C-signals with their counterparts in forbeside C (**6**) and F (**21**).

23 Forbeside H:

$$Qu\,i \xrightarrow{1\text{-}2} Xy\,l \xrightarrow{1\text{-}3} Qu\,i \xrightarrow{1\text{-}6} A$$

Occurrence: *Asterias forbesi* (*48*).
Physical data: m.p. 188°C; $[\alpha]_D = -4.7°$ (H_2O), FAB-MS (− ve ion), m/z 935 [$M_{SO_3^-}$].

The structure was derived by ^1H-^1H COSY data, which allowed assignments of ^1H chemical shifts for each sugar unit, followed by a HETCOR experiment, which permitted assignment of ^{13}C signals in the trisaccharide chain and revealed the location of interglycosidic linkages (*48*).

The saponins **21**–**23** are shortened glycosides possessing a tetrasaccharide (**21**) and a trisaccharide (**22**–**23**) chain. Compounds **21** and **22** are related to OA-I (**6**), the saponin containing the same 6-deoxy-*xylo*-hex-4-ulopyranosyl unit which is one major component of *A. forbesi*, whereas compound **23** is apparently related to versicoside A (**13**), which is also a major saponin from *A. forbesi*, by loss of the terminal trisaccharide residue attached at C-4 of the branched xylose. As in the case of the previous tetrasaccharide **20**, it is uncertain whether these shortened glycosides originate by enzymatic hydrolysis of the major penta- and hexa-glycosides or are their biological precursors.

24 Acanthaglycoside A:

$$Fuc \xrightarrow{1-2} Qui \xrightarrow{1-4} Xyl \xrightarrow{1-3} Qui \xrightarrow{1-6} B$$

$$\uparrow 1-2$$

$$Qui$$

OCCURRENCE: *Acanthaster planci* (*86*).
ACTIVITY: Inhibits cell division in fertilized sea urchin eggs; (ED_{50} 50 μg/ml; LD_{99} 50 μg/ml) (*34*).
PHYSICAL DATA: m.p. 208°C; $[\alpha]_D = -8.7°$ (H_2O); FAB-MS (+ ve ion), m/z 1265 $[M_{SO_3K} + H]^+$, 1249 $[M_{SO_3Na} + H]^+$.

The Δ^{24}, 23-keto structure of the aglycose was indicated by the olefinic singlet at δ 6.03 and the olefinic methyl protons at δ 1.92 and 2.16 in the ^1H-NMR spectrum, and confirmed by the CD spectrum, which showed a negative maximum $[\Theta]_{326} = -1050$, due to the $\pi \rightarrow \pi^*$ transition of C = O. The structure of the saccharide chain was determined by combining methylation analysis, ^{13}C-NMR spectroscopy and partial enzymatic hydrolysis, the latter affording a tetrasaccharide resulting from the loss of the terminal fucose along with the trisaccharide Qui(1 → 2)Xyl(1 → 3)Qui—B. The ^{13}C-NMR spectrum was examined by measuring the spin-lattice relaxation times (T_1), and FAB mass spectrometry was successfully used for confirmation of the structure of the natural oligoglycoside sulphate (*86*). The same saccharide chain was also foun in maculatoside **8** from *Luidia maculata*.

25 Acanthaglycoside B:

$$Fuc \xrightarrow{1-2} Glc \xrightarrow{1-4} Xyl \xrightarrow{1-3} Qui \xrightarrow{1-6} B$$

$$\uparrow 1-2$$

$$Qui$$

OCCURRENCE: *Acanthaster planci* (*80*).
PHYSICAL DATA: m.p. 230°C (dec.); $[\alpha]_D = +9°$ (CH_3OH); FAB-MS (+ ve ion), m/z 1303 $[M_{SO_3Na} + K]^+$, 1287 $[M_{SO_3Na} + Na]^+$, 1281 $[M_{SO_3K} + H]^+$.

The saccharide chain is identical with that of acanthaglycoside C (**7**), also isolated from *A. planci.*

26 Acanthaglycoside D:

$$Fuc \xrightarrow{1-2} Gal \xrightarrow{1-4} Xyl \xrightarrow{1-3} Qui \xrightarrow{1-6} B$$
$$\uparrow 1-2$$
$$Qui$$

OCCURRENCE: *Acanthaster planci (80).*
ACTIVITY: Inhibits cell division in fertilized sea urchin eggs; (ED_{50} 15 μg/ml; LD_{99} 5 μg/ml) (*34*).
PHYSICAL DATA: m.p. 226–227°C (dec.); $[\alpha]_D = +1.3°$ (H_2O); FAB-MS (+ ve ion), m/z 1303 $[M_{SO_3Na} + K]^+$, 1287 $[M_{SO_3Na} + Na]^+$.
The saccharide chain is identical with that of thornasteroside A (**4**).

27 Marthasteroside B:

$$Fuc \xrightarrow{1-2} Fuc \xrightarrow{1-4} Qui \xrightarrow{1-3} Glc \xrightarrow{1-6} C$$
$$\uparrow 1-2$$
$$Qui$$

OCCURRENCE: *Marthasterias glacialis (82, 83), Luidia maculata (66), Coscinasterias tenuispina (42).*
ACTIVITY: Inhibits cell division in fertilized sea urchin eggs (75% inhibition at 10^{-7} M); antifungal activity (minimum inhibitory amount against *Cladosporium cucumerinum*, 20 μg) (*35*).
PHYSICAL DATA: $[\alpha]_D = +9.0°$ (CH_3OH); FAB-MS (+ ve ion), m/z 1301 $[M_{SO_3Na} + K]^+$, 1285 $[M_{SO_3Na} + Na]^+$, 1279 $[M_{SO_3K} + H]^+$, 1263 $[M_{SO_3Na} + H]^+$.
Partial characterization of the asterosaponins from *Marthasterias glacialis* was described in 1970 by MACKIE and TURNER (*87*), who separated the steroidal glycosides into two components, designated M_1

and M_2, glycoside M_2 being the most active in eliciting the avoidance reaction of the mollusc *Buccinum undatum*. The same group in Aberdeen continued work on saponins from *M. glacialis* and determined the structure of the principal aglycones obtained by hydrolysis of the crude saponin mixture. Prolonged acid hydrolysis yielded marthasterone (desulphated C) and dihydromarthasterone (desulphated D) (*88*). Partial hydrolysis gave marthasterone-6-O-β-D-glucopyranoside (*89*). More recently we have reported the isolation of four principal saponins, marthasteroside A_1 (**14**), A_2 (**15**), B (**27**) and C (**29**) from the starfish *M. glacialis* (*82*), and determined their structures (*83*). The structures of the genuine aglycones were established by ^{1}H- and ^{13}C-NMR spectra of the intact saponins. Marthasteroside B (**27**) and C (**29**) are pentaglycosides containing the same oligosaccharide moiety. On permethylation followed by methanolysis and *p*-bromobenzoylation, both **27** and **29** gave permethylated methyl fucoside and permethylated methyl quinovoside, thus indicating the presence of one branching point, and methyl 2,4-di-O-(*p*-bromobenzoyl)-3-O-methyl-α-D-quinovopyranoside, methyl 2-O-(*p*-bromobenzoyl)-3,4-di-O-methyl-α-D-fucopyranoside and methyl 3-O-(*p*-bromobenzoyl)-2,4,-6-tri-O-methyl-α-D-glucopyranoside. Thus the interglycosidic linkages were determined. Partial enzymatic hydrolysis established the sequence of the sugars. The ^{13}C-NMR assignments, based on comparison with the references and known glycosidation shifts as well as FAB mass spectral fragments, were consistent with the proposed structures.

28 Luidiaglycoside C:

$$\text{Fuc} \xrightarrow{1\text{-}2} \text{Qui} \xrightarrow{1\text{-}4} \text{Qui} \xrightarrow{1\text{-}3} \text{Glc} \xrightarrow{1\text{-}6} \textbf{C}$$

$$\uparrow 1\text{-}2$$
$$\text{Qui}$$

OCCURRENCE: *Luidia maculata* (*90*).
ACTIVITY: Inhibits cell division in fertilized sea urchin eggs; (ED_{50} 10 μg/ml; LD_{99} 5 μg/ml) (*34*).
PHYSICAL DATA: $[\alpha]_D = +13.0°$ (CH$_3$OH); FAB-MS (+ ve ion), m/z 1280 $[M_{SO_3NH_4} + K]^+$.

The structure was derived as usual by methylation analysis, partial enzymatic hydrolysis, FAB-MS and ^{13}C-NMR spectroscopy. It differs from marthasteroside B (**27**) only in that the internal fucose is here replaced by quinovose.

29 Marthasteroside C:

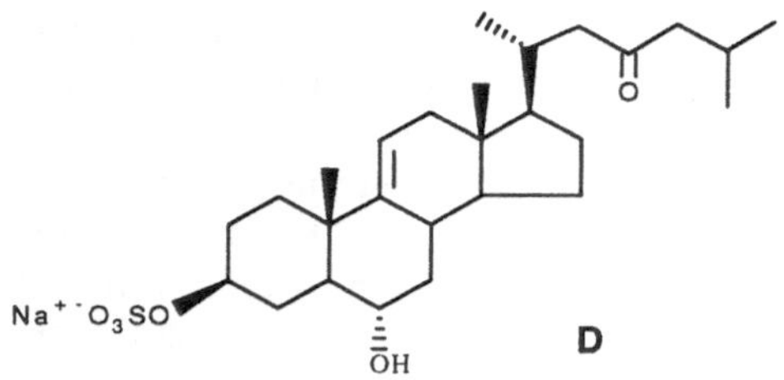
Na⁺ ⁻O₃SO
OH
O
D

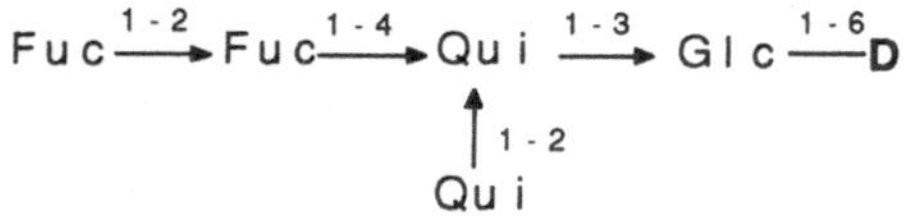
Fuc 1-2 Fuc 1-4 Qui 1-3 Glc 1-6 D
1-2
Qui

OCCURRENCE: *Marthasterias glacialis* (*82, 83*), *Luidia maculata* (*66*), *Coscinasterias tenuispina* (*42*).
ACTIVITY: Cytotoxic against turbinate cells from bovine fetus (10 μg/ml) (*36*); antifungal (minimum inhibitory amount against *Cladosporium cucumerinum*, 20 μg); also inhibits cell division in fertilized sea urchin eggs: (50% inhibition at 10^{-7} M) (*35*).
PHYSICAL DATA: $[\alpha]_D = +13.3°$ (CH_3OH); FAB-MS (+ve ion), m/z 1303 $[M_{SO_3Na} + K]^+$, 1287 $[M_{SO_3Na} + Na]^+$, 1281 $[M_{SO_3K} + H]^+$, 1265 $[M_{SO_3Na} + H]^+$.

This is the 24,25-dihydroderivative of marthasterone B (**27**). The structure of the aglycone was established by SMITH, TURNER and MACKIE (*88*).

30 Luidiaglycoside D:

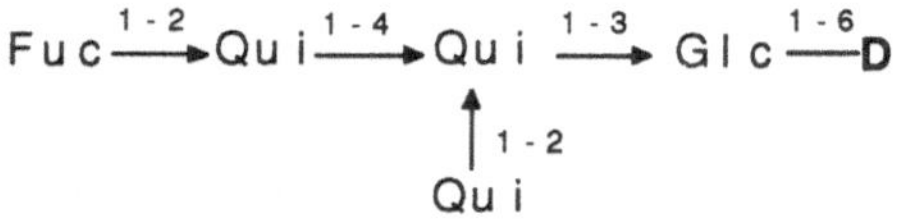
Fuc 1-2 Qui 1-4 Qui 1-3 Glc 1-6 D
1-2
Qui

OCCURRENCE: *Luidia maculata* (*90*).
ACTIVITY: Inhibits cell division in fertilized sea urchin eggs; (ED_{50} 5 μg/ml; LD_{99} 5 μg/ml) (34).
PHYSICAL DATA: $[\alpha]_D = +14.5°$(CH_3OH); FAB-MS (+ve ion), m/z 1282 $[M_{SO_3NH_4} + K]^+$.

The saccharide chain is as in luidiaglycoside C (**28**).

31 Ophidianoside B:

$$Fuc \xrightarrow{1-2} Gal \xrightarrow{1-4} Xyl \xrightarrow{1-3} Qui \xrightarrow{1-6} E$$
$$\uparrow^{1-2}$$
$$Qui$$

OCCURRENCE: *Ophidiaster ophidianus* (*67*), *Hacelia attenuata* (*67*).
PHYSICAL DATA: $[\alpha]_D = +2.8°$ (MeOH); FAB-MS (+ ve ion), m/z 1275 $[M_{SO_3Na} + Na]^+$.

The saccharide chain is identical with that of thornasteroside A (**4**); the 24-nor thornasterol A structure of the aglycone was determined by ^{1}H and ^{13}C-NMR spectroscopy of the intact saponin and comparison with thornasteroside A (**4**).

32 Ophidianoside C:

$$Fuc \xrightarrow{1-2} Xyl \xrightarrow{1-4} Xyl \xrightarrow{1-3} Qui \xrightarrow{1-6} E$$
$$\uparrow^{1-2}$$
$$Qui$$

OCCURRENCE: *Ophidiaster ophidianus* (*67*), *Hacelia attenuata* (*67*).
PHYSICAL DATA: $[\alpha]_D = -2.9°$ (MeOH); FAB-MS (+ ve ion), m/z 1245 $[M_{SO_3Na} + Na]^+$;

The saccharide chain is identical with that of ophidianoside F (**9**) and the structure was derived by following the usual steps: methylation analysis, enzymatic hydrolysis with *Charonia lampas* glycosidase mixture, affording the usual triglycoside Qui (1 → 2)Xyl (1 → 3)Qui—E along with the tetraglycoside originating by loss of the terminal fucose, and ^{13}C-NMR spectroscopy and FAB mass spectral fragmentation pattern.

Ophidianosides B (**31**) and C (**32**) are the major components of the saponin mixture of both *O. ophidianus* and *H. attenuata*. Sterols with a

24-norcholestane (or 26, 27-bis-nor-ergostane) skeleton represent a widespread class of marine sterols of dietary origin and have been found as minor sterol components in every marine invertebrate phylum. The discovery of a C-26 steroidal glycoside in two starfishes of the same family (Ophidiasteridae) as the major components of the saponin mixture is noteworthy.

33 Tenuispinoside A:

$$Fuc \xrightarrow{1-2} Gal \xrightarrow{1-4} Xyl \xrightarrow{1-3} Qui \xrightarrow{1-6} F$$
$$\uparrow 1-2$$
$$Qui$$

OCCURRENCE: *Coscinasterias tenuispina (42)*.
PHYSICAL DATA: $[\alpha]_D = -56.6°$ (MeOH); FAB-MS (+ve ion), m/z 1289 $[M_{SO_3Na} + Na]^+$, FAB-MS (− ve ion), m/z 1243 $[M_{SO_3^-}]$.

This saponin is isomeric with thornasteroside A (**4**); the saccharide chain is identical in both compounds and the only difference is in the steroid side chain in that *33* contains a 22,23-epoxy functionality instead of the common carbonyl group at C-23. Two one proton ¹H-NMR signals at δ 2.94 (td, J = 6, 2.5 Hz, 23-H) and 2.76 (d, J = 2.5 Hz. 22-H) coupled to each other, were indicative of the presence of a 22,23-epoxy group. The ¹³C-NMR spectrum supported this assumption, showing two signals characteristic of methine protons at 65.7 and 53.5 ppm. The *trans* configuration 22R, 23S was based on ¹³C-NMR data and comparison with model compounds.

34 Tenuispinoside B:

$$Fuc \xrightarrow{1-2} Fuc \xrightarrow{1-4} Xyl \xrightarrow{1-3} Qui \xrightarrow{1-6} F$$
$$\uparrow 1-2$$
$$Qui$$

OCCURRENCE: *Coscinasterias tenuispina (42)*.

PHYSICAL DATA: $[\alpha]_D = -11.9°$ (MeOH); FAB-MS (+ ve ion), m/z 1273 $[M_{SO_3Na} + Na]^+$, 1267 $[M_{SO_3K} + H]^+$; FAB-MS (− ve ion), m/z 1227 $[M_{SO_3^-}]$.

This substance is isomeric with the thornasterol A-containing regularoside B (11), a minor saponin isolated from *Halityle regularis (69)*, also co-occurring in *C. tenuispina*. The saccharide chain is identical in both compounds, while the aglycone of 34 has the 22,23-epoxy functionality of tenuispinoside A.

35 Asteroside A:

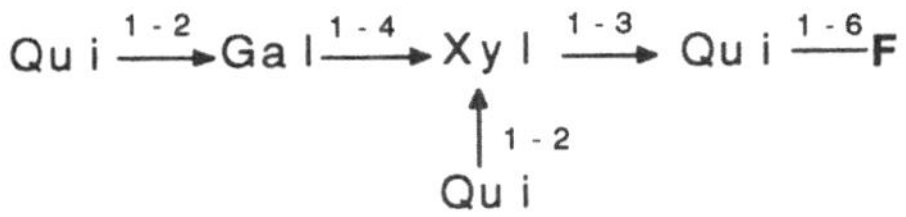

$$\text{Qu i} \xrightarrow{1-2} \text{Ga l} \xrightarrow{1-4} \text{Xy l} \xrightarrow{1-3} \text{Qu i} \xrightarrow{1-6} \text{F}$$

$$\uparrow 1-2$$

$$\text{Qu i}$$

OCCURRENCE: *Asterias amurensis (43)*.

PHYSICAL DATA: $[\alpha]_D = +5.8°$ (MeOH); FAB-MS (− ve ion), m/z 1243 $[M_{SO_3^-}]$.

This substance represents the third example among asterosaponins containing a (20R,22R,23S)-22-epoxy cholesten-9(11)-ene-3β,6α,20-triol 3β-sulphated aglycone. The saccharide chain is identical with that of glycoside B$_2$ (5), one of the major component of *Asterias amurensis* (43).

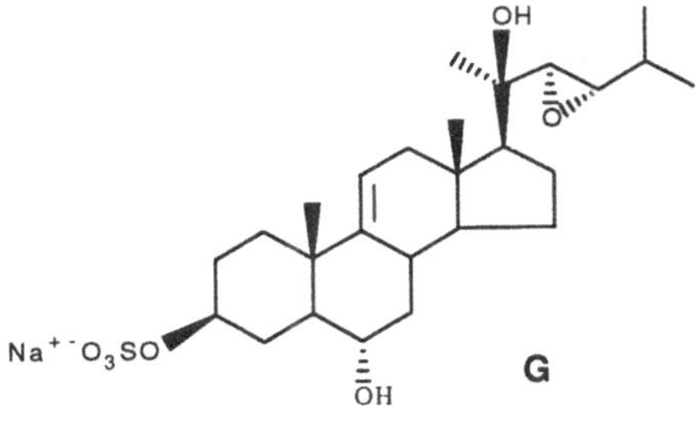

G

36 Asteroside B:

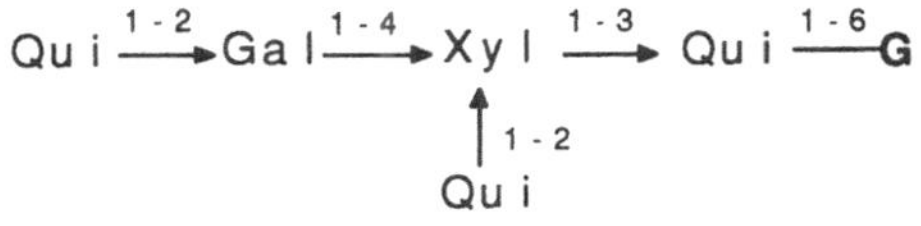

$$\text{Qu i} \xrightarrow{1-2} \text{Ga l} \xrightarrow{1-4} \text{Xy l} \xrightarrow{1-3} \text{Qu i} \xrightarrow{1-6} \text{G}$$

$$\uparrow 1-2$$

$$\text{Qu i}$$

Occurrence: *Asterias amurensis* (*43*).

Physical data: $[\alpha]_D = -9.2°$ (MeOH); FAB-MS (− ve ion), m/z 1229 $[M_{SO_3^-}]$.

The saccharide chain is identical with that of glycoside B_2 (**5**) and asteroside A (**35**). The FAB mass spectrum showed a molecular anion peak at m/z 1229 and the fragment ions corresponding to sequential loss of the sugar units, all shifted by fourteen mass units less relative to asteroside A (**35**, m/z 1243). The ^{1}H-NMR spectrum showed a one proton doublet (J = 2.2 Hz) at δ 2.83 coupled with a one proton double doublet (J = 2.2, 7.5 Hz) at δ 2.57, indicative of an epoxide functionality, and the isopropyl methyl signals were shifted to δ 0.98 and 1.04 ppm. The remaining signals were identical with those of asteroside A (**35**). Based on these observation the 22,23-epoxy-24-nor-5a-cholest-9(11)-ene-3β,6α,20-triol 3β-sulphated structure for the aglycone has been assumed. The ^{13}C-NMR supported this assumption and indicated the *trans* 22R,23R-stereochemistry.

37 Protoreasteroside:

Occurrence: *Protoreaster nodosus* and *Pentaceraster alveolatus* (*68*).

Physical data: $[\alpha]_D = +3.8°$ (MeOH); FAB-MS (+ ve ion), m/z 1289 $[M_{SO_3}Na + K]^+$, 1273 $[M_{SO_3Na} + Na]^+$, 1267 $[M_{SO_3K} + H]^+$, 1251 $[M_{SO_3Na} + H]^+$.

The saccharide chain is identical with those of maculatoside (**8**) and acanthaglycoside A(**24**). The elucidation of the structure of the aglycone

was pursued on the intact saponin. Differences in the shifts of the 18-
(δ 0.85 vs. 0.81) and 21-methyl (δ 1.25 vs. 1.37) protons on going from **37**
to the corresponding thornasterol A saponin **8** indicated perturbations
around C-22. In addition to the signals assigned to the saccharide chain
and those for C-3, C-6 and C-20 of the aglycone, the ^{13}C-NMR spectrum
contained one more hydroxymethine signal at 77.8 ppm, which confirmed the presence of an hydroxyl group at C-22. The 20R,22S configuration in protoreasteroside, the only known asterosaponin containing a
5α-cholesta-9(11),24(25)-dien-3β,6α,20,22-tetraol aglycone, was based on
the chemical shift of the C-21 methyl protons signal of the intact saponin
in d$_5$-pyridine (δ 1.64) and of the derived acetonide in d$_4$-methanol
(δ 1.37), in comparison with the reported data for 5α-cholesta-3β,20,22-
triol models (*91*).

38 Ovarian asterosaponin-4 (= Co-Aris III):

$$\text{Qu i} \xrightarrow{1\text{-}2} \text{G l c} \xrightarrow{1\text{-}4} \text{Xy l} \xrightarrow{1\text{-}3} \text{Qu i} \xrightarrow{1\text{-}6} \text{I}$$

$$\uparrow {}^{1\text{-}2}$$

$$\text{Qu i}$$

OCCURRENCE: *Asterias amurensis* (*25,43,92*).
ACTIVITY: Suppresses at 125 µg/ml spontaneous oocyte maturation of the
starfish (*92*) and induces the acrosome reaction (*25*).
PHYSICAL DATA: m.p. 189 − 201°C; $[\alpha]_D$ = + 3.7° (MeOH); FAB-MS (
− ve ion), m/z 1227 [M$_{SO_3^-}$].

Hydrolysis of **38** with naringinase gave an aglycone whose structure
was established by ^{1}H- and ^{13}C-NMR spectroscopy and by comparison
with thornasterol-A sulphate. Permethylation of the desulfated saponin
followed by analysis of the derived alditol acetates gave the interglycosidic linkages. Hydrolysis of **38** with a glycosidase mixture gave
three prosapogenols characterized as 6α-quinovopyranoside, 6α-

quinovopyranosyl-(1 → 2)-β-xylopyranoside, and 6α-quinovopyranosyl-(1 → 2)-β-xylopyranosyl-(1 → 3)-β-quinovopyranoside, of the desulphated steroid **I**, respectively. Thus, the saccharide chain of ovarian asterosaponin-4 could be defined as indicated in **38** (*92*).

39 Solasteroside A:

$$Fuc \xrightarrow{1\text{-}2} Fuc \xrightarrow{1\text{-}4} Xyl \xrightarrow{1\text{-}3} Qui \xrightarrow{1\text{-}6} I$$
$$\uparrow^{1\text{-}2}$$
$$Qui$$

OCCURRENCE: *Solaster borealis* (*93*).
PHYSICAL DATA: $[\alpha]_D = +1.1°$ (MeOH); FAB-MS (− ve ion), m/z 1211 $[M_{SO_3^-}]$.

The structure of **39** was derived by ^{1}H- and ^{13}C-NMR spectroscopy and by comparison with the ovarian asterosaponin-4 (**38**), which has the same aglycone, and regularoside B (**11**) and tenuispinoside B (**34**), which have the same saccharide chain.

40 Co-Aris II:

$$Fuc \xrightarrow{1\text{-}2} Fuc \xrightarrow{1\text{-}4} Qui$$
$$\uparrow^{1\text{-}2}$$
$$Qui$$

OCCURRENCE: *Asterias amurensis* (*25*).
PHYSICAL DATA: m.p. 192–200°C; FAB-MS (− ve ion), m/z 1311 $[M_{SO_3^-}$ + glycerin], 1219 $[M_{SO_3^-}]$ (keto form).

The saccharide chain is identical with that of Co-Aris I (**6**); the structure elucidation of the aglycone was pursued on the intact saponin.

Characteristic ^{1}H-NMR signals at δ 1.70 (21-CH$_3$), 1.90 and 2.11 (26-,27-CH$_3$) and 6.16 (24-H) led to the location of double bonds at the 17(20)- and 24(25)-positions. A saponin with 17(20), 24-dien-23-one side chain is unprecedented.

41 Tenuispinoside C:

$$\text{Fuc} \xrightarrow{1-2} \text{Gal} \xrightarrow{1-4} \text{Xyl} \xrightarrow{1-3} \text{Qui} \xrightarrow{1-6} \text{K}$$
$$\uparrow 1-2$$
$$\text{Qui}$$

OCCURRENCE: *Coscinasterias tenuispina (42)*.
PHYSICAL DATA: $[\alpha]_D = +37.5°$ (CH$_3$OH); FAB-MS (+ ve ion), m/z 1305 [M$_{SO_3Na}$ + Na]$^+$, 1299 [M$_{SO_3K}$ + H]$^+$; FAB-MS (− ve ion), m/z 1259 [M$_{SO_3^-}$].

Tenuispinoside C (**41**) represents a major departure from the previous aglycones, differing from the common thornasteroside A (**4**) by the presence of an extra hydroxyl group in the tetracyclic nucleus, as revealed by the FAB mass spectrum which showed the molecular anion peak (m/z 1259) and the fragment ions corresponding to the sequential loss of sugar units, sixteen mass units larger than those relative to thornasteroside A (m/z 1243). The ^{1}H-NMR spectrum exhibited the 12β-H signal as a doublet at δ 3.94 coupled with the signal of the 11-H proton (δ 5.57) by 5 Hz, indicative of its axial (α) orientation.

42 Asteroside D:

$$Qu\,i \xrightarrow{1-2} Gl\,c \xrightarrow{1-4} Xy\,l \xrightarrow{1-3} Qu\,i \xrightarrow{1-6} L$$
$$\uparrow 1-2$$
$$Qu\,i$$

OCCURRENCE: *Asterias amurensis (43)*.
PHYSICAL DATA: $[\alpha]_D = +5.0°$ (MeOH); FAB-MS (− ve ion), m/z 1349 $[M_{SO_3^-} + $ thioglycerol] (100%), 1241 $[M_{SO_3^-}]$ (50%).

This saponin is related to the ovarian asterosaponin-4 (**38**) by having an exo-methylene group at C-24 in the steroid side chain, as indicated by two one-proton olefinic signals at δ 4.72 and 4.78 and by the signals for 26- and 27-CH$_3$ which are shifted to δ 1.07 (6H,d,J = 7.5 Hz) ppm.

43 Versicoside C (= Thornasteroside B):

$$Fuc \xrightarrow{1-2} Ga\,l \xrightarrow{1-4} Xy\,l \xrightarrow{1-3} Qu\,i \xrightarrow{1-6} M$$
$$\uparrow 1-2$$
$$Qu\,i$$

OCCURRENCE: *Asterias amurensis* [cf.] *versicolor (77)*; *Coscinasterias tenuispina (42)*.
ACTIVITY: Inhibits cell division in fertilized sea urchin eggs; (ED$_{50}$ 5 µg/ml; LD$_{99}$ 2 µg/ml) (*34*); in this biological assay it is the most active compound among the asterosaponins.
PHYSICAL DATA: m.p. 224 − 226°C; $[\alpha]_D = -9.6°$ (H$_2$O); FAB-MS (+ ve ion), m/z 1319 $[M_{SO_3K} + Na]^+$, 1297 $[M_{SO_3K} + H]^+$, FAB-MS (− ve ion), m/z 1257 $[M_{SO_3^-}]$; CD (c = 0.917, water): $[\Theta]_{287} = -6084$.

Versicoside C is the 24R-methyl analogue of thornasteroside A (**4**). The 24R configuration was assigned by stereoselective synthesis of both

References, pp. 297–308

(20S,24R)-thornasterol B and (20S,24S)-thornasterol B (75). Differentiation between the two epimers is possible by the different molecular ellipticity [Θ] in the CD spectra, which is much higher in the 24R isomer: values reported are [Θ]$_{283}$ = $-$ 5780 for the 24R isomer and [Θ]$_{277}$ = $-$ 631 for the 24S isomer. Small but significative differences are also observed between the ^{13}C-NMR spectra of the two epimeric model compounds; the resonances for carbons 24 and 28 appeared in the spectrum of the 24R isomer at 54.8 and 11.9 ppm and in that of the 24S isomer at 54.1 and 12.5 ppm.

44 Versicoside B:

$$\text{Gal} \xrightarrow{1\cdot 3} \text{Fuc} \xrightarrow{1\cdot 2} \text{Gal} \xrightarrow{1\cdot 4} \text{Xyl} \xrightarrow{1\cdot 3} \text{Qui} \xrightarrow{1\cdot 6} \text{M}$$
$$\uparrow{\scriptstyle 1\cdot 2}$$
$$\text{Qui}$$

OCCURRENCE: *Asterias amurensis* [cf.] *versicolor* (*77*).

ACTIVITY: Inhibits cell division in fertilized sea urchin eggs: (ED$_{50}$ 10 µg/ml; LD$_{99}$ 5 µg/ml) (*34*).

PHYSICAL DATA: m.p. 206.5–208°C; [α]$_D$ = $-$ 8.4°(H$_2$O); FAB-MS (+ ve ion), m/z 1465 [M$_{SO_3Na}$ + Na]$^+$, 1459 [M$_{SO_3K}$ + H]$^+$, FAB-MS ($-$ ve ion), m/z 1419 [M$_{SO_3^-}$]; CD (c = 0.45,H$_2$O): [Θ]$_{284}$ = $-$ 6028.

Versicoside B is the 24R-methyl analogue of the hexaglycoside versicoside A(**13**).

45 Asteroside C:

$$\text{Qui} \xrightarrow{1\cdot 2} \text{Gal} \xrightarrow{1\cdot 4} \text{Xyl} \xrightarrow{1\cdot 3} \text{Qui} \xrightarrow{1\cdot 6} \text{M}$$
$$\uparrow{\scriptstyle 1\cdot 2}$$
$$\text{Qui}$$

OCCURRENCE: *Asterias amurensis* (*43*).

ACTIVITY: Inhibits cell division in fertilized sea urchin eggs (75% inhibition at 10^{-7} M) (*35*).

PHYSICAL DATA: [α]$_D$ = $-$ 2.0°(MeOH); FAB-MS ($-$ ve ion), m/z 1257 [M$_{SO_3^-}$].

Asteroside C is the 24-methyl analogue of glycoside B$_2$ (**5**).

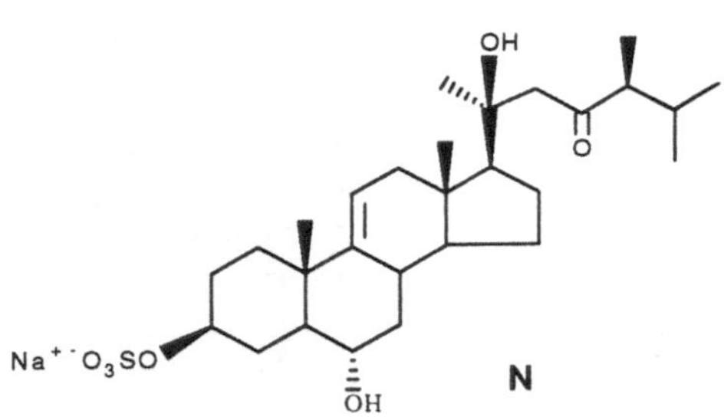

N

46 Acanthaglycoside F:

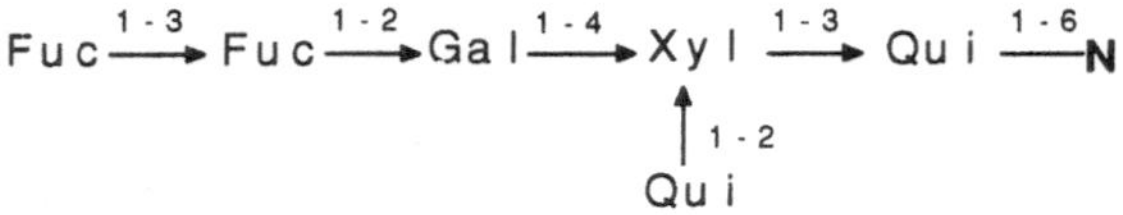

$$Fuc \xrightarrow{1-3} Fuc \xrightarrow{1-2} Gal \xrightarrow{1-4} Xyl \xrightarrow{1-3} Qui \xrightarrow{1-6} N$$
$$\uparrow 1-2$$
$$Qui$$

Occurrence: *Acanthaster planci (80)*.
Activity: Inhibits cell division in fertilized sea urchin eggs; (ED_{50} 5 µg/ml; LD_{99} 2 µg/ml) *(34)*.
Physical data: m.p. 233–228°C, $[\alpha]_D = +11.3°(H_2O)$, FAB-MS (− ve ion), m/z 1403 $[M_{SO_3^-}]^-$, CD (c = 0.900, H_2O): $[\Theta]_{275} = -685$.

Acanthaglycoside F is the 24S-methyl analogue of marthasteroside A_1 **(14)**.

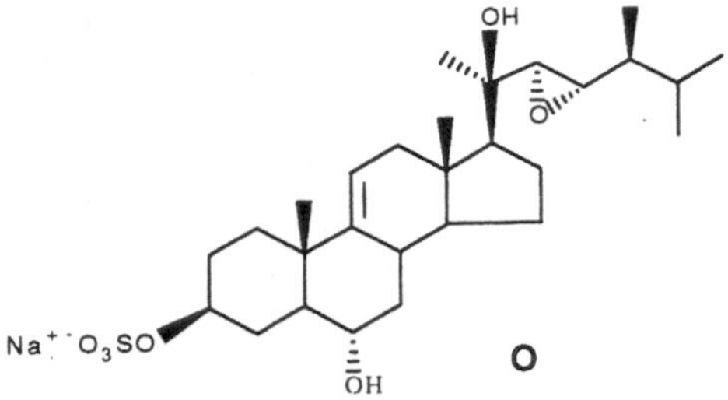

O

47 Regularoside A:

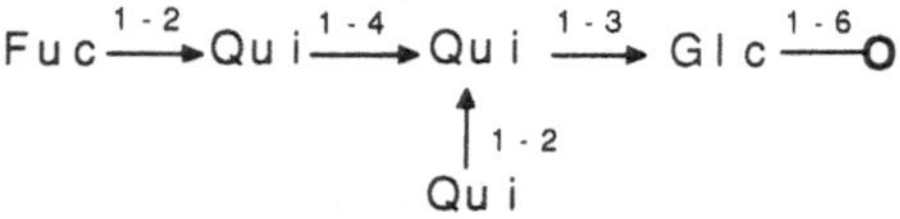

$$Fuc \xrightarrow{1-2} Qui \xrightarrow{1-4} Qui \xrightarrow{1-3} Glc \xrightarrow{1-6} O$$
$$\uparrow 1-2$$
$$Qui$$

Occurrence: *Halityle regularis (69)*.
Physical data: $[\alpha]_D = +12.3°(MeOH)$; FAB-MS (+ ve ion), m/z 1333 $[M_{SO_3^-Na} + K]^+$, 1317 $[M_{SO_3^-Na} + Na]^+$.

Regularoside A is the first reported asterosaponin with the 24-methyl-22,23-epoxycholestane aglycone. In addition to signals reminiscent of the spectra of asterosaponins containing a $\Delta^{9(11)}$-3β,6α,20-trihydroxysteroidal aglycone, the ^{1}H-NMR spectrum of **47** contained two one proton signals at δ 2.75 (d, $J = 7.5$ Hz) and 2.78 (dd, $J = 7.5$ Hz, 2.4 Hz), assigned to the epoxymethine protons, and three methyl doublet signals at δ 0.93, 0.97 and 1.02 assigned to 26-, 27- and 28-methyl protons. Besides the signals for the tetracyclic nucleus carbons already observed in the spectra of the many asterosaponins, the ^{13}C-NMR spectrum showed nine more signals including two doublets shifted downfield to 58.8 and 65.0 ppm, in agreement with the presence of a 24-methyl 22,23-epoxy functionality. The configurations at C-22, C-23 and C-24 were assigned by comparing the ^{1}H- and ^{13}C-NMR-spectra of **47** with those of the four possible *trans*-epoxy-22,23 cholestanes, the *cis*-stereochemistry being eliminated because of the resonance at low field of the C-24 signal in the spectrum of **47**. In the model 22,23(*cis*)-epoxycholestanes the resonances of C-20 and C-24 are shifted upfield by *ca.* 5.0 ppm relative to the *trans*-models because of the γ-gauche interaction. The saccharide chain of **47** is identical to that assigned to luidiaglycoside C (**28**) and D (**30**), minor saponins from *Luidia maculata* (*90*).

48 Pectinioside B:

$$\text{Ga l} \xrightarrow{1\text{-}4} \text{G l c} \xrightarrow{1\text{-}4} \text{Xy l} \xrightarrow{1\text{-}3} \text{Qu i} \xrightarrow{1\text{-}6} \text{O}$$

$$\underset{\text{Fu c}}{\overset{\text{1-2}}{\uparrow}} \qquad \underset{\text{Qu i}}{\overset{\text{1-2}}{\uparrow}}$$

OCCURRENCE: *Asterina pectinifera* (*46*).
ACTIVITY: No antitumor activity below 20 µg/ml (L 1210 and KB cells) (*37*).
PHYSICAL DATA: $[\alpha]_D = +5.5°(H_2O)$; FAB-MS ($-$ ve ion), m/z 1419 $[M_{SO_3^-}]$.

Pectinioside B is the first reported asterosaponin containing a hexasaccharide chain with two branches. Analysis of the terminal sugars indicated the presence of two branches. Partial hydrolysis using *Charonia lampas* glycosidase gave two prosapogenols, one pentaglycoside wherein the terminal galactose is missing, and a trisaccharide possessing two units of quinovose and one of xylose. ^{13}C-NMR studies of pectinioside B itself and of the derived shortened glycosides finally established the complete structure of the unique hexaglycoside.

49 Patirioside A:

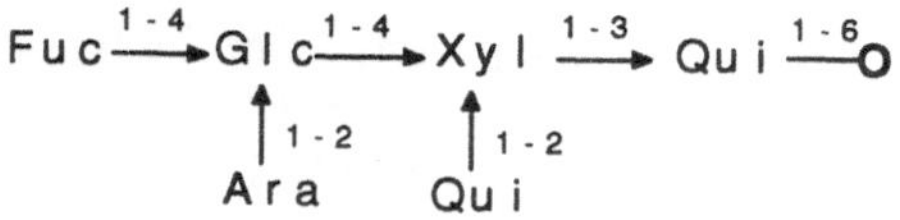

$$Fuc \xrightarrow{1-4} Glc \xrightarrow{1-4} Xyl \xrightarrow{1-3} Qui \xrightarrow{1-6} O$$

with branches:

Fuc←(1-2)Ara, Xyl←(1-2)Qui

OCCURRENCE: *Patiria miniata* (*64*).
PHYSICAL DATA: $[\alpha]_D = +5.8°$ (MeOH); FAB-MS (− ve ion), m/z 1389 $[M_{SO_3^-}]$.

The structure was derived by following the same steps used for the structure elucidation of pectinioside G (**18**) and patirioside B (**19**).

50 Henricioside A:

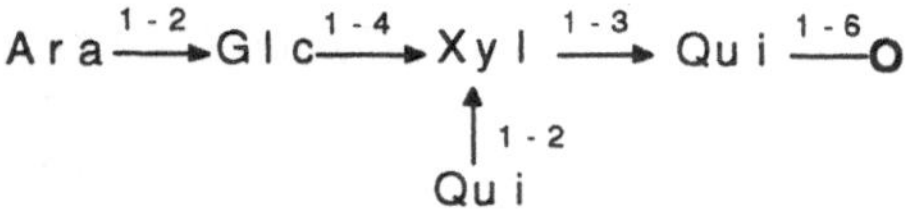

$$Ara \xrightarrow{1-2} Glc \xrightarrow{1-4} Xyl \xrightarrow{1-3} Qui \xrightarrow{1-6} O$$

with branch:

Xyl←(1-2)Qui

OCCURRENCE: *Henricia laeviuscula* (*94*).
ACTIVITY: Antifungal (minimum inhibitory amount against *Cladosporium cucumerinum*, 20 µg) (*35*).
PHYSICAL DATA: FAB-MS (− ve ion), m/z 1243 $[M_{SO_3^-}]$.

The structure was derived entirely by FAB-MS fragmentation pattern (sequence) and ^{13}C-NMR spectroscopy (interglycosidic linkages).

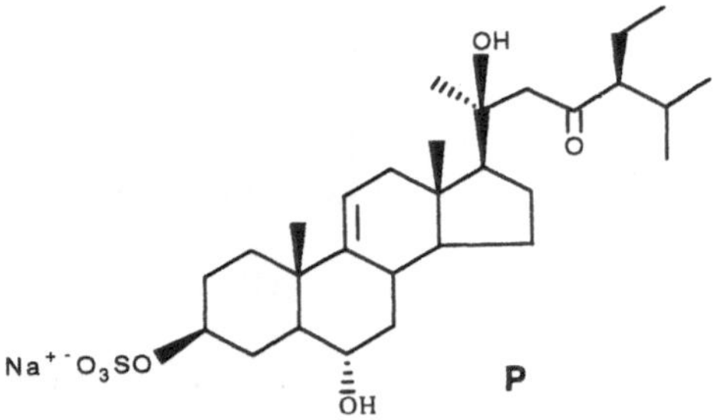

P

51 Pectinioside C:

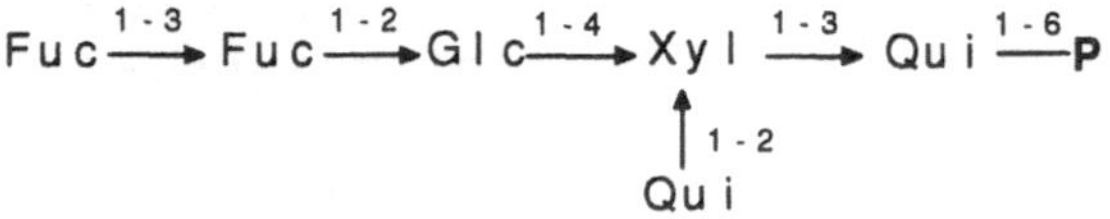

$$Fuc \xrightarrow{1-3} Fuc \xrightarrow{1-2} Glc \xrightarrow{1-4} Xyl \xrightarrow{1-3} Qui \xrightarrow{1-6} P$$

with branch:

Xyl←(1-2)Qui

OCCURRENCE: *Asterina pectinifera* (*47*).
ACTIVITY: Cytotoxic (IC_{50} 11.0 µg/ml against L 1210 and 11.5 µg/ml against KB cells (*37*).
PHYSICAL DATA: $[\alpha]_D = + 5.4°$(MeOH); FAB-MS (− ve ion), m/z 1417 $[M_{SO_3^-}]$.

The interglycosidic linkages of **51** were established by combining ^{13}C-NMR spectroscopy and methylation analysis. Partial enzymatic hydrolysis with *Charonia lampas* glycosidase mixture of **51** gave two prosapogenins, one triglycoside possessing two quinovosyl units and one xylosyl moiety, and one pentaglycoside resulting from the loss of the terminal fucose, which showed ^{13}C-NMR signals due to the sugar moiety superimposable on those of acanthaglycoside C (*7*). The structure of the 24-ethylthornasterol A for the aglycone was derived from spectral data. The 24S configuration was determined by comparing the physical data (^{1}H- and ^{13}C-NMR) of (20S,24S)- and (20S,24R)-3,6-di-O-acetyl-24-ethylthornasterols, stereoselectively synthesized, with those of the aglycone of **51**, obtained after enzymatic hydrolysis and subsequent solvolysis of the glycoside (*95*).

52 Forbeside E:

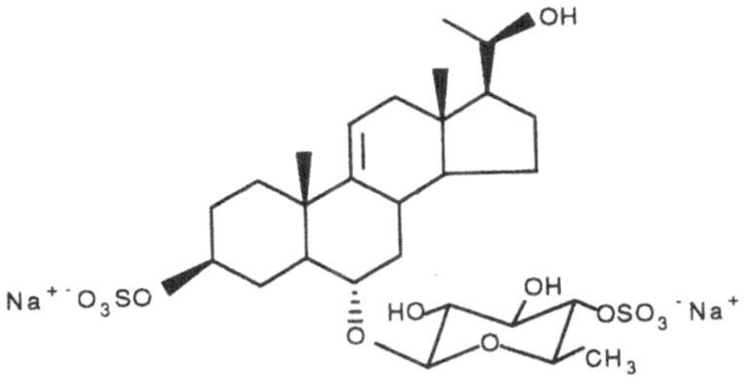

OCCURRENCE: *Asterias forbesi* (*96*).
PHYSICAL DATA: m.p. 238°C; $[\alpha]_D = + 9.5°$ (H_2O); FAB-MS (+ ve ion), m/z 706 $[M + Na - H]^+$.

Forbeside E is a minor constituent of the saponin fraction of the starfish *Asterias forbesi*, whose structure was determined principally by 2D-COSY, DQ COSY, HETCOR and DEPT experiments. We note that the aglycone of **52**, asterogenol, was also isolated as a hydrolysis product of the saponins from *Asterias vulgaris* (*97*), a finding which suggests that at last some of the saponins present in starfish species contain a reduced progesterone side chain.

3.1 Spectroscopy

The structure determination of the asterosaponins has involved the use of spectral techniques such as ^1H and ^{13}C-NMR spectroscopy, by which the structures of the native aglycones can be derived without degradation of the molecule, and by FAB mass spectrometry, which successfully gives the molecular weight of the underivatized sulphated saponins together with useful information on the saccharide sequence. The nature of the glycosidic linkages in the saccharide chain has generally been determined by interpretation of ^{13}C-NMR spectral data, a technique of increasing importance as a larger number of completely assigned model compounds became available.

3.1.1 ^1H-and ^{13}C-NMR spectra of the steroidal aglycones

The aglycones **E, F, G, H, J, K, L** and **O,** have never been obtained as free steroids by hydrolysis of the native glycosides and their structures are derived from ^1H and ^{13}C-NMR spectral analysis of the intact saponins.

Examination of the ^1H-NMR spectra of asterosaponins allows a quick identification of the aglycone structure (Table 2).

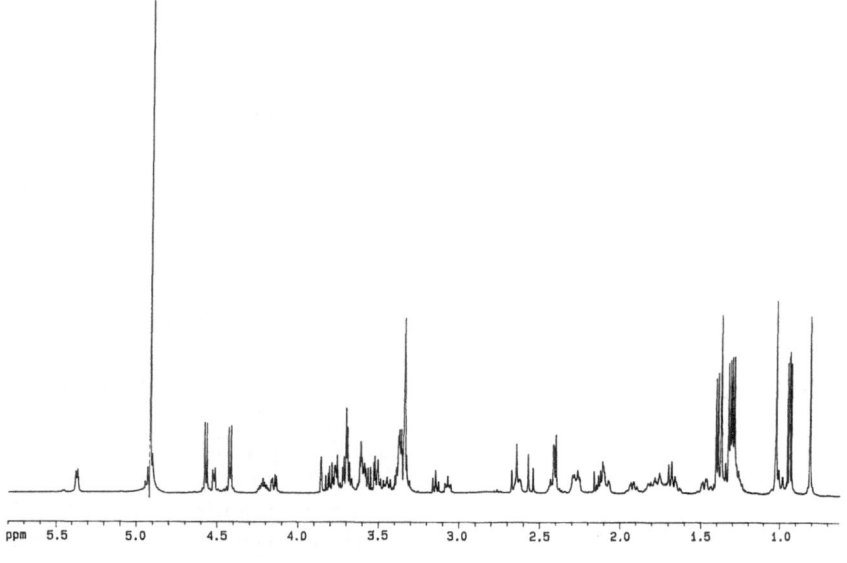

Fig. 1. ^1H NMR spectrum (500 MHz, CD_3OD) of thornasteroside A (**4**)

References, pp. 297–308

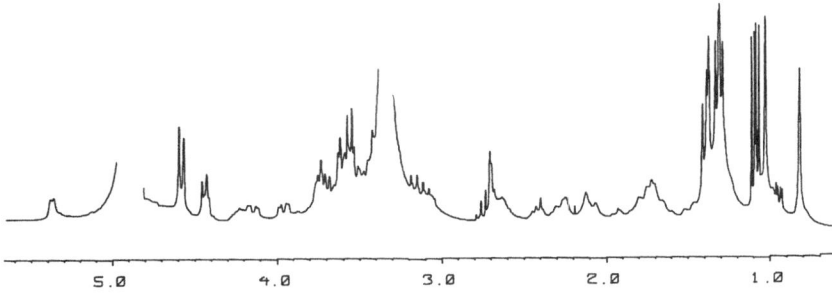

Fig. 2. ^{1}H NMR spectrum (250 MHz, CD$_3$OD) of ophidianoside C (**32**)

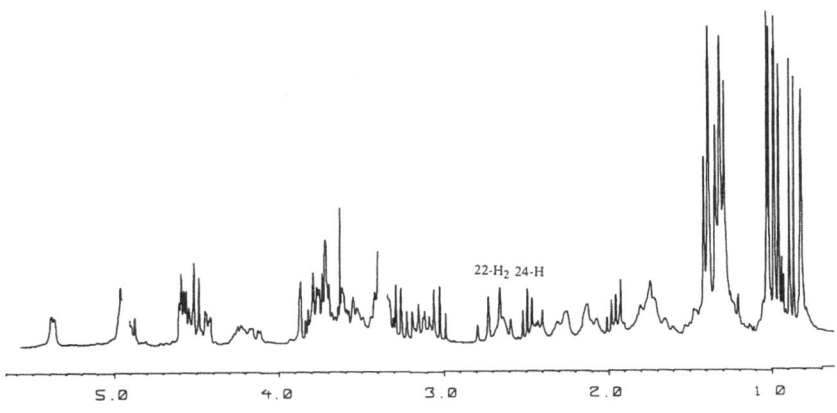

Fig. 3. ^{1}H NMR spectrum (250 MHz, CD$_3$OD) of thornasteroside B (**43**)

The ^{1}H-NMR spectra of asterosaponins are dominated in the downfield region (*ca.* δ 3.0–5.0 ppm) by the resonances of the hydroxymethine and hydroxymethylene protons of the sugar residues (see for example Figs. 1–4). Two isolated signals, appearing in this low field region at constant chemical shifts of δ 4.22 ppm (with the typical shape (W$_{1/2}$ = 22 Hz) of the 3α-proton in a 5α-cholestane skeleton) and of δ 5.37 ppm (br d, $J = 5.5$ Hz), are assigned to 3-H and 11-H of the aglycone. The resonance of the 3α-proton, shifted downfield from its usual position to δ4.22, also establishes the location of the sulphate residue at C-3, The signal for 6-H overlaps with the sugar signals and hence its chemical shift cannot be determined even at 500 MHz. In the spectrum of tenuispinoside C (**41**), which contains a hydroxyl group

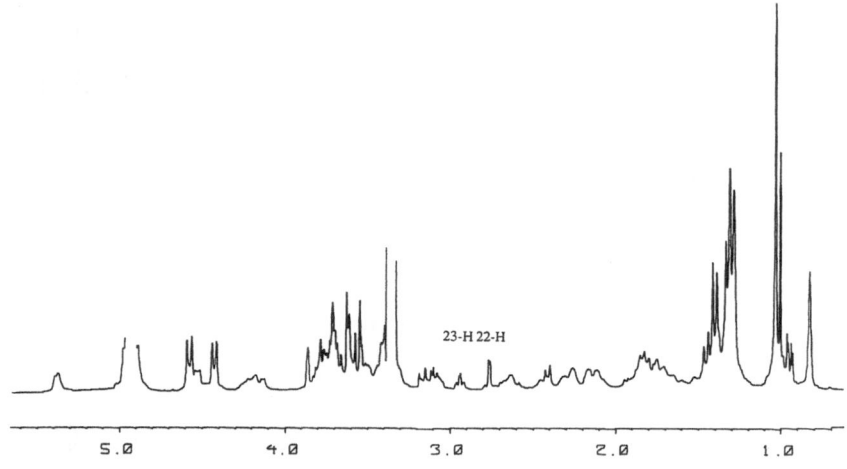

Fig. 4. ¹H NMR spectrum (250 MHz, CD₃OD) of tenuuispinoside A (**33**)

located at C-12α, the resonance of the olefinic proton at C-11 exhibits a
distinct downfield shift to α 5.57. The presence of this "extra" hydroxyl
group is also revealed by the appearance of the 12β-H signal as a doublet
at δ 3.94 ($J = 5$ Hz) whose coupling constant with the 11-H proton is
indicative of its axial (α) orientation (*42*).

In the high field region of the spectra, which is dominated by the
aglycone proton resonances, the methyl shifts and some of the resolved
signals of the side chain protons are very useful for structural assignment.
Also visible in this region of the spectrum are methyl doublets ranging
from δ 1.25 to 1.48 ppm, indicative of the presence of 6-deoxysugars.

The common 20-hydroxy functionality is clearly recognized as a
result of the shift of the C-21 methyl singlet which ranges from δ 1.37 to
δ 1.25, and also indicates the 20S configuration [*cfr.* (20S)- and (20R)-20-
hydroxycholesterol: δ 1.28 and 1.13, respectively], and by downfield shift
of the C-18 methyl singlet to *ca.* δ 0.82 ppm. In 20-hydroxy-23-
oxosteroids the C-21 methyl singlet is observed at a constant chemical
shift of δ 1.37; this value varies slightly in the other 20-hydroxy steroids
depending the functionalization of the side chain. All saponins containing thornasterol A (**A**) show characteristic C-22 and C-24 methylene
signals at *ca.* δ 2.60, an AB quartet with $J = 15$ Hz, and at *ca.* δ 2.40, a
doublet with $J = 6.5$ Hz (Fig. 1). Any change of these signals clearly
points to a different side chain. In ophidianoside B (**31**) and C (**32**),
asterosaponins containing the 24-nor-thornasterol A aglycone (**E**), the C-
22 and C-24 proton signals appear as a broad singlet at δ 2.70 (22-H₂)

References, pp. 297–308

and a multiplet at δ 2.73 (24-H) ppm (Fig. 2). Particularly characteristic in those steroids are also the isopropyl methyl doublet signals shifted downfield to δ 1.08 and 1.10 (*67*). In saponins containing the 24R-methyl thornasterol A (**M**) aglycone the signals C-22 methylene protons are observed as two well separated doublets at δ 2.62 and 2.76 and that of the C-24 proton appears as a clear quintet at 2.52 ppm (Fig. 3). The ¹H-NMR spectrum of acantaglycoside F (**46**), a saponin containing the 24S-methylthornasterol A aglycone (**N**), has not been described, but the ¹H-NMR spectra in CDCl₃ of the synthetic 3,6-diacetyl (20S,24R)- and (20S,24S)-24-methylthornasterols A have been reported and are super-imposable (*75*); differentiation between (24R)- and (24S)-thornasterol B (**M** and **N**) is made possible by the different molecular ellipticity [Θ] in the respective CD spectra, which is much higher in the 24R isomer: values reported are $[\Theta]_{288} = -5780$ for the 24R isomer and $[\Theta]_{277} = -631$ for the 24S isomer (*75*). Saponins containing 22,23-epoxysteroids are clearly recognized by a narrow doublet due to the 22-H, which exhibits an almost constant chemical shift in all compounds (*ca.* δ 2.80, *J* = 2.5 Hz) and by a 23-H signal, whose shape and chemical shift depend on the substitution pattern at C-24: a dt at δ 2.94 (*J* = 2.5, 6.0 Hz) in **33**, **34** and **35** (aglycone **F**; Fig. 4); a dd at δ 2.76 (*J* = 2.2, 7.5 Hz) in **36** (aglycone **G**) and a dd at δ 2.78 (*J* = 2.5, 7.5 Hz) in saponins **47–50** containing the 24-methyl-22, 23-epoxysteroidal aglycone **O**. The 22R, 23S configuration was suggested by the ¹³C-NMR spectral data and con-firmed by comparing the spectra with those of appropriate 22,23-(*trans*)-epoxy steroidal models. In protoreasteroside (**37**) (aglycone **H**), a distinc-tion between aglycones **H** and **I** by ¹H-NMR spectral data must rely solely on the chemical shifts of the C-21 methyls, because the sugar signals and the C-22 hydroxymethine resonance overlap. The C-21 methyl resonates at δ 1.25 in the aglycone **H** containing protoreastero-side (**37**) (*68*) and at δ 1.31 in saponins OA-4 (**38**) (*43*) and in solasteroside A (**39**) (93) which contains aglycone **I**. The 20R,22S configuration in protoreasteroside (**37**), the only known asterosaponin containing the 5α-cholesta-9(11), 24(25)-dien-3β,6α,20,22-tetraol aglycone (**H**), was based on spectral comparison with 5α-cholesta-3β, 20,22-triol models (*91*). The presence of a C-18 methyl signal at δ 0.70, shifted to higher field relative to the same signal in 20-hydroxysteroids, immediately denotes the presence of marthasterone (**C**) or dihydromarthasterone (**D**) type agly-cones lacking the 20-hydroxyl function. In marthasteroside C (**29**) (agly-cone **D**) the three side chain methyl doublets overlap at δ 0.94 whereas in marthasteroside B (**27**) (aglycone **C**) the resonances corresponding to the 26- and 27-methyl groups appear as two singlets at δ 1.94 and 2.11 and the 24-H olefinic singlet is seen at δ 6.20 ppm (*82*). The aglycone (**J**) found in CO-ARIS II (**40**) is characterized by the downfield shifts of the methyl signals to δ 1.70, 1.90 and 2.11 (21-, 26- and 27-H_3) and the olefinic singlet at δ 6.16 (24-H).

Carbon-13 NMR spectroscopy is a good tool to substantiate identification of the steroidal aglycones of asterosaponins. The chemical shifts of the aglycone carbons in representative saponins are compiled in Table 3. The assignments were made by information obtained through DEPT pulse sequences, comparison of the spectra with reference steroids and comparison within this series of aglycones, taking into account structure variations and known chemical shift rules. Recently Findlay *et al.* have applied modern high field 2D NMR techniques to structure elucidation of saponins from starfish *Asterias forbesi*, *e.g.* forbesides A (= versicoside A, **13**), B (= glycoside B_2, **5**) and C (= ovarian asterosaponin-1, **6**), all containing thornasterol A (**A**) as the aglycone, and have established the connectivity of all proton resonances in the aglycone using 2D-COSY, J-resolved, nOe and selected decoupling experiments, whereas chemical shifts assignments for all carbons belonging to the aglycone were derived by DEPT analysis and HETCOR experiments (*76*). The ^{13}C-NMR data for forbesides A, B and C were in excellent agreement with those reported for versicoside A (**13**), glycoside B_2 (**5**) and ovarian asterosaponin-1 (**6**), which were only based on chemical shift arguments. ^{13}C-NMR spectroscopy is also a very good tool for detecting the stereochemistry at C-22 and C-23 of 22,23-epoxysteroid aglycones The resonance of C-24 at relatively low field (41.4 ppm) in tenuispinoside A (**33**) (aglycone **F**) ruled out the epoxy-*cis*-stereochemistry, since in the model (*cis*)-22,23-epoxycholesta-5,7-dien-3β-ol the C-20 and C-24 signals are shifted significantly upfield (*e.g.* C-24: 38.1 ppm), relative to the *trans*-models. On the other hand the C-17 resonance at 59.8 ppm in tenuispinoside A (**33**), virtually unshifted relative to the same signal in thornasteroside A (**4**), indicated the 22R,23S-configuration (*trans*-β-epoxide). Indeed, analysis of model compounds showed that in (22R,23R)-22,23-epoxycholestan-3β-ol (*trans*-α-epoxide) the resonance of C-17 is shifted upfield (54.7 ppm) relative to 5α-cholestan-3β-ol (56.1 ppm), whereas in the (22S,23S)-stereoisomer (*trans*-β-epoxide) it remains at 56.2 ppm (*42*). [We note that the specification of configuration at C-22 changes on going from 22,23-epoxysterols to the 20-hydroxy-22,23-epoxysteroids in the asterosaponins (*e.g.* **33**)]. In a similar manner the same 22R,23S-configuration was assigned to the 22,23-epoxy-24-nor-5α-cholest-9(11)-ene-3β,6α,20-triol-3β-sulphated aglycone **G** of asteroside B (**36**) (*43*). Finally, the configuration at C-22, C-23 and C-24 in regularoside A (**47**), whose aglycone **O**, possess, beside the 22,23-epoxide function, a methyl group at the usual C-24 position, was made by comparing its ^{13}C-NMR spectrum with the spectra of the four stereo-isomeric models containing a (*trans*)-22,23-epoxy-24-methyl side chain. The (*cis*)-epoxy stereochemistry had been ruled out, as was stated earlier, by the paramagnetic shift of the C-24 signal (δ_C 42.0 ppm).

3.1.2 ^{1}H and ^{13}C-NMR spectra of the saccharide chains

The ^{1}H-NMR resonances of the sugar moieties overlap and form complex signals ranging from *ca.* δ 3.0 to 4.0 ppm in d_4-methanol. The signals of the anomeric protons usually appear as doublets with a *J* of *ca.* 7.5 Hz (indicative of a β-linkage; α for arabinose) at somewhat lower field ranging from δ 4.3 to δ 4.7 ppm, whereas the methyl doublets due to the 6-deoxyhexoses quinovose and fucose clearly emerge in the upfield region at δ 1.25–1.48 ppm.

In the spectra of the many asterosaponins with the β-D-quinovopyranosyl-(1 → 2)-β-D-xylopyranosyl-(1 → 3)-β-D-quinovo-pyranosyl moiety attached to C-6 of the aglycone and further substitution at C-4 of the xylose residue, a double doublet (*J* = 12, 4 Hz) at 4.14 ppm for 5-H$_{eq}$ of xylose, is observed. This signal, which does not overlap with anything else, is an excellent lead for the detection of 4-O-substituted xylose, the signal of 5-H$_{eq}$ being observed at δ 3.92 ppm in xylose units non-substituted at C-4. In spectra of the shortened triglycos-ides, obtained by partial hydrolysis with *Charonia lampas* glycosidase mixture which removes the sugar residues at C-4 of the branched xylose unit, this signal is in fact shifted to somewhat higher field at *ca.* δ 3.90 (dd, *J* = 12, 4 Hz). Further examples are provided by the spectra of ophidian-osides C (**32**) (Fig. 2) and F (**9**), which contain two double doublets at δ 4.13 (*J* = 11.5 and 4.5 Hz) and 3.95 (*J* = 11.1 and 4.6 Hz) indicative of the presence of a second xylose unit non-substituted at C-4.

In the second major series of asterosaponins with a β-D-quinovopyranosyl-(1 → 2)-β-D-quinovopyranosyl-(1 → 3)-β-D-quinovo-pyranosyl moiety attached at C-6 of the aglycone and further substitu-tion at C-4 of the central quinovose residue, (*e.g.* **10, 12, 16, 18, 19**), a methyl doublet is observed at significantly lower field, i.e. δ 1.48 ppm, and assigned to the methyl protons of the branched 2,4-disubstituted quinovose residue. In the spectra of the shortened triglycosides, obtained by enzymatic removal of the sugar residues at C-4 of the branched quinovose, this signal moved upfield to δ 1.40 ppm. Thus, the methyl doublet at *ca.* δ 1.48 is indicative of a 4-substituted quinovose. This doublet, which is clearly visible in a non-crowded region of the spectrum, has also been observed in the spectra of marthasteroside B (**27**) (*83*), marthasteroside C (**29**) (*83*), ovarian asterosaponin-1 (**6**) (*43*) and regularoside A (**47**) (*69*), all of them saponins having the branched 2,4-disubstituted quinovose moiety linked to C-3 of glucose.

Although ^{1}H-NMR spectroscopy has been an important source of information for studies dealing with structure determination of oligosaccharides, the potential of ^{13}C-NMR spectroscopy has also been recognized for a long time, especially because of the increased spectral dispersion it affords relative to ^{1}H-NMR. Assignments of the ^{13}C-NMR resonances of the oligosaccharide carbons are made possible by comparison with known assignments of the constituent monosaccharides. The effect of glycosidation is a major determinant of ^{13}C-NMR chemical shifts and consequently has been widely studied. It has been found that glycosidation shift effects, which describe the difference in chemical shift for a given carbon in a substituted monosaccharide relative to that in the free monosaccharide, are determined by the nearest neighbour residue in an oligosaccharide sequence. The chemical shifts of some asterosaponins representative of the twentytwo different saccharide chains encountered until now are compiled in Table 4. The spectra were recorded in d_5-pyridine, sometimes with the addition of a few drops of D_2O to improve solubility. The assignments of the sugar carbon atoms were based upon (a) comparison with the spectra of the appropriate methyl glycosides (*98, 99*), (b) the known glycosidation shifts (*100, 101*), (c) DEPT measurements and (d) sometimes supported by measuring the spin-lattice relaxation times (T_1) (*46, 79, 80, 86, 90*). The recently developed advanced NMR techniques, such as 2D heterocorrelated spectroscopy, have now been applied to the study of asterosaponins and allowed the structures of forbeside A (= versicoside **A, 13**), forbeside B (= glycoside B_2, **5**) and forbeside C (= ovarian asterosaponin-1, **6**) to be deduced entirely using two conceptually different types of NMR experiments (*76, 78*). First 2D-COSY corroborated by RECSY experiments provides an unambiguous assignment of proton signals belonging to each saccharide unit. Then, the sequence of monosaccharide units is deduced using experiments which measure through space inter-residue nuclear Overhauser effects by irradiation to the frequency of each anomeric signal. Corroboration as to the location of the interglycosidic linkages comes from ^{13}C shift data obtained by a hydrogen-carbon correlation experiment (HETCOR), which permits an unambiguous assignment of ^{13}C shifts. The ^{13}C-NMR data in Table 4 are consistent with the proposed structures and we note that the data for forbeside A (**13**), B (**5**) and C (**6**), obtained by using 2D heterocorrelated spectroscopy are in excellent agreement with those reported for versicoside A (**13**), glycoside B_2 (**5**) and ovarian asterosaponin-1 (**6**), whose assignments have been made mainly by comparison with reference compounds (*65, 61* and *59*).

References, pp. 297–308

A few further remarks may be added here. The unusual downfield position of the substituted carbon-3 of the monosaccharide unit directly attached to the aglycone is noteworthy. Its resonance at δ_c 90–91 ppm, at lower field than what expected for glycosidated C-3 in a β-glucopyranosyl or a β-quinovopyranosyl units which are reported to resonate at δ_c 87–88.5 ppm, appears as a distinctive feature of the ^{13}C-NMR spectra of all asterosaponins. In the spectra of the asterosaponins containing the *xylo*-hexos-4-ulopyranosyl unit (*e.g.* ovarian asterosaponin-1, **6**; forbeside F and G, **21** and **22**; Co-ARIS II, **40**) a signal for a fully substituted carbon, assigned to the gem-diol C-4 carbon of the above sugar unit, appears at *ca.* 92.7 ppm after equilibration with pyridine-d$_5$/water. The presence of the *xylo*-hexos-4-ulopyranosyl unit is also indicated by the peculiar high field resonance of the methyl carbon at C-5, δ_C 13.4 ppm, due to its β-position relative to the *gem*-diol function. The resonances of the methyl carbons of the quinovose units range from 17.7 to 18.3 ppm, while those of the fucose units are shifted upfield to *ca.* 16.6–17.0 ppm. Thus, ^{13}C-NMR spectroscopy is also a good tool for distinguishing between fucose and quinovose units. The signal of the C-5 methylene carbon of the xylosyl units is easily detected in a non-crowded region of the ^{13}C-NMR spectrum by a DEPT pulse sequence and its chemical shift is diagnostic for substitution at C-4. While appearing at *ca.* 67.0 ppm in xylosyl units non-substituted at C-4, the C-5 resonance moves to *ca.* 64.0 ppm as a consequence of glycosidation at C-4. Finally we would draw attention to the downfield shifts observed for C-2 of the first branching sugar starting from the aglycone and C-1 of the terminal quinovosyl unit attached to it upon desulphation [85.0 and 107.5 ppm *vs* 83.0 and 105.2 ppm in marthasteroside B (**27**)] (*83*). These shifts are common to the spectra of all asterosaponins subjected to desulphation (see Table 4). We would also note that a comparison of the spectra of the native saponins with those of the desulphated ones further confirms the location of the sulphate group at C-3; in the spectra of the desulphated saponins C-3 is shifted upfield by *ca.* 7.0 ppm and C-2 and C-4 are shifted downfield by *ca.* 3.0 ppm relative to the sulphated compounds.

3.1.3 FAB-Mass spectra

Fast atom bombardment (FAB) is the most useful mass spectrometric technique for the asterosaponins (*102*). It successfully gives the molecular weight of the underivatized sulphate saponins together with useful information on the saccharide sequence. Positive ion FAB-MS gives molecular ions in the protonated and cationized forms, $[M + H]^+$ and

$[M + C]^+$ where M is the molecular weight of the intact sodium (or potassium) salt and C is the mass or the cation (Na^+ or K^+). The relative intensity of this group of peaks is variable. For example in one spectrum of marthasteroside A_2 (15), the ion at m/z 1419 $[M_{SO_3Na} + Na]^+$ was the only one observed (Fig. 5); the more usual four peaks, *i.e.* $[M_{SO_3Na} + K]^+$, $[M_{SO_3Na} + Na]^+$ $[M_{SO_3K} + H]^+$, $[M_{SO_3Na} + H]^+$, were observed in spectra of the same sample recorded at different times as it occurred for marthasteroside A_1 (14), whose spectrum is shown in Fig. 6 (*83*). The spectra also contain a series of fragment ions. The principal fragmentations are considered to be due to the cleavage of the glycosidic bonds and two fragmentation pathways are apparent. In addition to the ions with the positive charge located on the aglycone-containing fragments, major ions are also produced with the positive charge located on the saccharide fragments. A comparison of the spectrum of marthasteroside B (27) with that of marthasteroside C (29) (Fig. 7) gives strong supporting evidence: the aglycone-containing fragments are shifted by two mass units on passing from 27 to 29, while the sugar fragments have identical masses. Cleavages can occur on both sides of glycosidic linkages with proton transfer. KOMORI *et al.* (*86*) have suggested that cleavages occur on one side of the glycosidic oxygen and that the series of doublets separated by 16 mass units is due to potassium- and sodium-cationized ions respectively. We think that cleavages on both sides of the glycosidic oxygen can also account for the presence of such doublets.

The isomeric nature of fucosyl and quinovosyl units causes ambiguity in the interpretation of the very complex spectral data and the saccharide sequence cannot be determined solely by mass spectrometry. Even so, some information can be extracted from the spectra. In the spectrum of marthasteroside C (Fig. 7), the peaks at *m/z* 623-607 resulting from the pentasaccharide fragment at *m/z* 785-769 by loss of 162 mass units (glucose) indicate that glucose is the sugar directly linked to the aglycone. The series of ions observed in the spectra of all four saponins, *viz. m/z* 625, 609, and 593 in the spectrum of marthasteroside A_1 (Fig. 6), *m/z* 609, 593, and 577 in the spectrum of marthasteroside A_2 (Fig. 5) and *m/z* 477, 461, and 445 in the spectra of the pentaglycosides (Fig. 7), suggest that the location of the branching point in each molecule is on the second monosaccharide unit starting from the aglycone.

The negative ions FAB-MS give an intense negatively charged molecular ion, $[M_{SO_3^-}]$, and a fragmentation arising from the cleavages of the glycosidic bonds from the terminal sugar moieties, with charge located on the aglycone-containing fragments (Figs. 8 and 9). In the spectra of saponins containing aglycones having the 20-hydroxy-23-oxoside chain, a major fragment corresponding to the C(20)–C(22) bond

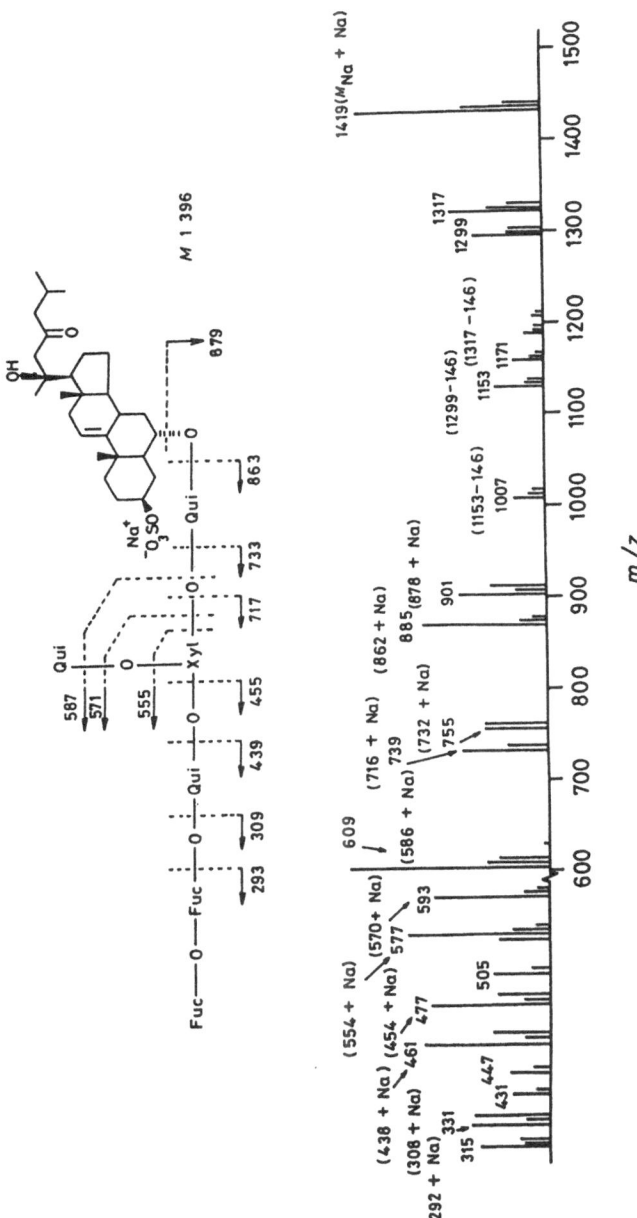

Fig. 5. (+ ve) FAB Mass spectrum of marthasteroside A₂ (15)

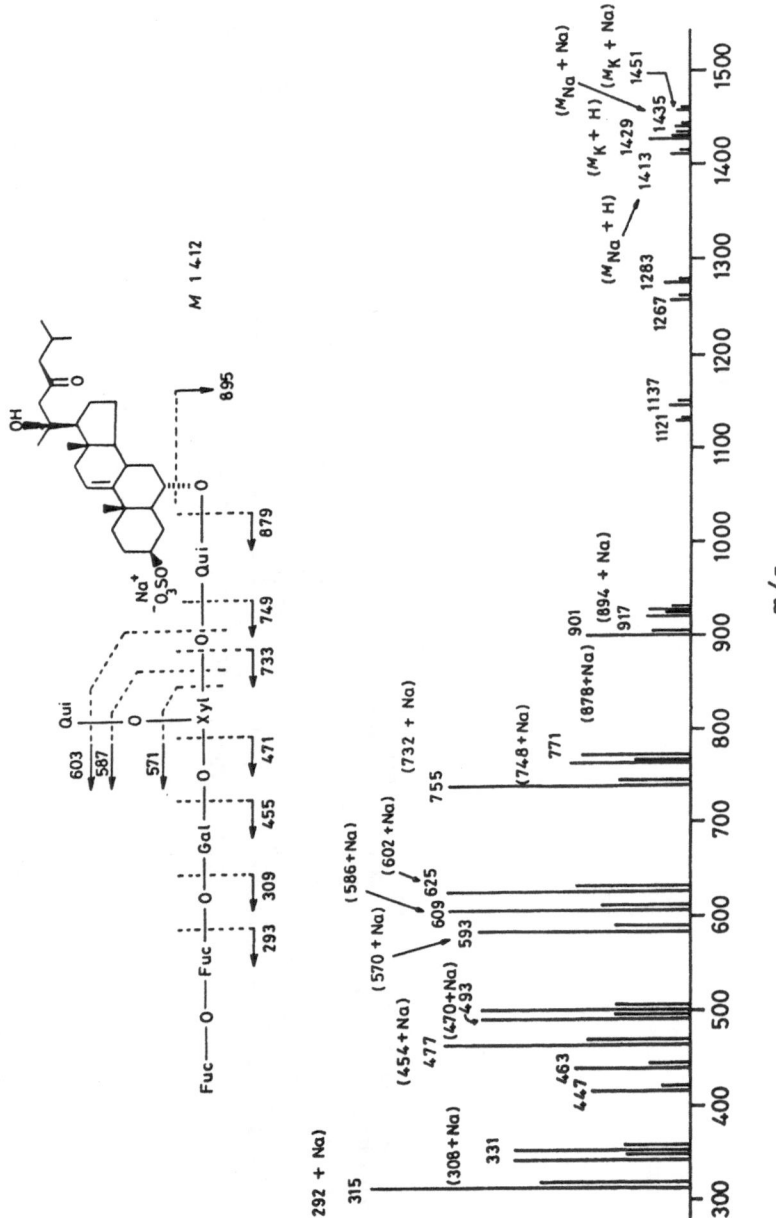

Fig. 6. (+ve) FAB Mass spectrum of marthasteroside A₁ (14)

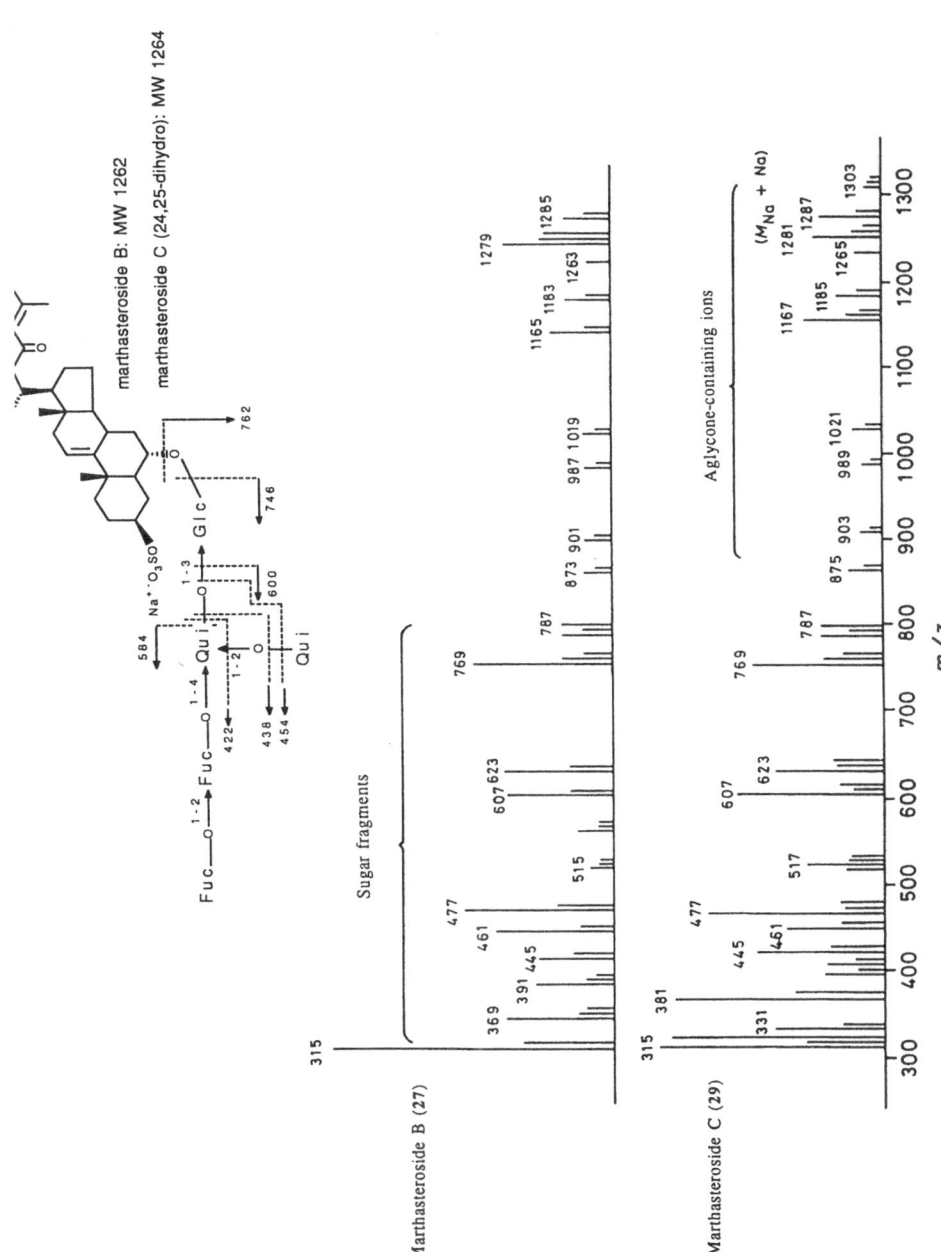

Fig. 7. (+ ve) FAB Mass spectrum of marthasteroside B (27) and C (29)

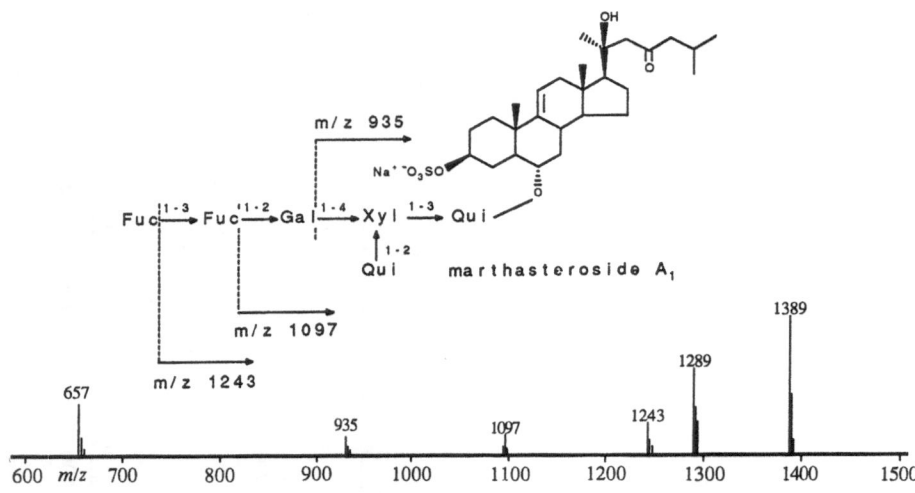

Fig. 8. (− ve) FAB Mass spectrum of marthasteroside A₁ (**14**)

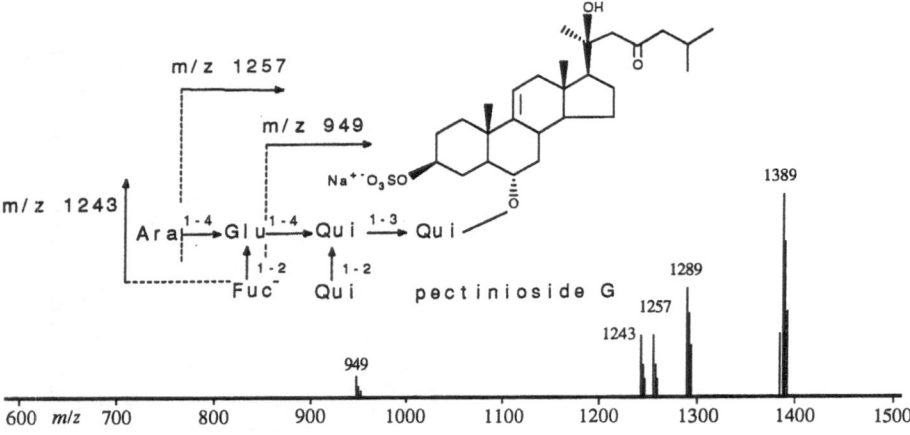

Fig. 9. (− ve) FAB Mass spectrum of pectinioside G (**18**)

cleavage and 1 H transfer (retro-aldol cleavage) is observed at m/z $[M_{SO_3^-} - 100]$ (pregnane type ion).

The negative ion FAB mass spectrum of pectinioside G (**18**), representative of the group of hexaglycosides with two branches, gave a molecular anion peak at m/z 1389 (100%) and major fragments at m/z 1289 $[M_{SO_3^-} - 100]$ (pregnane type ion by retroaldol cleavage of the

C(20)–C(22) bond and 1 H transfer), at m/z 1257 $[M_{SO_3^-}-132]$ and 1243 $[M_{SO_3^-}-146]$, corresponding to separate loss of the arabinosyl and of a deoxyhexosyl residues from $[M_{SO_3^-}]$, respectively (Fig. 9). A smaller fragment at m/z 949 was interpreted as due to the loss of arabinose, a deoxyhexose and glucose. These observations, together with results of methylation analysis indicating fucose, quinovose and arabinose as terminal sugars, suggested that a second branching of the saccharide chain was located on the glucose unit. Further information on the sugar sequence were obtained by enzymatic hydrolysis of pectinioside G (**18**) with *Charonia lampas* glycosidase mixture, which gave a pentaglycoside arising from loss of the terminal fucose and a triglycoside of thornasterol A sulphate containing three quinovose residues. The FAB mass spectrum of the pentaglycoside showed a molecular anion peak at m/z 1243 $[M_{SO_3^-}]$ and fragmentation ions at m/z 1111 and 949, ascribed to the sequential loss of arabinosyl and glucosyl residues. Taken together, data from FAB-MS, methylation analysis and partial enzymatic hydrolysis showed that the sugar sequence in the intact pectinioside G (**18**) must be fucose-(arabinose)-glucose-(quinovose)-quinovose-quinovose-thornasteryl A 3β-sulphate. The ^{13}C-NMR studies of pectinioside G itself and of the shortened glycosides obtained by partial hydrolysis established the nature of the interglycosidic linkages and allowed the complete structure to be defined (*63*).

Thus FAB mass spectrometry, preferably both in the negative and positive mode, offers an elegant and sound approach to the solution of the sugar sequence in natural asterosaponins.

We would draw the attention to the fact that in our experience the glycosidase mixture from *Charonia lampas*, which has been generally used to cleave the oligosaccharide chain of asterosaponins and has been an important tool for determining the sugar sequence, easily removes the sugars linked up to C-4 of the first branched unit starting from the aglycone, leaving a triglycoside which is more resistant to further enzymatic hydrolysis. For example, after 5 h hydrolysis of pectinioside G (**18**) the shortened pentaglycoside was almost the only detectable prosapogenin, while, when the hydrolysis was pursued for 24 h, the pentaglycoside completely disappeared and the triglycoside became the unique prosapogenin in the reaction medium.

4. Cyclic Steroidal Glycosides

Toxic saponins of a completely different structural type have been discovered in two species of the genus *Echinaster*. They have a number of unusual features when compared with the more common asterosaponins: there is no sulphate group and charge is due to a glucuronic acid unit in the saccharide moiety, the Δ^7-3β,6β-dihydroxysteroidal nucleus is unprecedented and, as the most remarkable feature, the trisaccharide chain is cyclized between C-3 and C-6 of the aglycone giving rise to a macrocyclic ring reminiscent of a crown ether. Sepositoside A (53), the major saponin from the Mediterranean starfish *Echinaster sepositus* (*103*), is accompanied by smaller amounts of three related oligoglycosides (54–56), which differ only in the structure of the side chain of the aglycones, all having a 22,23-epoxy functionality (*104*). A further representative of this class of glycosides was isolated from a Pacific starfish of the same genus, *Echinaster luzonicus*, and accordingly named luzonicoside A (57) (*105*). Because of these findings and because of the absence of asterosaponins in both *Echinaster* species we attributed a chemotaxonomic significance to the cyclic steroidal glycosides. This view is now questioned by the results of an investigation of a third species of *Echinaster*, *E. brasiliensis*, collected at Grand Bahama Island (Caribbean Sea). The cyclic steroidal glycosides were completely absent from polar extracts of this species, while typical asterosaponins, among which is the known marthasteroside A_1 (14), were isolated in relatively large amounts (*106*).

Structures and data on cyclic steroidal glycosides are presented below in the usual standardized form.

53 Sepositoside A

Occurrence: *Echinaster sepositus* (*103*).
Activity: Moderately toxic, $LD_{50} = 43$ mg Kg^{-1} by injection in mouse (*103*); cytotoxic towards bovine turbinate cells up to a level of 1 μg ml^{-1} (*36*); inhibits cell division of fertilized sea urchin eggs at 10^{-5} M (*ca.*30% inhibition) (*35*) and possesses antifungal activity (minimum inhibitory amount against *Cladosporium cucumerinum*, 5 μg) (*35*).
Physical data: $[\alpha]_D = -68.5°$ (H_2O); EI ms, m/z 1024 [M]$^+$ (permethylated sepositoside A).

The UV spectrum of sepositoside A (53) showed no significant absorption maximum above 210 nm. The inconsistency of the UV spectral properties of the natural glycoside when compared with those of sapogenin (λ_{max} 248 nm) made it clear that the structure 53a (Fig. 10) (*107*) could not be the structure of the intact glycoside prior to acid hydrolysis. The ^{13}C-NMR spectrum of the intact saponin showed the presence of only one double bond, δ_C 142.2 and 118.2 ppm, thus leading to the conclusion that the diene system in the genin (53a) was formed by

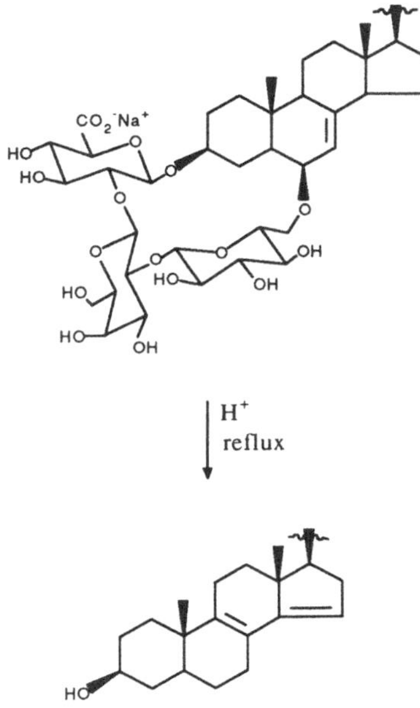

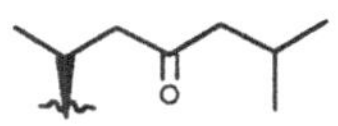

53, sepositoside A

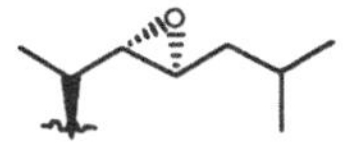

55

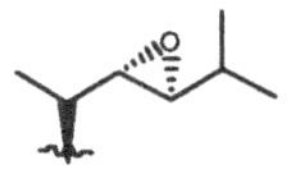

54

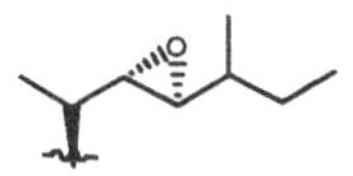

56

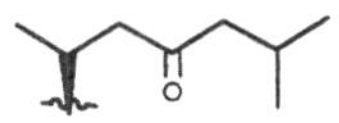

H⁺
reflux

53a

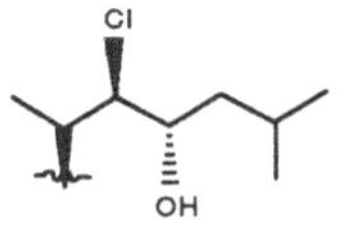

55a

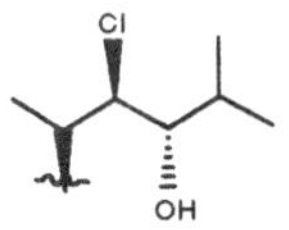

54a

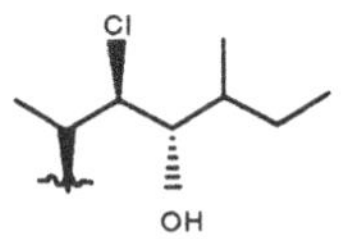

56a

53 (sepositoside A) $\xrightarrow{H^+, \text{ r.t.}}$ **53b**

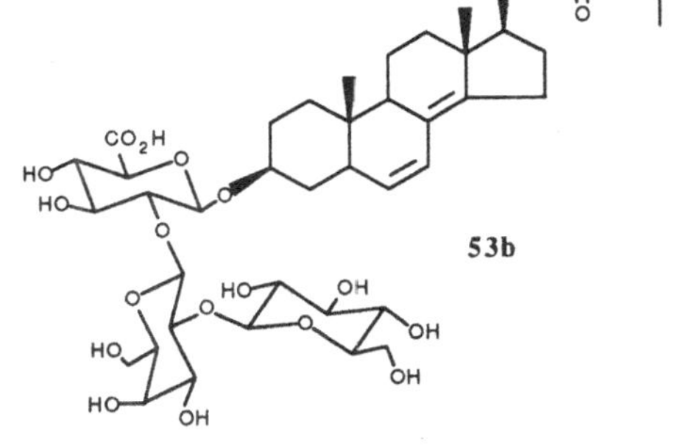

Fig. 10. Cyclic steroidal glycosides from the starfish *Echinaster sepositus*

dehydration and, probably, double bond migration during acid hydrolysis. The key step during the structural work was a mild acid hydrolysis
which resulted in opening of the macrocyclic ring, and gave rise to the
formation of the UV active glycoside **53b**, still containing glucose,
galactose and glucuronic acid. The sequence of sugars in **53b** was
determined by permethylation and electron impact mass spectral data of
the permethylated material, whereas the interglycosidic linkages were
determined by ^{13}C-NMR spectroscopy on the opened glycoside **53b**. The
facile formation of a steroidal $\Delta^{6,8(14)}$ diene suggested a Δ^7, 6-O-steroidal
structure for the intact saponin. This was confirmed by comparison with
model compounds which also indicated the 6β-stereochemistry. Finally,
a comparison of chemical and spectral properties of the intact sepositoside A (**53**) (m/z 1024 in EI-ms of the permethylated derivative; formation
of 2,3,4-tri-O-methyl glucose on acid hydrolysis of the permethylated
derivative; only one CH_2OH signal in the ^{13}C-NMR spectrum of **53** at
61.7 ppm) with those of the opened one (**53b**) (m/z 1038 in EI-ms of the
permethylated derivative, formation of 2,3,4,6-tetra-O-methyl glucose on
acid hydrolysis of the permethylated derivative; two CH_2OH signals in
the ^{13}C-NMR spectrum of **53b** at 62.6 and 62.0 ppm) gave the cyclic
structure **53** for the natural compound, in which the $HOCH_2$ of the
glucopyranosyl unit is attached to C-6 of the aglycone forming a 6β-O-
ethereal linkage.

54

OCCURRENCE: *Echinaster sepositus* (*104*).
PHYSICAL DATA: $[\alpha]_D = -56.5°$ (H_2O).

The structural assignment was made on the basis of spectral data
(^{1}H- and ^{13}C-NMR and chemical behaviour (formation of the opened
glycoside on very mild acid treatment) and comparison with the major
sepositoside A (**53**). On prolonged acid hydrolysis **54** yielded the chloro-
hydrin **54a** (Fig. 10) (*108*). The 22α,23α-epoxy-*trans*-stereochemistry was
assigned by comparing the ^{13}C-NMR shifts of the side chain carbons
with those of the 22,23-epoxysteroidal models.

55-56

OCCURRENCE: *Echinaster sepositus* (*104*).

As the two epoxy saponins **55–56** resisted attempts at separation,
structural studies were carried out on the mixture of the two isomers. On
prolonged acid hydrolysis they yielded the chlorohydrins **55a** and **56a**,
which were separated by reverse phase hplc (*108*).

57 Luzonicoside A

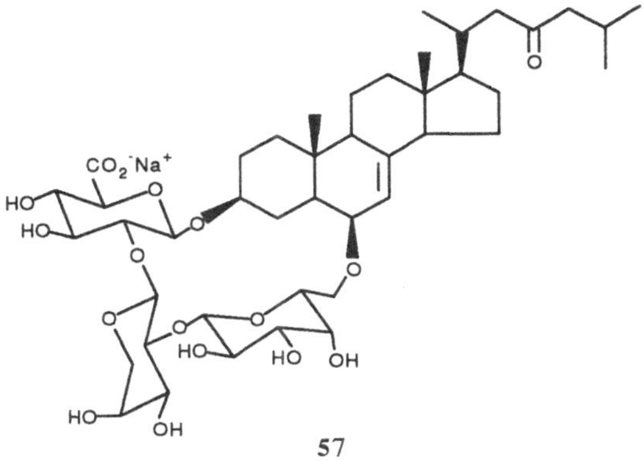

57

OCCURRENCE: *Echinaster luzonicus* (*104*).
PHYSICAL DATA: $[\alpha]_D = -66.0°$ (H_2O).

Comparison of ^1H- and ^{13}C-NMR spectra of **57** with those of sepositoside A (**53**) established that the steroidal parts of both molecules were identical; differences were observed in the sugar carbon region. The structure of the saccharide chain was determined by methylation analysis on the corresponding opened glycoside and confirmed by ^{13}C-NMR spectroscopy.

5. Glycosides of Polyhydroxysteroids

This third group of steroidal glycosides from starfishes shows a much larger degree of structural variability. Most of these compounds usually occur in minute amounts and are as widespread among starfishes as the asterosaponins, having been found, usually as complex mixtures, in almost all the species we have investigated (Table 1). They are composed of a polyhydroxysteroidal aglycone and a carbohydrate portion made up of one or two monosaccharide units, usually linked to each other and glycosidically attached at C-3 or C-24 of the aglycone it was. Only very recently we isolated from the New Caledonian species *Fromia monilis* cytotoxic triglycosides (**131**, **132**) (*109*), which constitute the only examples of triglycosides among more than one hundred different mono- and diglycosides of polyhydroxysteroids isolated so far. The most common monosaccharides are D-xylopyranose, often methylated at position 2 and/or 4 and occasionally at position 3, and L-arabinose, found in its furanose form. Rare examples of galactofuranosides (**121, 164, 165**), xylofuranosides (**84, 85, 107**) and fucofuranosides (**105, 106**) have also been found. Sulphated forms are quite common, with the sulphate group located on the steroidal moiety, at position 3β, 6α or 15α, or on the saccharide portion.

The first representative of such compounds, the cytotoxic nodososide (**146**), was initially isolated from Pacific *Protoreaster nodosus* (*110*) and later from other Valvatida species. Followed in succeeding years structures of more than one hundred different glycosides of polyhydroxylated steroids were determined. Structural variations are due to different hydroxylation patterns of the steroidal tetracyclic nucleus, the functionalization of the side-chain, the presence of sulphate and the nature and location of the saccharide moiety. Beside the invariable 3β-hydroxylation, other hydroxyl groups are commonly found at positions 6(α or β), 8, 15(α or β) and 24 of the aglycone with additional hydroxyl group(s) at one or more of the positions 4β, 7α and 16β.

All compounds are described on the following pages in the standardized form which includes sources, reported biological activities, selected spectral data and a brief account of the structure elucidation; the reported ^{1}H-NMR data are taken from solutions in d_4-methanol, unless otherwise stated.

Compounds **61–81** are representative of the subgroup characterized by steroidal aglycones with the basic 3β,6α,15α-hydroxylation pattern. Among this subgroup the amurensosides (**61–63**) from *Asterias amurensis* (43) are uncommon examples of steroidal glycosides lacking the hydroxyl group at C-8 of the aglycone. A second subgroup of compounds (**82–133**) contains aglycones with the basic 3β,6α,8,15β-hydroxylation pattern; the (24S) 5α-cholestane-3β,6α,8,15β,24-pentaol aglycone, first encountered in attenuatoside A II (**82**) from *Hacelia attenuata* (130), is the most common steroidal aglycone found in glycosides from starfishes. Compounds **134** to **165** represent the subgroup of glycosides whose aglycones possess the basic 3β,6β,8,15α-hydroxylation pattern; this group of compounds comprises examples of those glycosides in which the two monosaccharides are not linked to each other. The smallest subgroup which comprises compounds **166** to **169** has the basic 3β,6β,8,15β-hydroxylation pattern in the steroidal nucleus.

Compounds **58–60** cannot be classified in any of the above subgroups; they contain aglycones with unsaturated tetracyclic nuclei, a quite unusual feature among glycosides of polyhydroxysteroids from starfishes and only found, in addition to **58–60**, in a set of compounds with a Δ^4-steroidal skeleton.

The glycosides of polyhydroxysteroids often occur as complex mixtures; in addition to the previously discussed case of *Coscinasterias tenuispina*, whose polar steroids have been resolved into nineteen constituents (*42*), a further illustrative example of the structural variety of steroidal glycosides co-occurring in the same organism is represented by the twelve steriodal glycosides isolated from the starfish *Henricia laeviuscola* (*94*), which include a sulphated steroidal bioside, laeviuscoloside A (**58**), six sulphated monosides (**149–154**) with four types of steroidal aglycones (cholestane; 24-hydroxymethyl-; 24-(β-hydroxyethyl)- and 24-methyl-26-hydroxy-cholestane), all with the same nuclear hydroxylation pattern; two non-sulphated biosides with the two monosaccharide residues attached at different positions of their aglycones (**143, 144**); two non-sulphated monosides (**140, 141**) and the asterosaponin henricioside A (**50**). The analysis of the polar extractives from the starfish *Halityle regularis* led to the isolation of eight steroidal diglycosides, the halitylosides **93, 102, 111, 125, 126, 127, 128**, and **166**, along with two polyhydroxysteroids (**180, 181**) (*112*) and three asterosaponins including the novel regularosides A (**47**) and B(**11**) and the known thornasteroside A (**4**) (*69*). A recent reinvestigation of the polar extractives from the starfish *Culcita novaeguineae* (*44*) has led to the isolation of eleven polyhydroxysteroidal glycosides (**92, 93, 101, 102, 112, 119, 120, 125, 126, 166, 167**) and five polyhydroxysteroids (**187, 199, 200, 227, 229**), thus providing a further example of the complexity of steroidal mixtures occurring in a single organism.

A more recent example of a very complex polar steroid mixture in starfishes has been encountered during the analysis of polar extracts from the starfish *Solaster borealis*, which led to the isolation of a new asterosaponin, solasteroside A (**39**), five glycosides of polyhydroxysteroids, the novel borealosides A-D (**69–71 and 75**) and the known amurensoside B (**62**), along with nine polyhydroxysteroids (Table 1) (*93*).

58 Laeviuscoloside A

OCCURRENCE: *Henricia laeviuscola* (*94*).
ACTIVITY: Inhibits cell division of fertilized sea urchin eggs at 10^{-7} M (50% inhibition) and possesses antifungal activity (minimum inhibitory amount against *Cladosporium cucumerinum*, 1 μg) (*94*).
PHYSICAL DATA: $[\alpha]_D = +9.3°$ (MeOH); FAB MS (− ve ion), m/z 803 $[M_{SO_3^-}]$, 641 $[M_{SO_3^-}-162]$, 479 $[M_{SO_3^-} - 2 \times 162]$; selected ^{1}H-NMR signals: δ 0.68 (s, 18-H$_3$), 1.08 (s, 19-H$_3$), 1.62 (s, 26- or 27-H$_3$), 1.69 (s, 27- or 26-H$_3$), 3.70 (m, 3α-H), 4.46 (m, W$_{1/2}$ = 25 Hz, 6β-H), 4.57 (d, $J = 7$ Hz, 1'-H), 4.62 (d, $J = 7$ Hz, 1''-H), 5.12 (t, $J = 5.0$ Hz, 24-H).

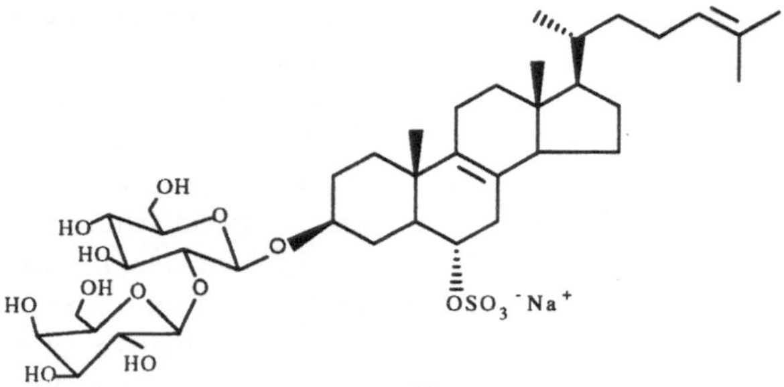

58

Laeviuscoloside A is the major steroidal glycoside obtained from the *Henricia laeviuscola* (40 mg from 3.3 kg fresh material). The sugar sequence Gal → Glc was confirmed by enzymatic hydrolysis with *Charonia lampas* glycosidase mixture affording, along with minor amount of the aglycone, a monoglycoside which gave methyl glucosides on acid methanolysis. The C-2 glycosidic linkage was evidenced by the ^{13}C-NMR spectrum, *i.e.* the high field shift exhibited by C-1, δ_C 101.9 ppm, and the downfield shift exhibited by C-2, δ_C 83.3 ppm, of the glucose residue. The sulphate group was located at C-6 by chemical shift considerations in comparison with the ^{1}H-NMR data of the corresponding desulphated analog (6-H: δ 4.46 *vs* 3.60 ppm) and accordingly the sugar moiety was located at C-3.

59 Moniloside A

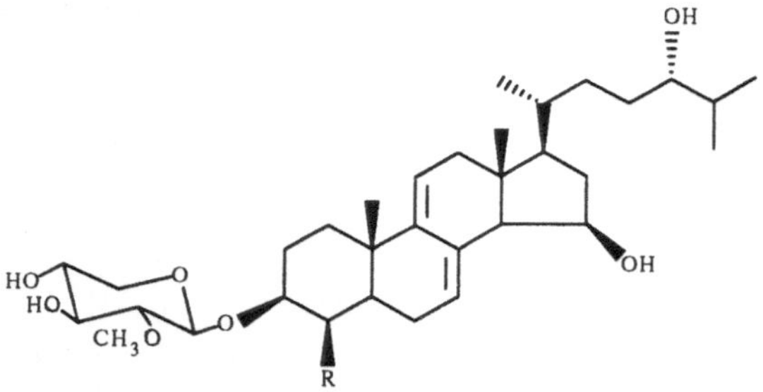

59, R=H

60, R=OH

Occurrence: *Fromia monilis* (*109*).
Activity: Cytotoxic on VERO cells at a concentration of 3 μg/5.10^4 cells.

PHYSICAL DATA: $[\alpha]_D = -30.2°$ (MeOH); UV, λ_{max} 235, 242 ($\epsilon = 12,300$) and 250 nm; FAB MS ($-$ ve ion), m/z 561 $[M-H]^-$; selected ^{1}H-NMR signals: δ 0.87 (s, 18-H$_3$), 1.00 (s, 19-H$_3$), 2.84 (dd, $J = 7, 9$ Hz, 2'-H), 3.26 (m, 24-H), 3.62 (m, 3α-H), 4.42 (d, $J = 7$ Hz, 1'-H), 4.52 (t, $J = 5.4$ Hz, 15α-H), 5.52 (br d, $J = 6.0$ Hz, 11-H), 5.79 (br d, $J = 5.0$ Hz, 7-H).

The structure was based entirely on spectral data. The chemical shift of the anomeric proton at δ 4.42 is a distinctive feature for location of the xylose at C-3 of the aglycone; in steroidal 24-O-xylopyranosides the signal is found systematically at δ 4.25 ppm. The 24S configuration is suggested by ^{13}C-NMR data and comparison with those of (24S)- and (24R)-24-hydroxycholesterol.

60 Moniloside B

OCCURRENCE: *Fromia monilis (109)*.
ACTIVITY: Cytotoxic on VERO cells at a concentration of 3 µg/5.10^4 cells.
PHYSICAL DATA: $[\alpha]_D = -24.2°$ (MeOH); UV (CH$_3$OH), λ_{max} 235, 242 ($\epsilon = 12,000$) and 250 nm; FAB MS ($-$ ve ion), m/z 577 $[M-H]^-$; selected ^{1}H-NMR signals: δ 0.87 (s, 18-H$_3$), 1.18 (s, 19-H$_3$), 1.95 (br t, $J = 15.5$ Hz, 6-H), 2.65 (br t, $J = 15.5$ Hz, 6-H), 3.62 (m, 3α-H), 3.96 (br s, 4α-H), 5.45 (br d, 11-H), 5.87 (br d, 7-H).

This substance is related to moniloside A by introduction of an additional hydroxyl group. The location of the "extra" hydroxyl group at C-4β was established by 2D-COSY experiments, which indicated that the 3-H signal (δ 3.62 m) is coupled to the 4-H signal at δ 3.96 and, continuing this J-connectivity path, it was possible to locate 5-H at δ 1.42 and H$_2$-6 at δ 1.95 and 2.65 ppm.

61 Amurensoside A

OCCURRENCE: *Asterias amurensis (43)*.
ACTIVITY: No activity in the sea urchin egg test below 10^{-5} M concentrations.
PHYSICAL DATA: $[\alpha]_D = +16.7°$ (MeOH); FAB MS ($-$ ve ion), m/z 647 $[M_{SO_3}^-]$; FAB MS ($+$ ve ion), m/z 693 $[M_{SO_3Na} + Na]^+$ (60%); 573 [693 - NaHSO$_4$]$^+$ (100%); selected ^{1}H-NMR signals: δ 0.77 (s, 18-H$_3$), 0.88 (s, 19-H$_3$), 0.96 (6H, d, $J = 7.0$ Hz, 26- and 27- H$_3$), 3.30 (2H, m, 24-H and 6β-H), 3.50 (m, 3α-H), 4.25 (d, $J = 7.5$ Hz, 1'-H), 4.50 (dt, $J = 2.5, 10.0$ Hz, 15β-H); ^{13}C-NMR spectrum of the steroid portion in Table 5.

The location of sulphate at C-15 is indicated by the ^{13}C-NMR frequencies of C-14 (δ_C 61.5 ppm), C-15 (δ_C 81.7 ppm) and C-16 (δ_C 38.7 ppm), which are shifted by -2.5, $+7.7$ and -1.6 ppm relative to 15α-

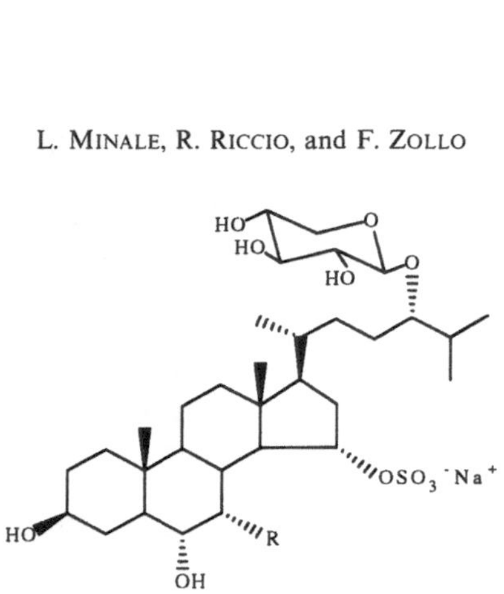

61, R=H

62, R=OH

cholestanol, and confirmed by solvolysis of **61** in pyridine-dioxane which afforded the desulphated derivative showing in the ^{1}H-NMR spectrum the H-15 signal shifted upfield to δ 3.87 (δ 4.50 in **61**) ppm. Consideration of the ^{13}C-NMR data established the site of glycosidation at C-24. The configuration at C-24 was determined on the derived 3β,6α,15α-trimethoxy-5α-cholestan-24-ol (methylation with CH$_3$I in DMF/NaH of the desulphated **61** followed by acid hydrolysis) after derivatization with chiral (α-methoxy-α-trifluoromethyl) phenylacetic acid (MTPA, Mosher's reagent) followed by ^{1}H-NMR measurements. The signals for the diastereotopic methyl protons at C-26 and C-27 were shifted upfield to δ 0.84 and 0.86 in the ^{1}H-NMR spectrum of the (+)MTPA ester and shifted downfield to δ 0.89 and 0.91 in that of the (−)MTPA ester in agreement with the 24S-configuration. The reverse occurs with the (24R) 24-hydroxysteroids (*crf.* section 9.1); the term (+) or (−) MTPA ester refers to an ester obtained by using the acid chloride prepared from R(+) − or S(−)-acid, respectively.

62 Amurensoside B

Occurrence: *Asterias amurensis (43).*
Activity: No activity in the sea urchin egg test below 10^{-5} M concentrations.
Physical data: $[\alpha]_D$ = + 12.5° (MeOH); FAB MS (− ve ion), m/z 663 [M$_{SO_3}$] (100%); selected ^{1}H-NMR signals: δ 4.21 (t, J = 2.5, 7β-H); ^{13}C-NMR spectrum of the steroid in Table 5.

This substance is related to amurensoside A by introduction of an additional hydroxyl group at position 7α of the aglycone, δ$_H$ 4.21 as an apparent triplet ($J = 2.5$ Hz), which sharpened on irradiation at δ 3.30 (6-H). Confirmation was obtained by ^{13}C-NMR spectroscopy. Taking **61** as starting structure and using the substituent effects that have been published for hydroxysteroids (*113, 114*), the calculated ^{13}C-NMR spectrum for the compound with 7α-hydroxy group is in excellent agreement with the experimental one, whereas the calculated spectrum for the compound with the the 7β-hydroxyl group was significantly different. Particularly significant for 7α-axial hydroxylation are the frequencies for C-5, C-9 and C-14 (γ-carbons) in the spectrum of **62**, which are shifted upfield by 8.3, 8.8 and 5.5 (γ-gauche effects) ppm, respectively relative to **61**.

63 Amurensoside C

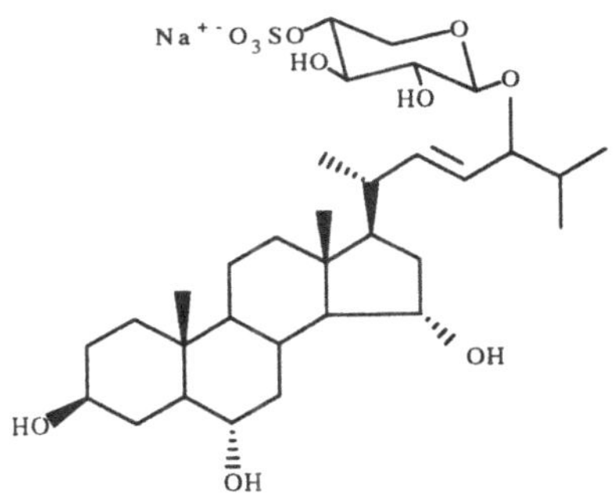

63

OCCURRENCE: *Asterias amurensis (43)*.

ACTIVITY: No activity in the sea urchin egg test below 10^{-5} M concentrations.

PHYSICAL DATA: $[\alpha]_D = + 6.8°$ (MeOH); FAB MS (− ve ion), m/z 645 $[M_{SO_3}]$; FAB MS (+ ve ion), m/z 691 $[M_{SO_3Na} + Na]^+$ (100%); 669 $[M_{SO_3Na} + H]^+$ (25%); selected ^{1}H NMR signals: δ 1.06 (d, $J = 7.0$ Hz, 21-H$_3$), 3.72 (dd, $J = 7.0$, 5.5 Hz, 24-H), 3.91 (dt, $J = 2.5$, 10.0 Hz, 15β-H), 5.37 (dd, $J = 15.0$, 6.0 Hz, 23-H), 5.47 (dd, $J = 15.0$, 7.0 Hz, 22-H); ^{13}C-NMR spectrum of the steroid portion in Table 5.

Assignment of the NMR signals of the xylopyranosyl 4-O-sulphated moiety.

Position	^{1}H	^{13}C
1	4.26 d(7.5 Hz)	104.2
2	3.27 dd(7.5, 9.5 Hz)	75.4

3	3.51 t(9.5 Hz)	76.2
4	4.19 m	77.6
5	3.30 t(10 Hz)	64.9
	4.15 dd(10, 5 Hz)	

The upfield shift of 4'-H from δ 4.19 in **63** to 3.50 ppm in the desulphated derivative showed that the sulphate is located at C-4' of xylose. This was confirmed by the ^{13}C-NMR spectrum in which the C-4' signal is shifted by 6.2 ppm to δ_C 77.6 ppm, whereas C-3' and C-5' are shifted upfield by 1.7 and 2.2 ppm to δ_C 76.2 and 64.9 ppm, respectively, relative to xyloside **61**.

64 Asterosaponin P-1

Occurrence: *Patiria pectinifera (115), Oreaster reticulatus, (116), Patiria miniata (117).*
Physical data: m.p. 191–192°C; $[\alpha_D]$ = + 3.0° (MeOH); FAB MS (+ ve ion), m/z 739 $[M_{SO_3Na} + K]^+$; 723 $[M_{SO_3Na} + Na]^+$; selected ^{1}H-NMR signals: δ 1.00 (s, 18-H$_3$), 1.06 (s, 19-H$_3$), 3.64 (dt, J = 3.5, 10.0 Hz, 6β-H), 4.25 (dt, J = 3.5, 9.5 Hz, 15β-H); ^{13}C-NMR spectrum of the steroid portion in Table 5.

Assignments of the NMR signals of the arabinofuranosyl moiety in **64** and **65**.

	^{1}H		^{13}C	
Position	**64**	**65**	**64**	**65**
1	4.97 s	5.02 s	109.4	108.4
2	4.07 d(J = 2 Hz)	4.27 m	81.2	82.3
3	3.62 dd(J = 7, 3 Hz)	4.51 dd(J = 5.5, 3 Hz)	89.2	83.7
4	4.15 m	4.27 m	81.7	82.8
5	4.22 m	3.76 dd(J = 12, 5 Hz)	68.3	62.6
		3.87 dd(J = 12, 3 Hz)		
OCH$_3$	3.60		57.8	

On acid hydrolysis this substance gave 3-O-methyl-L-arabinose, affording arabinose on demethylation with BF$_3$, and peaks in the mass spectrum of the alditol peracetate derivative corresponding to a 3-O-methylpentose. Comparative ^{13}C-NMR data of **64** and its desulphated derivative established the structural characteristics of the carbohydrate moiety such as the α-configuration, the furanose form and the location of sulphate at C-5'.

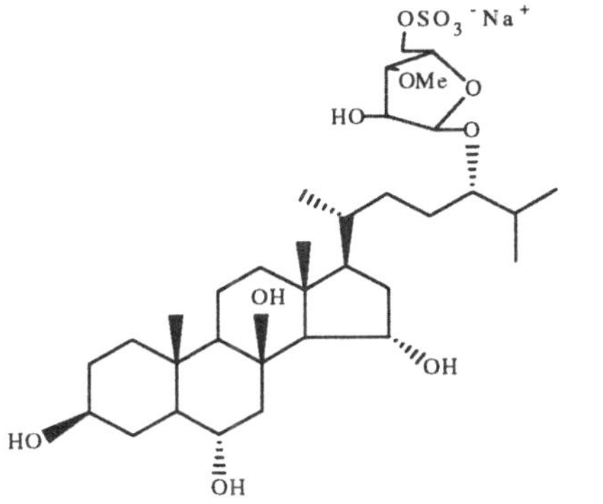

64

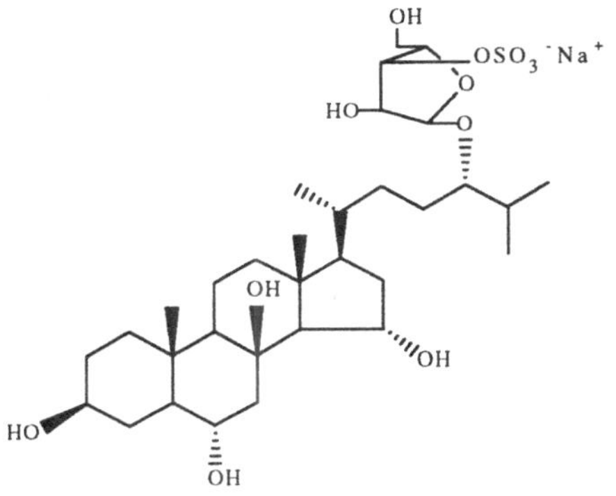

65

Acid hydrolysis gave a set of sapogenols, which interfered with the attempt to establish the structure of the genuine aglycone. High temperature catalytic reduction (Pd/CaCO$_3$ at 330°C) afforded 5α-cholestane. A comparison of ^{1}H-NMR spectra for desulphated **64** and the corresponding 3,6,15-triketone and 6,15-diketone, obtained by oxidation with CrO$_3$/pyridine and CrO$_3$/CH$_3$CO$_2$H, respectively, allowed identification of the proton at δ 3.57 associated with 24-hydroxy group, which is the site of glycosidation (115). The stereochemistry was tentatively assigned as 24S by analogy with nodososide (146) and because the ^{13}C-NMR shifts of the side chain carbons were virtually identical in both spectra (116). Recently X-ray crystallographic analysis of olesulphated asterosaponin P-1 has confirmed the 24-S configuration of the aglycone (193).

The steroidal aglycone of asterosaponin P-1 is common among the glycosides of starfishes and was later found in glycoside **65** from *Oreaster reticulatus* (*116*), in miniatoside B (**66**) co-occurring with the previous **64** and **65** in *Patiria miniata* (*117*), in crossasteroside B (**67**) and C (**68**) isolated from *Crossaster papposus* (*118*) and in borealosides A-C (**69–71**) from *Solaster borealis* (*93*). As non-glycosidated steroid, this aglycone has been isolated from *Asterina pectinifera* (*119*).

65

OCCURRENCE: *Oreaster reticulatus* (*116*), *Patiria miniata* (*117*).
PHYSICAL DATA: $[\alpha]_D = +0.2°$ (MeOH); FAB MS (+ ve ion), m/z 709 $[M_{SO_3Na} + Na]^+$.

Comparative ^{1}H-NMR data of **65** and its desulphated derivative established the location of the sulphate at C-3′ of the arabinofuranosyl residue; desulphation produced changes in the shifts of 3′-H (δ 4.51 *vs.* 3.87), and also in those of 2′-H and 4′-H (δ 4.27 *vs.* 4.02).

66 Miniatoside B

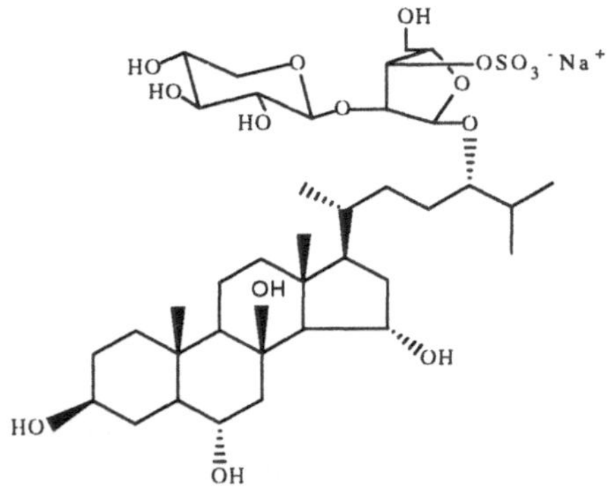

66

OCCURRENCE: *Patiria miniata* (*117*).
PHYSICAL DATA: $[\alpha]_D = -12.3°$ (MeOH); FAB MS (− ve ion), m/z 795 $[M_{SO_3^-}]$; selected ^{1}H-NMR signals: δ 4.65 (dd, $J = 5.8, 3.0$ Hz, 3′-H), 5.18 (br s, 1′-H).

This substance is the 2′-β-D-xylopyranosyl derivative of the previous glycoside **65** which also occurs in *Patiria miniata*. A major fragment in the FAB MS at m/z 663 corresponding to the loss of the xylosyl residue (132 mass units) indicated the sugar sequence. Analysis of the ^{13}C-NMR data and comparison with those of **65** established that the β-xylopyranosyl residue was attached at C-2′ of the (3′-O-sulphate)-arabinofuranose (C-2′: 90.3 *vs.* 82.3 ppm). Moreover, since the anomeric proton of the arabinofuranosyl unit always exhibits a downfield shift to δ 5.10-5.18 in 2′-substituted arabinofuranosides, the chemical shift at δ 5.18 for 1′-H in **66** was additional evidence for the glycosidation at C-2′.

67 Crossasteroside B

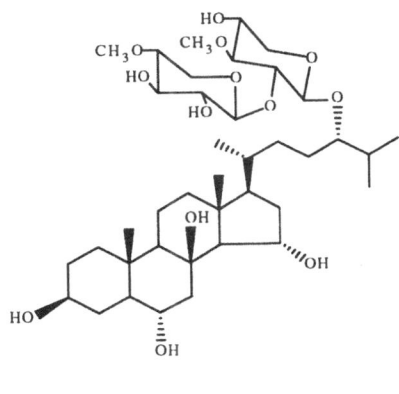

67

OCCURRENCE: *Crossaster papposus* (*118*).
ACTIVITY: Antiviral against herpes virus, SHV-1 at a dose of 1 μg/ml; also shows very feeble cytotoxic and antibacterial activity.
PHYSICAL DATA: $[\alpha]_D = -5.0°$ (MeOH); FAB MS (+ ve ion), m/z 767 $[M]^+$.
  The structure of the carbohydrate moiety was established by ^{1}H and ^{13}C-NMR spectral data and comparison with those of the major crossasteroside A (**72**).

68 Crossasteroside C

OCCURRENCE: *Crossaster papposus* (*118*).
PHYSICAL DATA: $[\alpha]_D = -8.0°$ (MeOH); FAB MS (+ ve ion), m/z 753 $[M + Na]^+$; for assignments of NMR signals of the disaccharide moiety see culcitoside C_5 (**101**).
  ^{1}H- and ^{13}C-NMR spectral data indicated that the substance contained an α-arabinosyl and a 4-O-methyl-β-xylopyranosyl unit [C-4′: 80.8 ppm, shifted downfield by 10.4 ppm and C-5′: 64.3 ppm, shifted

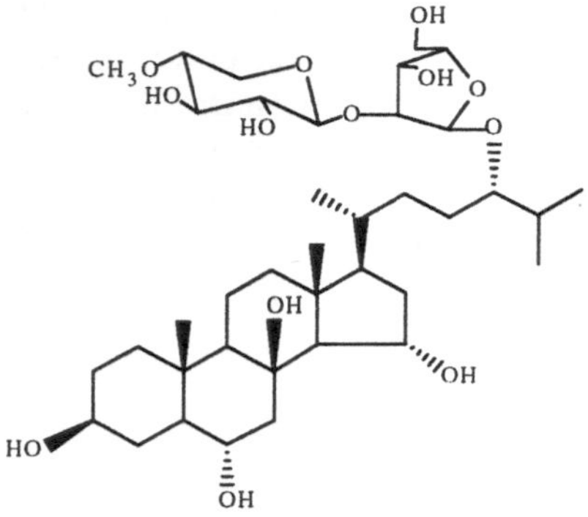

68

upfield by 2.0 ppm relative to methyl-β-D-xylopyranoside]. The chemical shifts observed for C-1′ and C-2′ of the arabinosyl moiety [107.8 and
92.7 ppm *vs.* 109.3 and 81.9 observed in methyl-α-D-arabinofuranoside]
established the location of the terminal 4-O-methylxylose at C-2′ of the
arabinose. Once again we note that the signal for 1′-H of arabinose is
shifted downfield to δ 5.11, thus indicating a substitution at C-2′.

69 Borealoside A

OCCURRENCE: *Solaster borealis* (*93*).
PHYSICAL DATA: $[\alpha]_D = \pm 0°$ (MeOH); FAB MS (− ve ion), m/z 809
$[M_{SO_3^-}]$; selected ^{1}H-NMR signals: δ 2.89 (dd, $J = 9.5, 7.5$ Hz, 2″-H), 4.15
(m, 5′-H_2), 5.11 (br s, 1′-H).

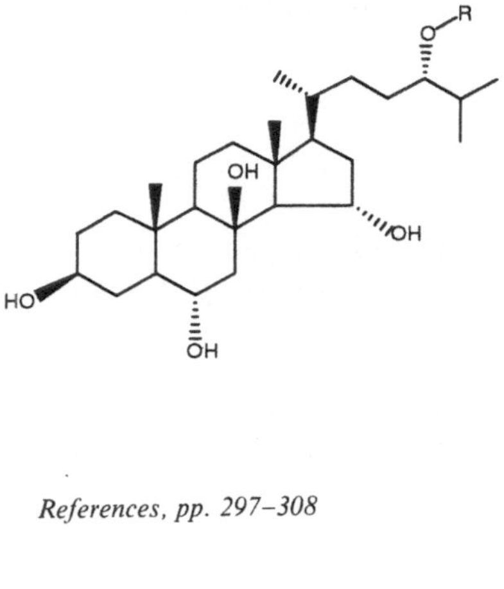

69, R=

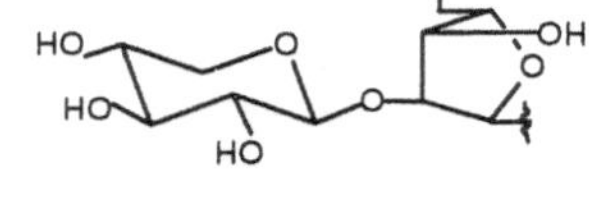

70, R=

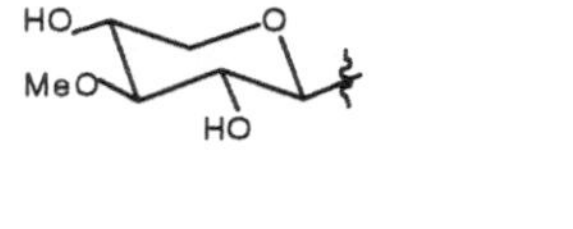

71, R=

Analysis of ^{1}H-and ^{13}C-NMR spectral data indicated the presence of a 2-O-methyl-β-xylopyranosyl [$\delta_{H-2''}$: 2.89 (dd, J = 9.5, 7.5 Hz); $\delta_{C-2''}$: 84.5 pp vs 74.0 in methyl-β-D-xylopyranoside] and a (5'-O-sulphated)-arabinofuranosyl [$\delta_{H_{2-5'}}$: 4.15 (m); $\delta_{C-5'}$: 68.1 ppm vs. 62.4 in methyl-α-L-arabinoside] unit. In the FAB MS the loss of 146 mass units from the anion peak corresponding to a methoxylated pentose unit indicated the sugar sequence as in **69**, whereas the ^{1}H- and ^{13}C-NMR shifts of the arabinosyl unit [C-1':107.6; C-2':92.0 ppm] established the interglycosidic (1 → 2) linkage. The 24S configuration is proposed on the basis of the ^{13}C-NMR frequencies of the side chain carbons which are virtually identical with those of nodososide (**146**) and other (24S)-24-O-α-L-arabinofuranosides (*120*).

70 Borealoside B

OCCURRENCE: *Solaster borealis* (*93*).
PHYSICAL DATA: [α]$_D$ = + 1.4° (MeOH); FAB MS (− ve ion), m/z 795 [$M_{SO_3^-}$] and 663 [$M_{SO_3^-}$ 132] $^-$; selected ^{1}H-NMR signals: δ 3.17: (dd, J = 9.5, 7.5 Hz, 2''-H), 4.15 (m, 5'-H$_2$), 5.10 (br s, 1'-H).

Structure **70** is based only on the close similarity of the ^{1}H-NMR spectrum to that of borealoside A (**69**), except for the lack of the methoxyl singlet and the downfield shift to δ 3.17 of the signal for 2''-H of the xylose residue. Borealoside B is isomeric with miniatoside B (**66**) the only difference being the sulphation at C-5' of arabinose instead of at C-3'.

71 Borealoside C

OCCURRENCE: *Solaster borealis* (*93*).
PHYSICAL DATA: [α$_D$] = + 6.7° (MeOH); FAB MS (− ve ion), m/z 597 [M-H]$^-$; selected ^{1}H-NMR signals: δ 3.03 (t, J = 9.5 Hz, 3'-H), 4.25 (d, J = 7.5 Hz, 1'-H).

The location of the methyl group at 3-OH of the xylose residue was suggested by the high field shift observed for 3'-H (δ 3.03 vs. 3.30 in steroidal xylopyranosides) and confirmed by the ^{13}C-NMR spectrum ($\delta_{C-3'}$: 87.6 vs. 77.7 ppm in steroidal xylopyranosides). The 24S configuration in borealoside C was derived from the ^{13}C-NMR shifts of the side chain carbons which were virtually identical with that in (24S)-24-O-β-D-xylopyranosyl steroids (*e.g.* amurensoside A, **61**), and confirmed by conversion of the derived 3β,6α,15α-trimethoxycholest-8(9)-en-24-ol into 24-O-(+)MTPA ester, and comparison of the ^{1}H-NMR spectrum to those of 24-O-(+)MTPA esters of known stereochemistry (*crf.* amurensoside A, **61** and section 9.1).

72 Crossasteroside A

OCCURRENCE: *Crossaster papposus* (*121*)
ACTIVITY: Causes 80% inhibition of the electrically induced contraction of the guinea-pig ileum at concentration of 100 μg/ml, and possess some cytotoxicity at a dose of 10 μg/ml and activity on human lymphoma cell (JURCAT) growth at a dose of 500 μg/ml.
PHYSICAL DATA: $[\alpha]_D = -19.5°$ (MeOH); FAB MS (+ ve ion), m/z 783 $[M + Na]^+$; selected ^{1}H-NMR signals: δ 1.01 (s, 18-H$_3$), 1.03 (s, 19-H$_3$), 3.60 (m, 3α-H), 3.68 (2H, m, 6- and 7-H), 4.33 (dt, $J = 9.0, 3.0$ Hz, 15β-H); ^{13}C-NMR spectrum of the steroid portion in Table 5.

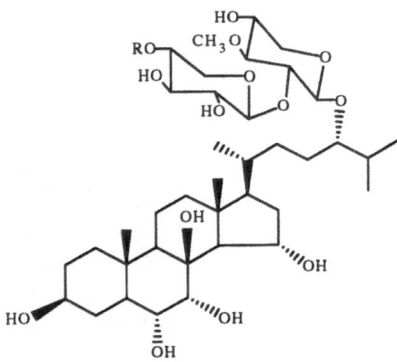

72, R=CH$_3$

73, R=H

^{13}C-NMR assignments of the disaccharide moiety of **72**

Position	4-O-methyl xylose	3-O-methyl xylose
1	105.0	103.3
2	75.0	79.1
3	76.4	86.6
4	81.0	71.1
5	64.3	66.6
OCH$_3$	58.9	60.6

Introduction of an additional hydroxyl group at position 7α of the aglycone is accompanied in the ^{1}H-NMR spectrum by appearance of a two protons signal at δ 3.78 due to the overlapping hydroxymethine protons at C-6 and C-7. In the spectrum of the derived 3,6,5',4',2'',3''-

hexaacetate the resonance of 7-H remained essentially unchanged at δ 3.62 (d, J = 3.0 Hz), while the signal for 6-H moved downfield to 5.17 (dd, J = 12.5,3.0 Hz). Decoupling proved that the two protons were coupled by 3.0 Hz. The structure of the saccharide moiety was determined by acid methanolysis followed by p-bromobenzoylation, which established the presence of a 3-O-methyl-xylopyranose and a 4-O-methyl-xylopyranose, and ¹H-NMR analysis of the derived hexaacetate, which indicated the 4-O-methylxylopyranosyl (1 → 2)-3-O-methylxylopyranosyl sequence. The ¹³C-NMR spectrum of **72** provided supporting evidence.

73 Crossasteroside D

OCCURRENCE: *Crossaster papposus (121)*
PHYSICAL DATA: $[\alpha]_D$ = − 9.2° (MeOH); FAB MS (+ ve ion), m/z 769 [M + Na]⁺.

This substance is related to the previous crossasteroside A (**72**) by lacking the methyl group at C-4″ of the terminal xylosyl unit; the structure was derived by comparison of the ¹H- and ¹³C-NMR spectra. The FAB mass spectrum showed a small peak at m/z 615 corresponding to the loss of xylose (132 mass units) from [M + Na]⁺, thus supporting the sugar sequence as indicated in **73**.

74 Attenuatoside C

OCCURRENCE: *Hacelia attenuata (122)*
PHYSICAL DATA: $[\alpha]_D$ = + 4.7° (MeOH); selected ¹H-NMR signals: δ 0.99 (s, 18-H₃), 1.21 (s, 19-H₃), 2.47 (dt, J = 4.5, 14.0 Hz, 7β-H), 3.46 (dt, J

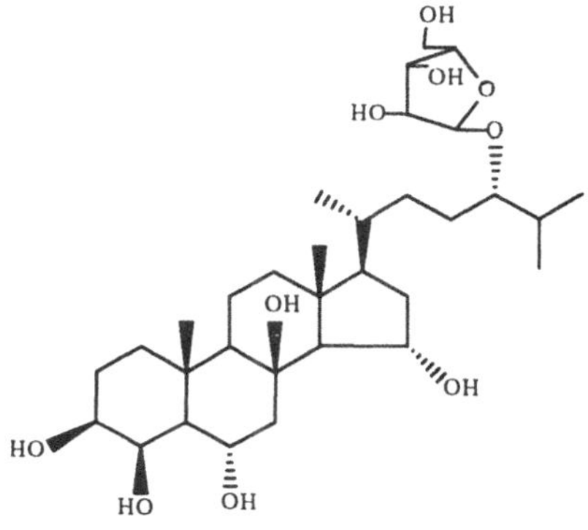

74

= 12.0, 4.5 Hz, 3α-H), 4.11 (dt, J = 4.5, 11.4 Hz, 6β-H), 4.24 (dt, J = 4.5, 11.0 Hz, 15β-H), 4.28 (br s, 4α-H).

The steroid aglycone of attenuatoside C is related to the common 5α-cholestane-3β,6α,8,15α,24-pentaol aglycone of asterosaponin P-1 (**64**) by introduction of an additional hydroxyl group at 4β-position. The resonance of the hydrogen atom at C-4 is observed at δ 4.28 (broad singlet), whereas the resonances of 6-H and 19-H$_3$ have experienced a distinct downfield shift to δ 4.11 (δ 3.64 in **64**) and 1.21 (δ 1.05 in **64**). A small upfield shift is also observed for 3-H. The upfield shifts exhibited by C-2 (5.3 ppm) and C-6 (2.6 ppm) and the downfield shift experienced by C-19 (3.9 ppm) relative to **64** are the significant features of the ^{13}C-NMR spectrum and confirm the location of the new hydroxyl group at C-4β (see also **75** in Table 5). Spectral comparison of attenuatoside C (**74**) with the isomeric attenuatoside B-II (**99**), which differs in the stereochemistry at C-15, was of value for the ^{13}C-NMR assignment. The arabinose was shown to belong to the L series by the large positive split CD curves (A = 120) of the methyl 2,3,4-tri-O-(p-bromobenzoyl)-β-arabinopyranoside (*123*).

75 Borealoside D

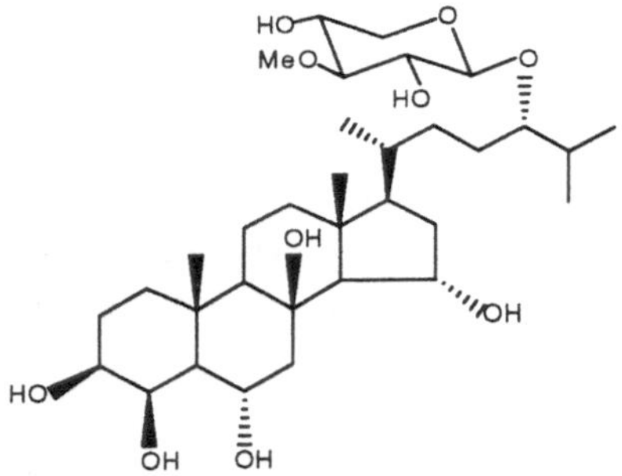

75

Occurrence: *Solaster borealis* (*93*).
Physical data: [α]$_D$ = + 15.5° (MeOH); FAB MS (− ve ion), m/z 613 [M-H]$^-$; ^{13}C-NMR of the steroid in Table 5.

This is the 4β-hydroxyderivative of borealoside C (**71**); the structure was derived by ^{1}H-and ^{13}C-NMR spectra and comparison with **71**.

76 6-*Epi*-nodososide

Occurrence: *Pentaceraster alveolatus* (*124*).
Physical data: [α]$_D$ = ± 0° (MeOH); FAB MS (+ ve ion), m/z 769 [M + Na]$^+$; selected ^{1}H-NMR signals: δ 1.00 (s, 18-H$_3$), 1.20 (s, 19-H$_3$), 3.88

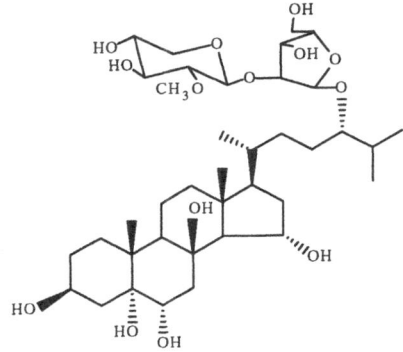

76

(dd, $J = 11.2, 4.5$ Hz, 6β-H), 4.00 (m, 3α-H), 4.24 (dt, $J = 3.0, 9.8$ Hz, 15β-H), 4.40 (d, $J = 7.5$ Hz, 1″-H), 5.11 (s, 1′-H); for spectral data of the saccharide moiety *cfr.* nodososide (**146**); ^{13}C-NMR of the steroid in Table 5.

This substance in the C-6 epimer of nodososide (**146**) and the structure was derived by spectral comparison.

77

OCCURRENCE: *Poraster superbus (149)*.
PHYSICAL DATA: $[\alpha]_D = + 10.4°$ (MeOH); FAB MS (+ ve ion), m/z 637 [M + Na]$^+$, 615 [M + H]$^+$; selected ^{1}H-NMR signals: δ 1.15 (s, 18-H$_3$), 1.05 (s, 19-H$_3$), 2.43 (dd, $J = 13.5, 4.2$ Hz, 7-H), 3.32 (dd, $J = 10.5, 5.0$ Hz, 26-H), 3.46 (dd, $J = 10.5, 6.0$ Hz, 26-H), 4.01 (dd, $J = 8.0, 2.5$ Hz, 16α-H), 4.08 (dd, $J = 11.0, 2.5$ Hz, 15β-H), 4.44 (d, $J = 7.5$ Hz, 1′-H).

77, R=H

78, R=OH

The steroid aglycone has been previously isolated as one of the major steroidal constituents from *Protoreaster nodosus* (125). The 16β-hydroxylation is common among the glycosides and polyhydroxysteroids from starfishes, whereas the 26-hydroxylation is only rarely encountered among glycosides, but very common among the polyhydroxysteroids. For assignment of the stereochemistry at C-25 *cfr.* section 9.2. The site of glycosidation at C-3 was identified by ^{13}C-NMR spectral comparison with the corresponding aglycone (173); C-3 is deshielded by 7.9 ppm (δ_C 80.1 *vs.* 72.2 ppm), whereas C-2 and C-4 are shielded by 1.5 (δ_C 30.1 *vs.* 31.5 ppm) and 3.1 (δ_C 29.3 *vs.* 32.4 ppm) ppm, respectively. The signal of the anomeric hydrogen is observed at δ 4.44, whereas in 24-O-xylopyranosides the same signal is seen at δ 4.25. Thus, this signal proved useful for assigning positions of xylose attachment.

78

Occurrence: *Poraster superbus* (149).
Physical data: $[\alpha]_D = +3.4°$ (MeOH); FAB MS (+ ve ion), m/z 653 [M + Na]$^+$, 631 [M + H]$^+$; selected ^{1}H-NMR signals: δ 3.82 (2H, m, 6- and 7-H), 4.03 (dd, J = 8, 2.5 Hz, 16α-H), 4.17 (dd, J = 11.0, 2.5 Hz, 15β-H), 4.44 (d, J = 7.5 Hz, 1'-H).

The steroid aglycone (174) has been previously isolated from *Protoreaster nodosus* (125).

79

Occurrence: *Patiria pectinifera* (126).
Physical data: m.p. 248–249°C; $[\alpha]_D = +3.8°$ (MeOH); EI ms, m/z 610 (0.3%).

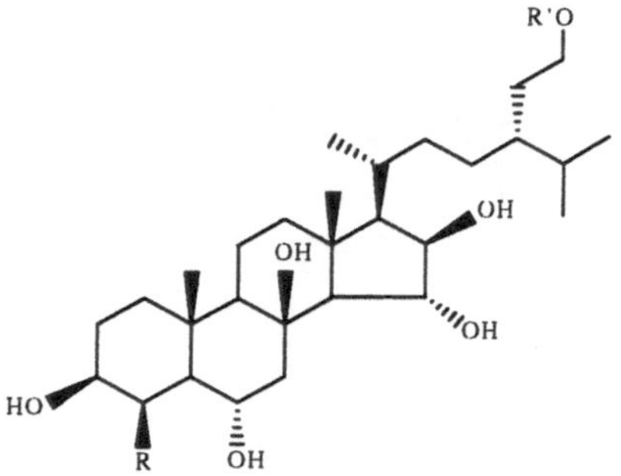

79, R=H, R'=

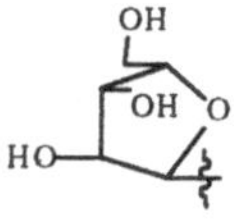

80, R=H, R'=

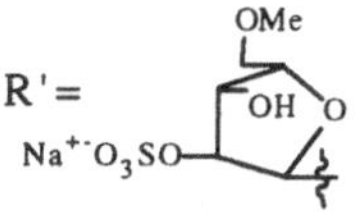

81, R=OH, R'=

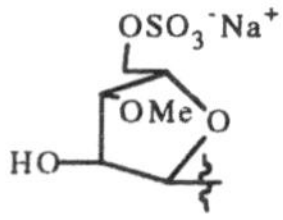

Glycoside **79** has been isolated from a glycosidic fraction from *Patiria pectinifera* after desulphation. **79** along with attenuatoside S-I (**157**) from *Hacelia attenuata* (*127*) and echinasteroside B (**153**) from *Echinaster sepositus* (*128*) were the first glycosides of the 29-hydroxystigmastane series detected in starfishes. Since then the structures of many glycosides of this series have been reported. Structure **79** was deduced mainly from spectroscopic data and comparison with the 5α-cholestane-3β,6α,8,-15α,16β-26-hexaol (**173**) found in the same starfish and previously in *Protoreaster nodosus* (*125*). Acetylation of desulphated **76** gave the 3, 6, 15, 2′, 3′, 5′-hexaacetate, in whose ¹H-NMR spectrum the signals for the side chain hydroxymethylene protons were barely shifted, thus indicating the site of glycosidation to be there. That a 29-β-hydroxyethyl cholestane side chain was present was deduced by comparing the experimental side chain carbon shifts with those calculated for different structure variants. The configuration at C-24 was left unassigned. We now propose the 24R configuration by analogy with the many 24-(β-hydroxyethyl) steroidal glycosides isolated from starfishes, for which the 24R configuration was assigned on the basis of ¹H- and ¹³C-NMR shifts of the isopropyl methyls (*cfr.* section 9.4).

80 Miniatoside A

OCCURRENCE: *Patiria miniata* (*117*).
PHYSICAL DATA: $[\alpha]_D = + 19.0°$ (MeOH); FAB MS (− ve ion), m/z 721 $[M_{SO_3^-}]$; selected ¹H-NMR signals; δ 0.88 ($J = 7.0$ Hz, 26- or 27-H_3), 0.90 ($J = 7.0$ Hz, 27- or 26-H_3), 1.05 (s, 19-H_3), 1.14 (s, 18-H_3), 2.44 (dd, $J = 12.5$, 3.5 Hz, 7-H), 3.65 (dt, $J = 3.5$, 9.0 Hz, 6β-H), 3.76 (m, 29-H_2), 4.01 (dd, $J = 8.2$, 2.5 Hz, 16α-H), 4.08 (dd, $J = 11.0$, 2.5 Hz, 15β-H); the ¹³C-NMR spectrum of the steroid portion in Table 5.

Assignments of the NMR signals of the 5-O-methyl-arabinosyl-2-sulphate moiety

Position	¹H	¹³C
1	5.17 bs	107.8
2	4.59 bd ($J = 3$ Hz)	89.1
3	4.04 dd ($J = 3.0$, 6.0 Hz)	83.5
4	4.06 m	78.4
5	3.63 dd ($J = 12.5$, 3.5 Hz)	74.0
	3.58 dd ($J = 12.5$, 5.0 Hz)	
OCH₃	3.42	59.4

This is the minor component (2 mg from 3.5 kg of fresh animal) among the fifteen isolated steroidal constituents from *P. miniata*. The structure of the aglycone was derived from spectral data and comparison with (24R)-24-ethyl-5α-cholestane-3β,6α,8,15α,16β,29-hexaol isolated as the 29-sulphate from the starfish *Poraster superbus* (*149*). The 5-O-methyl-2-O-sulphate-α-arabinofuranosyl structure for the sugar moiety was derived both from ^{13}C-NMR and 500 MHz ^{1}H-NMR spectrometers and double resonance experiments, which permitted unequivocal assignments of signals and connetivity. The paramagnetic shift of 2'-H (δ 4.59 *vs.* 3.98 in α-L-arabinofuranoside) confirmed the attachment of sulphate at C-2'. The 24R configuration is based on the chemical shifts of the isopropyl methyls, $δ_H$ 0.88 d, 0.90 d, $δ_C$ 19.9, 19.3 ppm, and comparison with steroids of known configuration (*cfr.* section 9.4).

81 Asterosaponin P-2

OCCURRENCE: *Patiria pectinifera* (*129*).

The sugar moiety of **81** has been already encountered in asterosaponin P-1 (**64**) whereas the steroid portion is the 4β-hydroxy derivative of the glycoside **79**. **64** and **79** co-occur with **81** in *P. pectinifera*.

82 Attenuatoside A-II

OCCURRENCE: *Hacelia attenuata* (*130*).

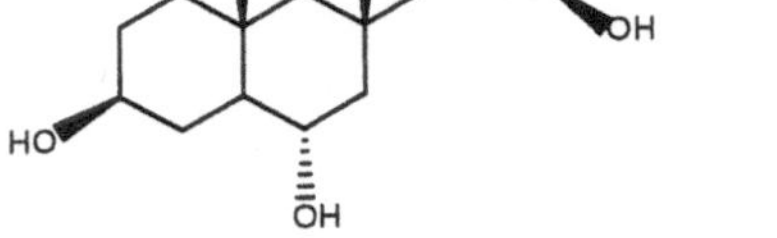

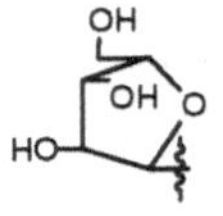

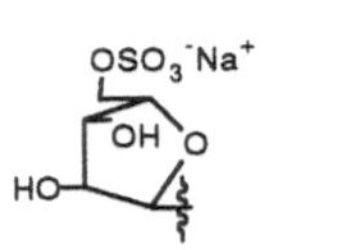

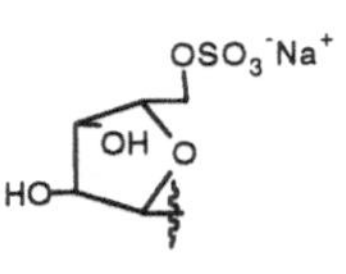

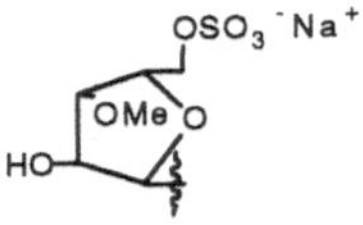

PHYSICAL DATA: $[\alpha]_D = -15.7°$ (MeOH); selected ^{1}H-NMR signals: δ 1.01 (s, 19-H$_3$), 1.30 (s, 18-H$_3$), 2.48 (dd, $J = 13.5, 4.5$ Hz, 7-H), 3.74 (dt, $J = 4.5, 11.0$ Hz, 6β-H), 4.46 (dt, $J = 1.6, 5.1$ Hz, 15α-H), 4.95 (br s, 1'-H).

This substance was isolated as very minor component from *H. attenuata* and the structure was derived by ^{1}H- and ^{13}C-NMR spectral comparison with the major component attenuatoside A-I (**91**), the first glycoside of the series with the (24S)-5α-cholestane-3β,6α,8,15β,24-pentaol aglycone (for ^{13}C-NMR data *cfr.* **85** in Table 5) detected in starfishes. The non-glycosidated steroid (**227**) was subsequently isolated from *Gomophia watsoni* (*131*).

83 Scoparioside A

OCCURRENCE: *Astropecten scoparius* (*84*).

PHYSICAL DATA: $[\alpha]_D = -8.5°$ (MeOH); FAB MS (− ve ion), m/z 663 $[M_{SO_3}^-]$; ^{1}H-NMR signals of the arabinosyl moiety below.

^{1}H-NMR spectrometry indicated the location of the sulphate at C-5' of the arabinose (δ 4.15 *vs.* 3.66 dd–3.78 dd in the desulphate derivative) and ^{13}C-NMR spectrometry indicated the attachment of the sugar moiety at C-24 of the aglycone. The 24S configuration is proposed by analogy with the (24S) 5α-cholestane-3β,6α,8,15β,24-pentaol 24-sulphate (**228**), co-occurring in the same organism, for which the configuration at C-24 was assigned by conversion of the 3β,6α,15β-trimethoxy-5α-cholestane-8,24-diol, obtained from **228** by methylation followed by solvolysis, into the 24-(+)MTPA ester and ^{1}H-NMR measurement (*cfr.* amurensoside A, **61** and section 9.1).

84 Scoparioside B

OCCURRENCE: *Astropecten scoparius* (*84*).

PHYSICAL DATA: $[\alpha]_D = \pm 0°$ (MeOH); FAB MS (− ve ion), m/z 663 $[M_{SO_3}^-]$.

^{1}H-NMR assignments of the sugar moieties in **83** and **84**.

Position	arabinofuranosyl (**83**)	xylofuranosyl (**84**)
1	4.94 d ($J = 1.5$ Hz)	4.96 d ($J = 1.5$ Hz)
2	3.99 dd ($J = 1.5, 3.7$ Hz)	4.04 dd ($J = 1.5, 3.7$ Hz)
3	3.90 dd ($J = 6.0, 3.7$ Hz)	4.07 dd ($J = 3.7, 5.0$ Hz)
4	4.15 m	4.42 m
5	4.15 m	4.20 dd ($J = 12.0, 5.0$ Hz)
		4.42 dd ($J = 12.0, 6.2$ Hz)

The major difference in the ^{1}H-NMR spectrum of **84**, when compared with that of the isomeric **83**, was the resonance of 4'-H of the sugar moiety at lower field (δ 4.42 compared with δ 4.15 in **83**), in agreement with a xylofuranosyl structure (1,3-*syn* interaction of H-4'/2'-OH as compared with the *anti* arrangement of the arabinofuranosyl structure). The upfield shift of 5'-H$_2$ from δ 4.42/4.20 in **84** to δ 3.90/3.79 ppm in the desulphated derivative confirmed the location of the sulphate. Scoparioside B was the third xylofuranosyl-containing glycoside to be found in starfishes, after indicoside B (**85**) and C (**103**) from *Astropecten indicus* (*132*).

85 Indicoside B

OCCURRENCE: *Astropecten indicus* (*132*).
PHYSICAL DATA: [α]$_D$ = − 3.0° (MeOH); FAB MS (− ve ion), m/z 677 [M$_{SO_3^-}$]. ^{1}H-NMR signals of the sugar moiety: δ 3.72 (dd, J = 1.5, 5.0 Hz, 3'-H), 4.11 (t, J = 1.5 Hz, 2'-H), 4.16 (dd, J = 7.5, 11.0 Hz, 5'-H), 4.33 (dd, J = 5.0, 11.0 Hz, 5'-H), 4.50 (m, 4'-H) 4.96, (d, J = 1.5 Hz, 1'-H); ^{13}C-NMR spectrum of the steroid portion in Table 5.

Detailed ^{1}H-NMR analysis including NOEDS and decoupling experiments revealed the presence of a xylofuranose structure. Full elucidation of the monosaccharide moiety was obtained by acid methanolysis followed by *p*-bromobenzoylation affording methyl 3-O-methyl-2,4-di-O-(*p*-bromobenzoyl)-α-D-xylopyranoside. Indicoside B (**85**) and indicoside C (**107**) represent the first two cases of steroidal xylofuranosides form to be found in starfishes.

86 Pycnopodioside A

OCCURRENCE: *Pycnopodia helianthoides* (*74*).
ACTIVITY: inhibits cell division of fertilized sea urchin eggs at a concentration of 10^{-5} M (50% inhibition) (*35*).
PHYSICAL DATA: [α]$_D$ = + 2.0° (MeOH); FAB MS (− ve ion), m/z 583 [M–H]$^-$; selected ^{1}H-NMR signals: δ 4.27 (d, J = 7.5 Hz, 1'-H); selected ^{13}C-NMR signals: δ 86.5 (C-24).

87 Pycnopodioside B

OCCURRENCE: *Pycnopodia helianthoides* (*74*).
PHYSICAL DATA: [α]$_D$ = + 1.8° (MeOH); FAB MS (− ve ion), m/z 663 [M$_{SO_3^-}$].

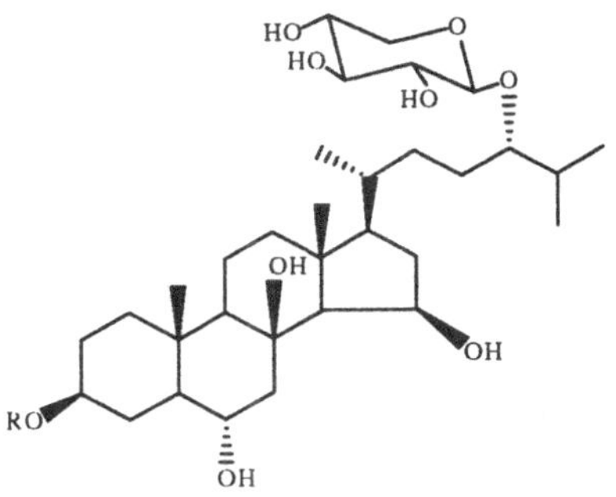

86, R=H

87, R=SO$_3^-$Na$^+$

^{1}H-($\delta_{H-1'}$: 4.27 d) and ^{13}C-NMR (δ_{C-24}: 86.5 ppm) indicated the attachment of the β-xylopyranosyl unit at C-24 in both compounds; downfield shifts of signals for 3-H (δ 4.24 m) and C-3 (80.0 ppm) in **87** relative to **86** (δ_H: 3.55 m; δ_C: 72.2 ppm) established the location of the sulphate there. The 24S-configuration was assigned on the basis of the ^{13}C-NMR shifts of the side chain carbons which were identical with those assigned to the corresponding carbons of the (24S) 24-O-xylopyranoside amurensoside A (**61**).

88 Pycnopodioside C

OCCURRENCE: *Pycnopodia helianthoides (74)*.
ACTIVITY: No activity in the sea urchin egg test below 10^{-5} M concentrations.

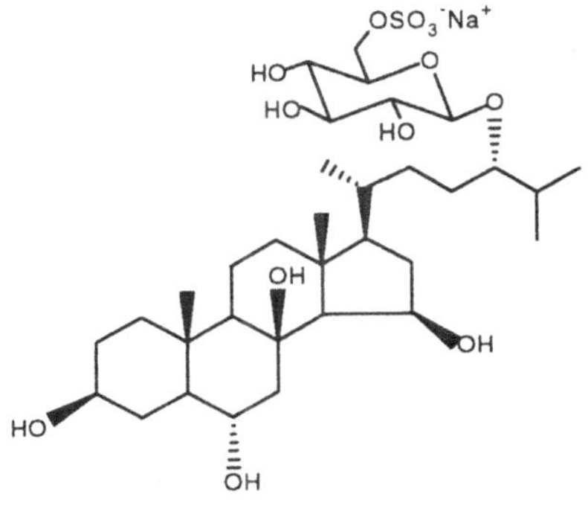

88

Physical data: $[\alpha]_D = +4.2°$ (MeOH); FAB MS (− ve ion), m/z 693 $[M_{SO_3^-}]$; selected ^{1}H-NMR signals: δ 4.18 (dd, $J = 5.0$, 12.0 Hz, 6′-H), 4.34 (dd, $J = 2.0$, 12.0 Hz, 6′-H).

The location of the sulphate at C-6 of glucose was indicated by the ^{1}H-NMR spectrum ($\delta_{H_3-6'}$: 41.18–4.34 vs. 3.72–3.90 in the desulphated one) and confirmed by ^{13}C-NMR spectrum ($\delta_{C-6'}$: 68.5 vs. 61.9 in methyl-β-glucopyranoside).

89 Glacialoside A

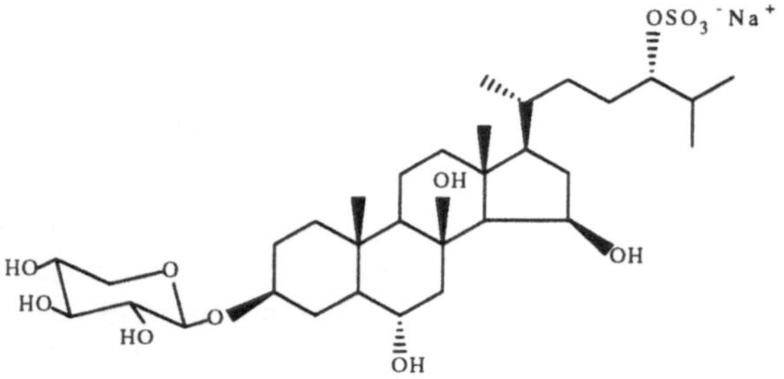

89

Occurrence: *Marthasterias glacialis* (*133*).
Physical data: $[\alpha]_D = +3.0°$ (MeOH); FAB MS (− ve ion), m/z 663 $[M_{SO_3^-}]$; selected ^{1}H-NMR signals: δ 0.95 (d, $J = 7.0$ Hz, 26- or 27-H$_3$), 0.96 (d, $J = 7.0$ Hz, 27- or 26-H$_3$), 3.55 (m, 3α-H), 4.14 (q, $J = 6.5$ Hz, 24-H), 4.39 (d, $J = 7.5$ Hz, 1′-H).

In addition to major amounts of the asterosaponins marthasterosides A$_1$ (**14**), A$_2$ (**15**), B (**27**) and C (**29**), the Mediterranean starfish *M. glacialis* also gave in minute amounts two sulphated steroidal monoglycosides, glacialoside A (**89**) and glacialoside B (**104**). The position of the sulphate at C-24 and of the β-xylopyranosyl unit at C-3 was shown by the ^{1}H- and ^{13}C-NMR spectra and comparison with the desulphated derivative. The paramagnetic shift of the signal of the anomeric proton (δ 4.39) was once again indicative of the location of the xylopyranosyl unit at C-3 (δ 4.27 in the isomeric 24-O-xylopyranoside, **87**). The stereochemistry at C-24 (24S) was determined on the desulphated analog as in amurensoside A (**61**).

90 Scoparioside C

Occurrence: *Astropecten scoparius* (*84*).
Physical data: $[\alpha]_D = +6.8°$ (MeOH); FAB MS (− ve ion), m/z 677 $[M_{SO_3^-}]$.

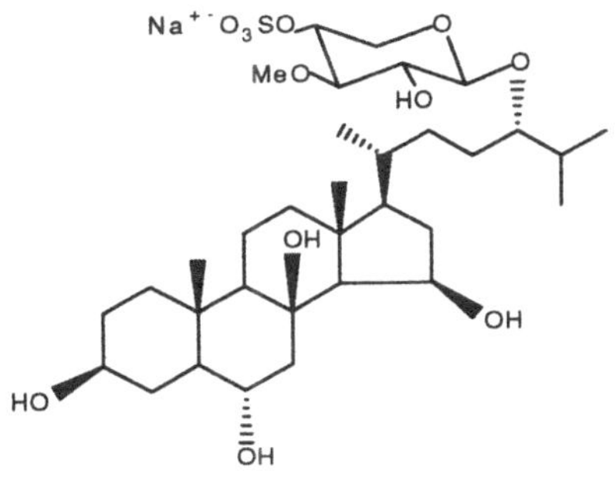

90

Assignments of the NMR signals of the 3-O-methyl-xylopyranosyl-4-sulphate moiety.

Position	^{1}H	^{13}C
1	4.30 d ($J = 7.5$ Hz)	104.7
2	3.28 dd ($J = 7.5, 9.0$ Hz)	74.6
3	3.35 t ($J = 9.0$ Hz)	85.1
4	4.28 m	77.6
5	4.25 dd ($J = 10.6, 5.0$ Hz)	64.7
	3.22 t ($J = 10.6$ Hz)	
OCH$_3$	3.65 s	60.1

The position of the sulphate and the methyl in the β-xylopyranosyl residue was established as being at C-4' and C-3', respectively, by ^{1}H- and ^{13}C-NMR spectrometry. Desulphation afforded an O-methylxyloside with proton signals for the sugar moiety identical with those observed in the 3-O-methylxylopyranosides (*cfr.* **71, 75**). Significant feature in the ^{1}H-NMR of the desulphated sample was the presence of the triplet for 3'-H shifted upfield to δ 3.03.

91 Attenuatoside A-I

OCCURRENCE: *Hacelia attenuata (111).*
PHYSICAL DATA: $[\alpha]_D = -20.6°$ (MeOH); FAB MS (− ve ion), m/z 753 [M + Na]$^+$; 500 MHz ^{1}H-NMR of the sugar moiety: xylose, δ 4.40 (d, $J = 7.8$ Hz, 1'-H), 2.84 (dd, $J = 9.1, 7.8$ Hz, 2'-H), 3.35 (under CD$_3$OD, 3'-H), 3.46 (ddd, $J = 10.2, 9.1, 5.6$ Hz, 4'-H), 3.13 (dd, $J = 11.3, 10.2$ Hz, 5'-H), 3.78 (dd, $J = 11.3, 5.6$ Hz, 5'-H); arabinose, δ 5.08 (d, $J = 1.1$ Hz, 1'-H), 4.06 (dd, $J = 4.0, 1.1$ Hz, 2'-H), 3.98 (dd, $J = 7.8, 4.0$ Hz, 3'-H), 3.93

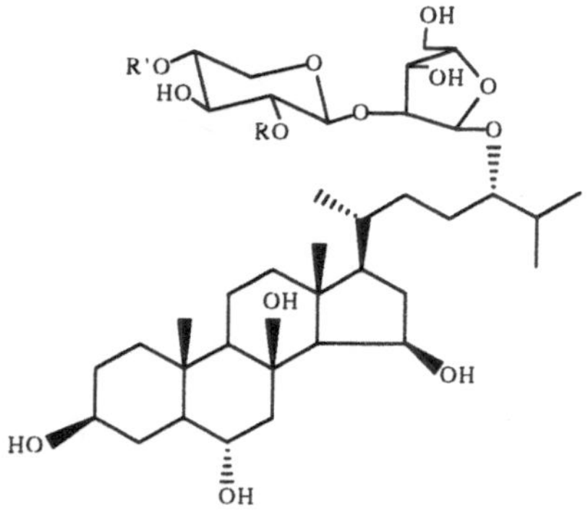

91, R=CH$_3$, R'=H

92, R=H, R'=CH$_3$

93, R=R'=CH$_3$

94, R=R'=CH$_3$, Δ^{22}

(ddd, $J = 7.8, 5.1, 2.9$ Hz, 4'-H), 3.75 (dd, $J = 12.5, 2.9$ Hz, 5'-H), 3.63 (dd, $J = 12.5, 5.1$ Hz, 5'-H); OCH$_3$, δ 3.52 (s).

^{13}C-NMR assignments of the disaccharide moiety

Position	2-O-methyl xylose	arabinose
1	105.1	107.6
2	84.3	93.1
3	77.8	77.6
4	71.1	85.0
5	67.1	62.5
OCH$_3$	60.6	

Attenuatoside A-I, the first glycoside of a series with 5α-cholestane-3β,6α,8,15β,24-pentaol as the aglycone to be found in starfishes, is the major component isolated from *Hacelia attenuata*. The presence of 2-O-methyl-β-D-xylopyranosyl (1 → 2)-α-L-arabinofuranosyl moiety, a common disaccharide unit among glycosides from starfishes, was suggested by 500 MHz ^{1}H- and ^{13}C-NMR data for **91** and supported by comparison with the data for the sugar moiety of nodososide (**146**). Acetylation and ^{1}H-NMR data of the derived heptaacetate confirmed the sequence and the interglycosidic linkage. The stereochemical assignment of the sugars as D-xylose and L-arabinose was assigned from exciton split CD data of the *p*-bromobenzoates of the monosaccharides (*123*), obtained by acid methanolysis followed by *p*-bromobenzoylation. The structure of the aglycone was established by the spectral data of the intact glycoside. The presence of the 15β-hydroxyl group was supported by formation of a 8,15-phenylboronate.

92 Culcitoside C_4

OCCURRENCE: *Culcita novaeguineae* (*44*).
PHYSICAL DATA: $[\alpha]_D = -20.3°$ (MeOH); FAB MS (− ve ion), m/z 729 [M–H]⁻; selected ^{1}H-NMR signals: δ 3.14 (1H, t, $J = 10.6$ Hz, 5''-H_2), 3.21 (m, 4''-H), 4.05 (1H, dd, $J = 10.6$, 5.0 Hz, 5''-H_2).

The location of the methoxy group at C-4 of the terminal xylopyranosyl unit was indicated by the upfield shift of the proton at C-4'' to δ 3.21 (*ca.* 3.50 in methyl β-xylopyranoside) and the downfield shift of the equatorial proton at C-5'' to δ 4.05 (*ca.* 3.80 in methyl β-xylopyranoside) and supported by ^{13}C-NMR data (C-4'': 80.9, C-5'': 64.6; 71.4 and 66.7 ppm in methyl β-D-xylopyranoside). For complete NMR assignments of the disaccharides moiety *cfr.* **101**. Culcitoside C_4, which is isomeric with the previous attenuatoside A-II (**91**) and differs only in the position of the methyl group in the saccharide moiety, co-occurs with ten other steroidal glycosides and five polyhydroxysteroids in *Culcita novaeguineae* (*44*).

93 Halityloside E

OCCURRENCE: *Halityle regularis* (*112*), *Nardoa gomophia* (*70*), *Sphaerodiscus placenta* (*134*), *Culcita novaeguineae* (*44*).
ACTIVITY: Inhibits cell division of fertilized sea urchin eggs at a concentration of 10^{-7} M (50% inhibition) and possesses antifungal activity (minimum inhibitory amount against *Cladosporium cucumerinum*, 10 μg) (35).
PHYSICAL DATA: $[\alpha]_D = -20.0°$ (MeOH); FAB MS (+ ve ion), m/z 767 [M + Na]⁺.

Assignments of NMR signals of the disaccharide moiety in **93**.

Position	xylose		arabinose	
	^{1}H	^{13}C	^{1}H	^{13}C
1	4.44 (d, $J = 7.6$ Hz)	105.1	5.11 (br s)	108.0
2	2.90 (dd, $J = 9.0$, 7.6 Hz)	84.9	4.08 (d, $J = 4.0$ Hz)	92.4
3	3.42 (t, $J = 9.0$ Hz)	76.6	4.02 (m)	77.9

4	3.20 (m)	81.0	3.98 (m)	84.2
5	3.14 (t, J = 10.6 Hz)	64.4	3.65 (dd, J = 12.5, 4.8 Hz)	62.8
	4.02 (dd, J = 10.6, 4.5 Hz)		3.79 (dd, J = 12.5, 3.0 Hz)	
OCH_3	3.50 (s)-3.61 (s)	61.0–59.0		

The structure was derived from ^{1}H- and ^{13}C-NMR data and comparison with halityloside D (**102**), the major component of the group of steroidal glycosides isolated from *Halityle regularis* (*112*).

94 22-Dehydrohalityloside E (= poranoside A)

OCCURRENCE: *Sphaerodiscus placenta* (*134*), *Porania pulvillus* (*135*).
PHYSICAL DATA: FAB MS (+ ve ion), m/z 765 [M + Na]$^+$, 743 [M + H]$^+$; selected ^{1}H-NMR signals: δ 1.05 (d, J = 7.0 Hz, 21-H$_3$), 1.32 (s, 18-H$_3$), 1.77 (m, 25-H), 2.21 (m, 20-H), 3.70 (t, J = 7.0 Hz, 24-H), 5.35 (dd, J = 15.0, 7.5 Hz, 23-H), 5.46 (dd, J = 15.0, 7.5 Hz, 22-H).
The FAB MS spectrum gave pseudomolecular ion peaks two mass units fewer than halityloside E (**93**). The ^{1}H-NMR spectrum with double resonance experiments established the sequence C-20/C-27.

95 Distolasteroside D$_1$

96 Distolasteroside D$_2$

97 Distolasteroside D$_3$

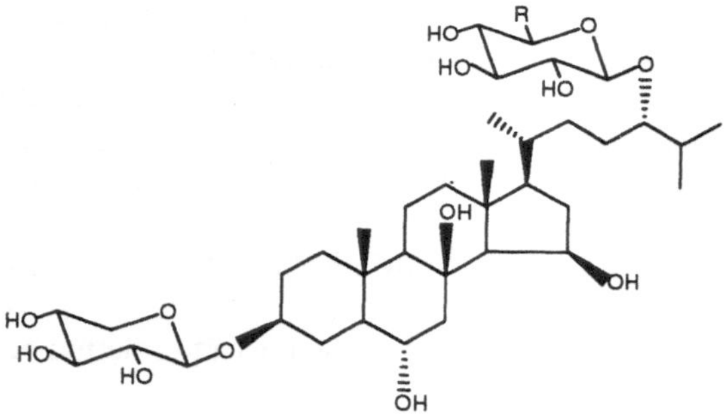

95, R=H

96, Δ^{22}, R=H

97, R=CH$_2$OH

OCCURRENCE: *Distolasterias nippon* (136).

Distolasterosides D_1-D_3 exemplify a group of steroid diglycosides from starfishes in which the two monosaccharides are not linked to each other (see for example 138).

98 Amurensoside D

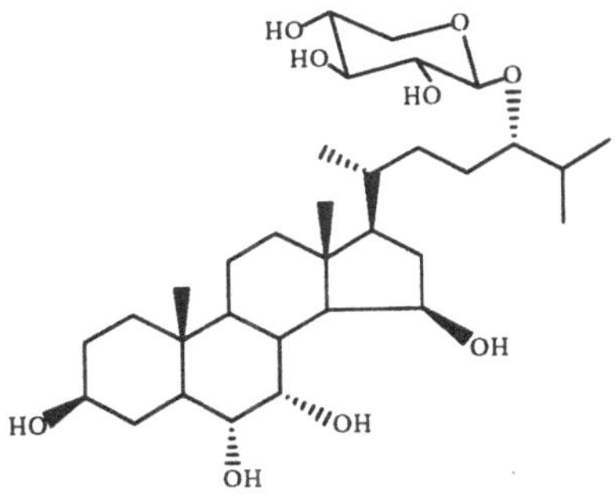

98

OCCURRENCE: *Asterias amurensis* (43).

PHYSICAL DATA: $[\alpha]_D = +6.8°$ (MeOH); FAB MS (− ve ion), m/z 583 [M−H]$^-$; selected ^{1}H-NMR signals: δ 0.97 (s, 18-H$_3$), 4.05 (t, $J = 2.0$ Hz, 7β-H), 4.34 (t, $J = 5.0$ Hz, 15α-H).

This is the only non-sulphated steroidal glycoside isolated from *A. amurensis* and is isomeric with the desulphated amurensoside B (62). The 15β-hydroxy stereochemistry in 98 was suggested by the broad triplet at δ 4.34 whose shape is indicative of a 15β-hydroxy group (137) and by the downfield shift of the C-18 methyl signal to δ 0.97 (δ 0.75 in the 15α-hydroxy analog) and was supported by ^{13}C-NMR spectral data and comparison with those published for hydroxysteroids (113, 114).

99 Attenuatoside B-2

OCCURRENCE: *Hacelia attenuata* (122).

PHYSICAL DATA: $[\alpha]_D = -9.0°$ (MeOH); selected ^{1}H-NMR signals: δ 1.18 (s, 19-H$_3$), 1.29 (s, 18-H$_3$), 4.29 (br s, 4α-H), 4.19 (dt, $J = 4.5, 11.4$ Hz, 6β-H), 4.45 (dd, $J = 6.0, 1.8$ Hz, 15α-H).

This substance is an epimer of attenuatoside C (74), co-occurs in the same organism, and differs from 74 in the stereochemistry at C-15, which in 99 is 15β-OH. The broad triplet at δ 4.24 ($J = 11.0, 4.5$ Hz) in 74 is

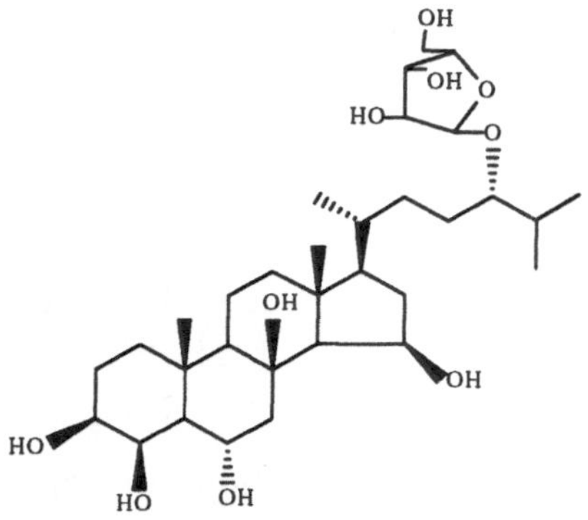

99

replaced in **99** by a broad triplet at δ 4.45, and the signal for C-18 methyl protons is shifted downfield to δ 1.29 (δ 0.99 in **74**).

100 Attenuatoside B-I

Occurrence: *Hacelia attenuata* (*122*).
Physical data: $[\alpha]_D = -12.0°$ (MeOH); FAB MS (+ ve ion), m/z 769 $[M + Na]^+$.

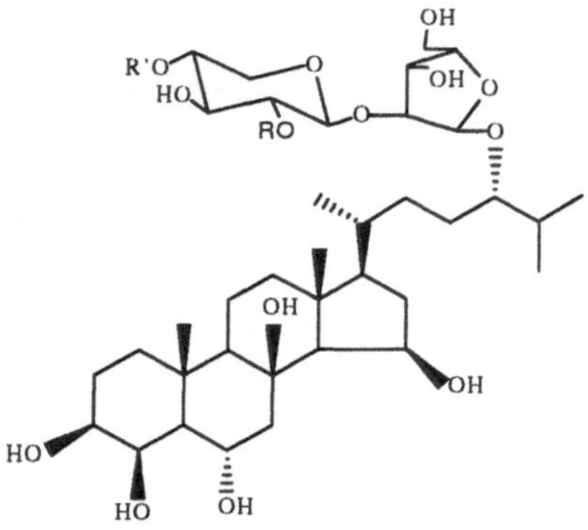

100, R=CH$_3$, R'=H

101, R=H, R'=CH$_3$

102, R=R'=CH$_3$

103, R=R'=CH$_3$, Δ^{22E}

Structure **100** was assigned to attenuatoside B-I by comparing its spectral data with those of the major constitutent attenuatoside A-I (**91**); attenuatoside B-I is related to attenuatoside A-I by containing an additional hydroxyl group at 4β-position. Acetylation of **100** gave an heptaacetate, which on oxidation with pyridinium dichromate provided a 4-keto derivative, whose ^{1}H-NMR spectrum was devoid of the H-4 signal and showed the 19-methyl signal shifted upfield to δ 1.00, consistent with the 4β-OH assignment in **100**. For the NMR data of the disaccharide moiety see attenuatoside A-I (**91**).

101 Culcitoside C_5:

OCCURRENCE: *Culcita novaeguineae (44)*.
PHYSICAL DATA: $[\alpha]_D = -22.4°$ (MeOH); FAB MS (− ve ion), m/z 745 [M−H]$^-$.
Assignments of the NMR signals of the disaccharide moiety in **101**.

Position	4-O-methyl-β-xylopyranosyl ^{1}H	^{13}C	α-L-arabinofuranosyl ^{1}H	^{13}C
1	4.37 (d, $J = 7.5$ Hz)	105.2	5.11 (d, $J = 1.5$ Hz)	107.8
2	3.18 (dd, $J = 7.5, 9.0$ Hz)	75.2	4.06 (dd, $J = 1.5, 4.0$ Hz)	93.0
3	3.40 (t, $J = 9.0$ Hz)	77.0	4.04 (m)	77.6
4	3.21 (m)	80.9	4.00 (m)	83.8
5	3.14 (t, $J = 10.6$ Hz)	64.6	3.81 (dd, $J = 11.2, 3.0$ Hz)	62.7
	4.05 (dd, $J = 10.6, 4.5$ Hz)		3.65 (dd, $J = 11.2, 4.2$ Hz)	
OCH$_3$	3.50 (s)	59.1		

Culcitoside C_5 (**101**) is related to culcitoside C_4 (**92**) in having an additional hydroxyl group at position 4β of the steroid moiety.

102 Halityloside D (= culcitoside C_1)

OCCURRENCE: *Halityle regularis (112)*, *Culcita novaeguineae (44, 138)*, *Nardoa gomophia* and *N. novaecaledoniae (70)*, *Linckia guildingi (138)*.
ACTIVITY: Inhibits cell division of fertilized sea urchin eggs at a concentration of 10^{-6} M (25% inhibition) and possesses some cytotoxic activity at 10 μg/ml against turbinate cells and some antibacterial activity against *S. aureus*.
PHYSICAL DATA: $[\alpha]_D = -13.7°$ (MeOH); FAB MS (+ ve ion), m/z 783 [M + Na]$^+$.

This is the second major glycoside component of *Halityle regularis (112)* and is related to attenuatoside B-I (**100**) in having a methoxy group at C-4 of the 2-O-methylxylose unit. Acid methanolysis of **101** followed by *p*-bromobenzoylation of the reaction mixture gave 2,4-di-O-methyl-3-O-(*p*-bromobenzoyl)-α-D-xylopyranoside identified by ^{1}H-NMR spectrometry, thus establishing the location of the methoxyl groups at C-2 and C-4 of the xylose unit. The 3β,4β-di-hydroxy assignment was confirmed by formation of an acetonide on heating **101** in dry acetone containing *p*-TsOH (*112*). For the NMR data of the sugar moiety *cfr.* **93**, whereas for data of the steroid *cfr.* **99**.

103 22-Dehydrohalityloside D

Occurrence: *Sphaerodiscus placenta* (*134*).
Physical data: FAB MS (+ ve ion), m/z 781 [M + Na]$^+$, 759 [M + H]$^+$.

The structure of **103**, a very minor component of *S. placenta*, was derived by comparing the spectral data and comparison with those of halityloside D (**102**).

104 Glacialoside B

Occurrence: *Marthasterias glacialis* (*133*).
Physical data: FAB MS (-ve ion), m/z 679 [M$_{SO_3^-}$]; selected ^{1}H-NMR signals: δ 4.20 (m, 3α-H), 4.63 (br s, 4α-H).

104

Glacialoside B, a minor glycoside from *M. glacialis*, is related to pycnopodioside B (**87**) in having an additional hydroxyl group at position 4β. The location of the sulphate at C-3 was indicated by the downfield shift of the 3-H and 4-H signals to δ4.20 and 4.63 respectively, while the ^{13}C NMR spectrum (δ_{C-24}: 86.5 ppm) confirmed the location of the sugar at C-24. All signals of the side chain carbons were identical with those of the corresponding carbons in amurensoside A (**61**).

105 Imbricatoside B

106 Imbricatoside A

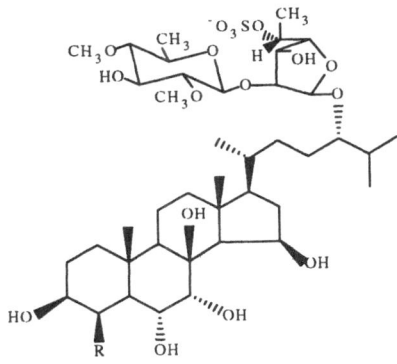

105, R=H

106, R=OH

OCCURRENCE: *Dermasterias imbricata* (*139*).

ACTIVITY: Both imbricatosides A and B inhibit cell division of fertilized sea urchin eggs at a concentration of 10^{-7} M (50% inhibition). Tested for antifungal activity they were inactive at a dose below 40 µg (*Cladosporium cucumerinum*) (*35*).

PHYSICAL DATA: Imbricatoside B: $[\alpha]_D = +2.0°$ (MeOH); FAB MS (− ve ion), m/z 867 [$M_{SO_3^-}$]; selected ^{1}H-NMR signals: δ 1.02 (s, 19-H$_3$), 3.85 (m, 6β-H), 3.88 (t, $J = 2.0$ Hz, 7β-H), 4.58 (br t, $J = 6.0$ Hz, 15α-H); Imbricatoside A: $[\alpha]_D = -27.5°$ (MeOH); FAB MS (− ve ion), m/z 883 [$M_{SO_3^-}$]; ^{13}C-NMR spectrum of the steroid portion in Table 5.

Assignments of the NMR signals of the disaccharide moiety in **105** and **106**.

	2,4-di-O-methyl-β-quinovopyranosyl		5-O-sulphate-β-fucofuranosyl	
Position	^{1}H	^{13}C	^{1}H	^{13}C
1	4.47 (d, $J = 7.5$ Hz)	104.0	5.12 (br s)	107.7
2	2.89 (dd, $J = 7.5, 9.0$ Hz)	85.4	4.10 (br d, $J = 3.0$ Hz)	92.1

3	3.44 (t. $J = 9.0$ Hz)	74.1	4.30 (dd, $J = 3.0, 6.0$ Hz)	77.2
4	2.76 (t, $J = 9.0$ Hz)	86.8	3.96 (dd, $J = 2.0, 6.0$ Hz)	86.8
5	3.30 (m)	72.3	4.65 (m)	77.4
CH_3	1.30 (d, $J = 7.0$ Hz)	17.3	1.46 (d, $J = 7.0$ Hz)	18.4
OCH_3	3.58 (s)	60.6		
	3.62 (s)	61.2		

Imbricatosides B (**105**) and A (**106**) are related by the presence of an "extra" hydroxyl group at C-4β in imbricatoside A, which is the major compound. The structure of the saccharide moiety in **106**, which includes the rare fucofuranosyl unit, never encountered previously among glycosides from echinoderms, is based on acid methanolysis which afforded methyl 2,4-di-O-methyl-quinovoside and methyl fucosides, and on ¹H- and ¹³C-NMR data. The ¹H-NMR spectrum and sequential decoupling permitted the assignment of signals corresponding to the β-fucofuranoside moiety. The sulphate was located at C-5′ in view of the signal for 5′-H, which in the desulphated derivative shifted from δ 4.65 to 3.85 ppm. The signal of C-2′, which is shifted downfield to 92.1 ppm (glycosidation shift), established the location of the terminal 2,4-di-O-methylquinovose there. The structures of the steroid aglycones in **105** and **106** were established by comparative analysis of their NMR spectra.

107 Indicoside C

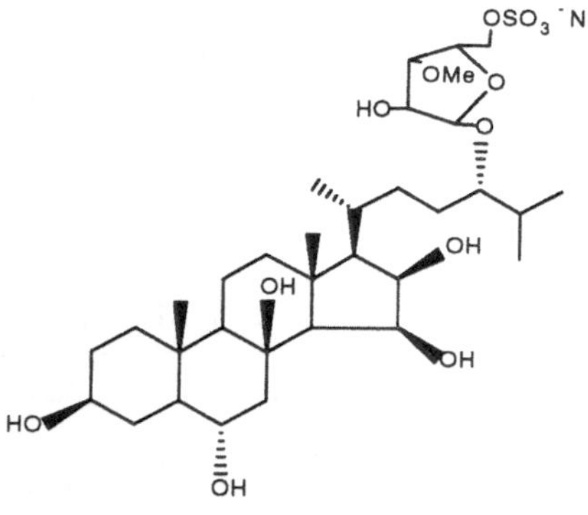

107

OCCURRENCE: *Astropecten indicus (132)*.
PHYSICAL DATA: $[\alpha]_D = -3.8°$ (MeOH); FAB MS (− ve ion), m/z 693 $[M_{SO_3^-}]$; selected ¹H-NMR signals: δ 1.27 (s, 18-H_3), 3.75 (dd, $J = 3.5, 12.0$ Hz, 6β-H), 4.30 (t, $J = 6.4$ Hz, 16α-H), 4.42 (br t, $J = 6.4$ Hz, 15α-H); ¹³C-NMR spectrum of the steroid portion in Table 5.

Indicoside C is related to the previous xylofuranoside indicoside B (85) by having an additional hydroxyl group at position 16β of the steroid portion. The structure of this minor glycoside was derived by comparing the spectral data with those of indicoside B (85) and halitylo-side A (126) and B (125), glycosides isolated earlier from starfishes with the steroidal 15β,16β-dihydroxylation pattern (112).

108 Coscinasteroside F

OCCURRENCE: *Coscinasterias tenuispina* (42).
PHYSICAL DATA: $[\alpha]_D = -8.8°$ (MeOH); FAB MS (− ve ion), m/z 599 [M–H]⁻; selected ¹H-NMR signals: δ 1.30 (s, 18-H₃), 4.45 (br t, J = 6.0 Hz, 15α-H).

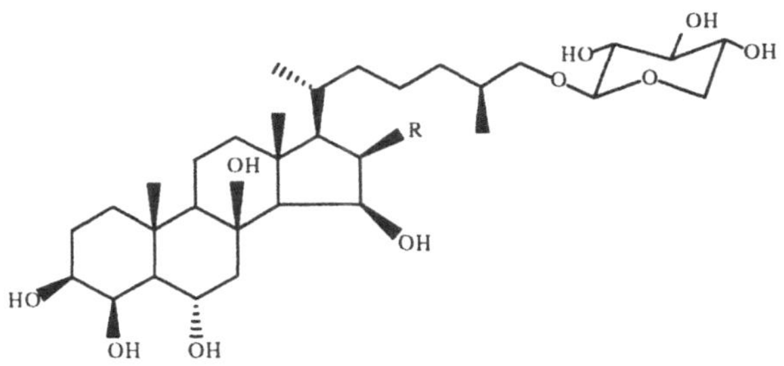

108, R=H

109, R=OH

109 Coscinasteroside E

OCCURRENCE: *Coscinasterias tenuispina* (42).
PHYSICAL DATA: $[\alpha]_D = -11.8°$ (MeOH); FAB MS (− ve ion), m/z 615 [M–H]⁻; selected ¹H-NMR signals: δ 1.27 (s, 18-H₃), 4.25 (t, J = 6.0 Hz, 16α-H), 4.40 (dd, J = 6.0, 5.0 Hz, 15α-H).

Coscinasterosides F (108) and E (109) are related in that 109 has an "extra" hydroxyl group at C-16β of the steroid portion. The steroid aglycone (181) of coscinasteroside E (109) was previously isolated from *Halityle regularis* (112), and also as 3-sulphated derivative from the same *C. tenuispina*. Comparison of the ¹³C-NMR spectrum of 109 with that of its aglycone identified the site of glycosidation as C-26. The C-25S configuration was suggested by analogy with 181 (*cfr.* section 9.2). Application of the method of molecular rotation difference showed that the sugar was D-xylose. The structure of coscinasteroside F (**108**) was derived from spectral data and comparison with those of **109**. Glycosidation at C-26 causes the resonance of one of the C-26 methylene protons to be shifted to δ 3.74 (δ 3.47 in the non-glycosidated steroid **181**).

110 Coscinasteroside A

110

Occurrence: *Coscinasterias tenuispina* (*42*).
Physical data: $[\alpha]_D = -1.4°$ (MeOH), FAB MS (− ve ion), m/z 707 [$M_{SO_3^-}$]; FAB MS (+ ve ion), m/z 753 [$M_{SO_3Na} + Na$]$^+$; selected ^{1}H-NMR signals: δ 0.98 (d, $J = 7.5$ Hz, 27-H$_3$), 1.02 (d, $J = 7.5$ Hz, 28-H$_3$), 1.42 (d, $J = 5.0$ Hz, 14-H), 3.78 (dd, $J = 10.0, 7.5$ Hz, 26-H), 3.86 (dd, $J = 10.0, 2.0$ Hz, 6β-H), 3.91 (d, $J = 2.0$ Hz, 7β-H), 4.08 (dd, $J = 10.0, 5.0$ Hz, 26-H), 4.23 (t, $J = 6.0$ Hz, 16α-H), 4.42 (d, $J = 7.5$ Hz, 1'-H), 4.45 (dd, $J = 5.0, 6.0$ Hz, 15α-H), 5.40 (dd, $J = 15.0, 7.0$ Hz, 23-H), 5.50 (dd, $J = 15.0, 6.0$ Hz, 22-H).

The 500 MHz ^{1}H-NMR spectrum and sequential decoupling experiments permitted assignments of signals of the steroid moiety from C-3 to C-7 and from C-14 to C-28. The two moieties were obviously connected through C-8 bearing an hydroxyl group (signal for a quaternary carbon in the ^{13}C-NMR at 79.3 ppm). Confirmation was obtained by NOE experiments indicating proximity of the protons at C-7 and C-15. The coupling between the C-6 and C-7 protons indicated the 6α,7α-dihydroxy stereochemistry, confirmed by enhancement of the H-6 signal when the C-19 methyl signal was irradiated. Upfield shifts of the C-26 protons from 3.78/4.08 ppm in **110** to 3.42/3.54 in the desulphated derivative indicated location of sulphate at C-26. The chemical shift of C-3 (δ_C 79.9 ppm *vs.* 72.3 in a non-glycosidated model steroid) established the attachment of xylose at C-3.

After synthesis of C-24, C-25 stereoisomeric model steroids (*cfr.* section 9.5) (*140*), the relative *threo*-stereochemistry can now be assigned to coscinasteroside A (110). The chemical shifts of the 27- and 28-methyl groups (^{1}H and ^{13}C) are used as indication of the relative stereochemistry. The absolute configuration C-24R, C-25S shown in formula 110, is suggested by analogy with echinasteroside A (152) (140).

111 Halityloside I

111

OCCURRENCE: *Halityle regularis* (*112*), *Nardoa gomophia* (*70*).
PHYSICAL DATA: FAB MS (+ ve ion), m/z 883 [M_{SO_3Na} + Na]$^+$; selected ^{1}H-NMR signals: δ 0.94 (d, J = 7.0 Hz, 27-H$_3$), 0.96 (d, J = 7.0 Hz, 28-H$_3$), 1.04 (d, J = 7.0 Hz, 21-H$_3$), 2.72 (dd, J = 12.0, 4.0 Hz, 7-H), 4.33 (br s, 4α-H), 4.95 (dt, J = 5.4, 10.5 Hz, 6β-H).

Halityloside I is a minor component among the glycosides from *H. regularis*. The structure of the saccharide moiety was derived from the examination of ^{1}H- and ^{13}C-NMR data (*cfr.* section 5.1) and comparison with the major constituent halityloside A (126). The downfield shift of the signal for 6-H in 111 indicates that the sulphate is located at C-6, while the chemical shift of the methylene carbon at C-26 (δ$_C$ 74.0 ppm) indicates that the sugar moiety is attached at C-26 (*112*). The stereochemistry at C-24 and C-25 of the side chain is now suggested by analogy with echinasteroside A (152) (140).

112 Culcitoside C$_8$

OCCURRENCE: *Culcita novaeguineae* (*44*).
PHYSICAL DATA: [α]$_D$ = − 1.4° (MeOH); FAB MS (− ve ion), m/z 757 [M–H]$^-$, 611 [M–H–146]$^-$; selected ^{1}H-NMR signals: δ 1.12 (d, J = 7.0 Hz, 27-H$_3$), 4.77 and 4.87 (br s, 28-H$_2$).

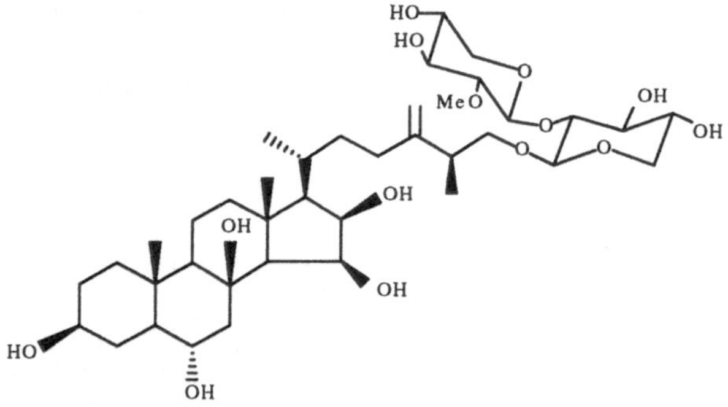

112

Culcitoside C$_8$ contains a saccharide moiety already encountered in previously isolated glycosides (*cfr.* halityloside A, **126**). The 3β,6α,8,15β,16β hydroxylation pattern of the steroid nucleus is commonly found in steroids from starfishes (*cfr.* halityloside A, **126**). The 24-methylene-26-hydroxy side chain was evident from the ^{1}H-NMR data. The chemical shift of the C-26 oxygenated methylene at 74.9 ppm and comparison with the corresponding signal at 67.6 ppm observed in the spectra of 24-methylene-26-hydroxy steroids confirmed the location of the sugar moiety at C-26. The 25S configuration is proposed by analogy with steroid **187**, isolated from the same organism.

113 Coscinasteroside D

Occurrence: *Coscinasterias tenuispina* (*42*).
Physical data: $[\alpha]_D = +11.6°$ (MeOH); FAB MS (+ ve ion), m/z 753 $[M_{SO_3Na} + Na]^+$; selected ^{1}H-NMR signals: δ 1.02 (d, $J = 7.0$ Hz, 27-

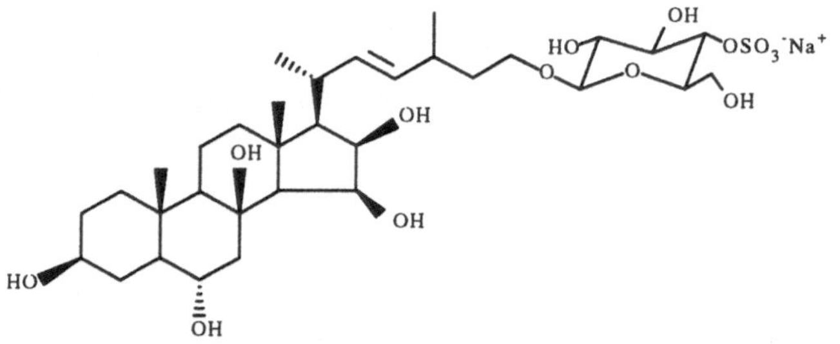

113

H$_3$), 1.06 (d, J = 7.0 Hz, 21-H$_3$), 3.56 (1H, m, 26-H$_2$), 3.96 (1H, m, 26-H$_2$), 4.13 (t, J = 9.0 Hz, 4'-H), 4.31 (d, J = 7.5 Hz, 1'-H), 5.35 (dd, J = 15.0, 8.0 Hz, 23-H), 5.57 (dd, J = 15.0, 7.5 Hz, 22-H).

The Δ^{22},27-nor-24-methyl-26-hydroxy side chain was evident from the ^{1}H-NMR spectrum and sequential decoupling.

114 Placentoside A

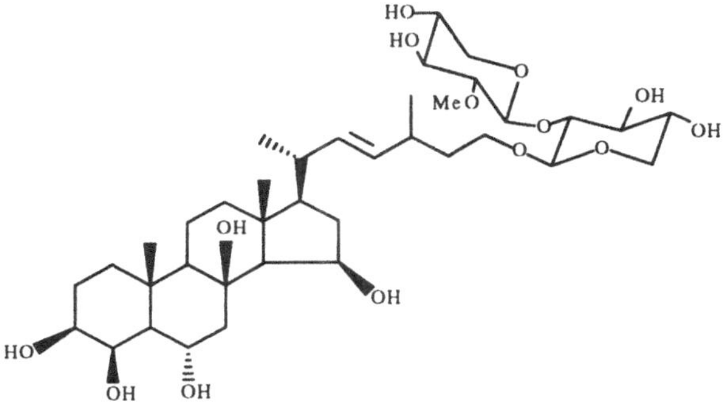

114

OCCURRENCE: *Sphaerodiscus placenta* (*134*).

PHYSICAL DATA: FAB MS (– ve ion), m/z 743 [M–H]$^-$; selected ^{1}H NMR signals: 3.58 (1H, m, 26-H$_2$), 3.96 (1H, m, 26-H$_2$), 5.17 (dd, J = 15.0, 7.5 Hz, 23-H), 5.30 (dd, J = 15.0, 7.5 Hz, 22-H).

Placentoside A contains the common 2-O-methyl-β-xylopyranosyl (1 → 2)-β-xylopyranosyl disaccharide moiety already encountered in glycosides from starfishes (*e.g.* halityloside A, **126**), while the aglycone contains the same steroidal nucleus as halityloside D (**102**) and other glycosides (for ^{1}H-NMR data *cfr.* **99**) and the same side chain as coscinasteroside D (**113**). The chemical shift of the olefinic protons is most sensitive to substitution on ring D. In **114** they are shifted upfield relative to **113**.

115 Coscinasteroside C

OCCURRENCE: *Coscinasterias tenuispina* (*42*).

PHYSICAL DATA: [α]$_D$ = + 3.6° (MeOH); FAB MS (+ ve ion), m/z 767 [M$_{SO_3Na}$ + Na]$^+$; selected ^{1}H-NMR signals: δ 0.88 (d, J = 7.0 Hz, 26- or 27-H$_3$), 0.94 (d, J = 7.0 Hz, 27- or 26-H$_3$), 1.08 (d, J = 7.0 Hz, 21-H$_3$), 1.68 (m, 25-H), 2.14 (m, 24-H), 3.48 (1H, t, J = 9.0 Hz, 28-H$_2$), 4.07 (1H,

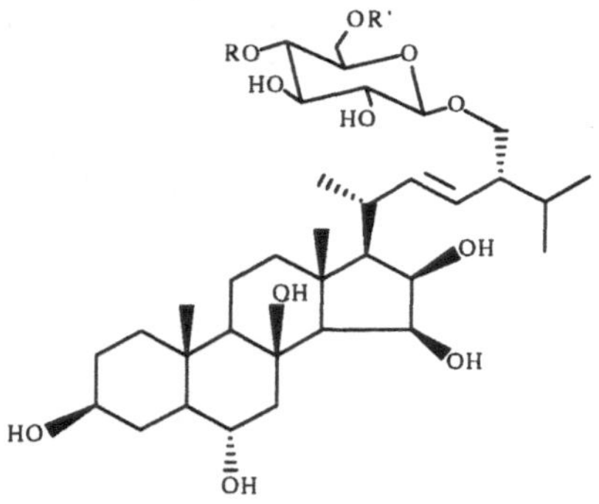

115, R=SO$_3^-$Na$^+$, R'=H

116, R=H, R'=SO$_3^-$Na$^+$

dd, $J = 5.0, 9.0$ Hz, 28-H$_2$), 4.13 (t, $J = 9.0$ Hz, 4'-H), 5.37 (dd, $J = 15.0$, 9.0 Hz, 23-H), 5.45 (dd, $J = 15.0, 7.0$ Hz, 22-H).

Coscinasteroside C is the first steroidal glycoside of the 24-hydroxy-methyl cholestane series to be isolated. The ^{1}H-NMR spectrum and sequential decoupling permitted assignment of the signals of the side chain protons. The recent stereoselective synthesis of (24S)- and (24R)-24-(hydroxymethyl)-cholesta-5,22(E)-dien-3β-ols and comparison of spectral data has confirmed the gross structure and has allowed the 24R configuration in 115 to be established (*cfr.* 116 and section 9.3).

116 Pisasteroside A

Occurrence: *Pisaster ochraceus* and *P. brevispinus (72).*
Physical data: [α]$_D$ = + 4.0° (MeOH); FAB MS (− ve ion), m/z 721 [M$_{SO_3^-}$].

Pisasteroside A is isomeric with coscinasteroside C (115) isolated from *C. tenuispina,* the only difference being the location of the sulphate at C-6' of the glycosyl moiety in 116 instead of at C-4' in 115. Enzymatic hydrolysis with *Charonia lampas* glycosidase mixture of desulphated 116 removed glucose to yield the steroid aglycone which was converted into the 3,6,28-(+)MTPA triester. The chemical shifts of the signals for the C-21 methyl (δ 1.06) and C-28 methylene protons (δ 4.37 d) were in excellent agreement with the corresponding signals in the spectrum of the 24R model and quite different from those of the 24S model (*cfr.* section 9.3).

117 Culcitoside C_3

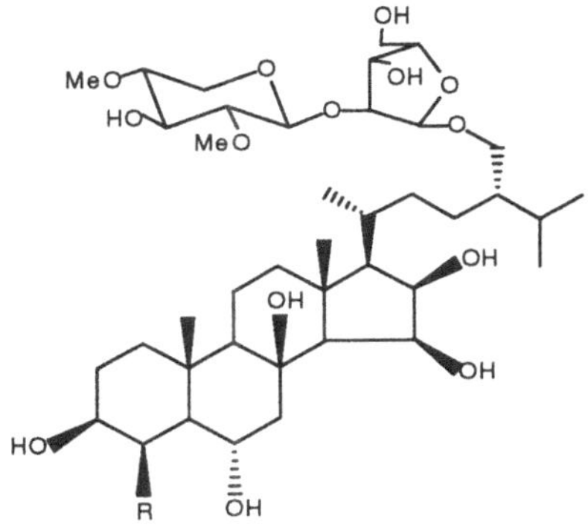

117, R = H

118, R = OH

118 Culcitoside C_2

OCCURRENCE: *Culcita novaeguineae (141)*.

The structures of **117** and **118** were proposed leaving the configuration at C-24 unassigned. We now suggest the 24S configuration by analogy with culcitoside C_6 (**119**).

119 Culcitoside C_6

OCCURRENCE: *Culcita novaeguineae (44)*.
PHYSICAL DATA: $[\alpha]_D = -23.2°$ (MeOH); FAB MS (− ve ion), m/z 773 $(M-H)^-$.

For the NMR data of the saccharide moiety *cfr.* **93**; for the NMR data of the pentahydroxycholestane nucleus *cfr.* **99**. The ^{1}H-NMR shift values for the 28-methylene protons [3.46 (dd, $J = 10.0$, 6.5 Hz) − 3.56 (dd, $J = 10.0$, 6.5 Hz)] in the spectrum of the derived steroid **119a**, virtually identical with the corresponding signals of the synthetic 24S isomer and different from those of the 24R isomer (2 H doublet at δ 3.52), were used to determine the 24S configuration in **119** (*cfr.* section 9.3; Fig. 16).

120 Culcitoside C_7

OCCURRENCE: *Culcita novaeguineae (44)*.
PHYSICAL DATA: $[\alpha]_D = -17.7°$ (MeOH); FAB MS (− ve ion), m/z 853 $[M_{SO_3^-}]$.

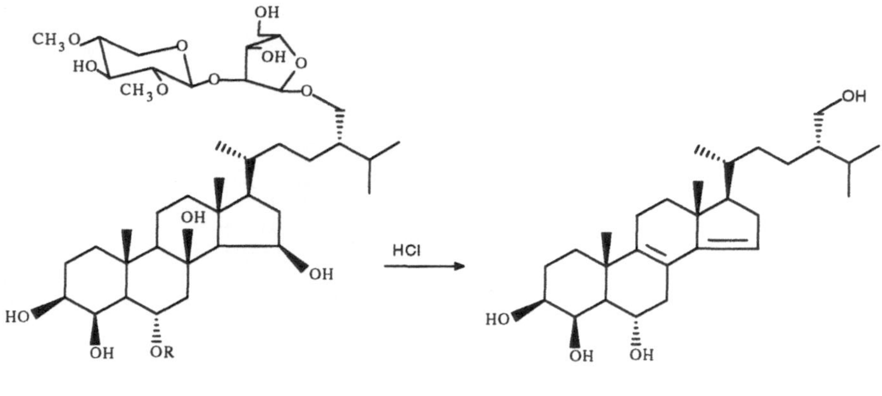

119, R = H

120, R = SO₃⁻Na⁺

119a

In the ^{1}H-NMR spectrum sulphation at C-6 is accompanied by downfield shifts of 6-H (δ 4.90 *vs.* 4.19 in **119**), 7β-H (δ 2.74 *vs.* 2.49 in **119**) and 19-H$_3$ (δ 1.26 *vs.* 1.19 in **119**).

121 Indicoside A

OCCURRENCE: *Astropecten indicus (142)*.
PHYSICAL DATA: $[\alpha]_D = -69.4°$ (MeOH); FAB MS ($-$ ve ion), m/z 673 [M–H]$^-$; selected ^{1}H-NMR signals: δ 3.36 (1H, d, $J = 12.0$ Hz, 28-H$_2$), 3.69 (1H, d, $J = 12.0$ Hz, 28-H$_2$); ^{13}C-NMR assignments of the sugar

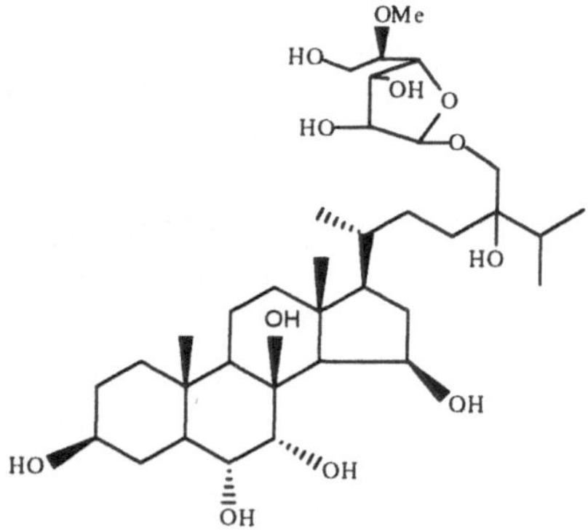

121

moiety: C-1: 109.8, C-2: 82.9, C-3: 79.0, C-4: 84.7, C-5: 83.4, C-6: 62.4, OCH$_3$: 59.5 ppm.[13]

^{1}H-NMR spectroscopy indicated the presence of a –CH$_2$O– group located on a quaternary carbon. Acetylation of **121** gave a pentaacetate (characterized as a 3,6,2',3',6'-pentaacetate) in whose ^{1}H-NMR spectrum the –CH$_2$O– proton signals were virtually unchanged, thus clarifying that the sugar is attached at C-28. The 5-O-methylgalactofuranosyl structure of the sugar moiety was established by detailed analysis of the ^{1}H-NMR spectrum of the acetate, which showed six resolved bands for the sugar protons that were correlated by sequential decoupling. NOEDS experiments confirmed the galactofuranose structure. Final supporting evidence was provided by the ^{13}C-NMR spectrum and comparison with the spectrum of methyl 5-O-methyl-β-O-galactofuranoside.

122 Gomophioside B

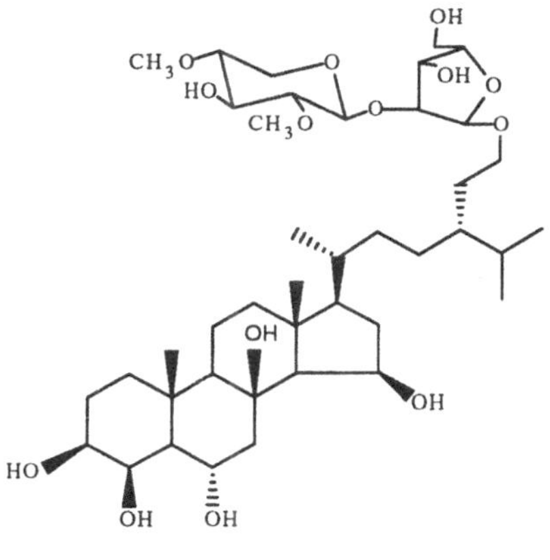

122

OCCURRENCE: *Gomophia watsoni (131)*.

PHYSICAL DATA: $[\alpha]_D = -21.5°$ (MeOH); FAB MS (+ ve ion), m/z 811 [M + Na]$^+$; selected ^{1}H-NMR signals: δ 0.87 (d, $J = 7.0$ Hz, 26- or 27-H$_3$), 0.90 (d, $J = 7.0$ Hz, 27- or 26-H$_3$); selected ^{13}C-NMR signals: δ 19.1, 20.0 (C-26, -27).

For NMR data of the saccharide moiety *cfr.* **93**; for NMR data of the 3β,4β,6α,8,15β-pentahydroxy steroidal nucleus *cfr.* **99**. Presence of the 24-(β-hydroxyethyl) cholestane side chain was confirmed by ^{13}C-NMR data which also indicated the attachment of the disaccharide moiety at

C-29. Chemical shift differences between the C-26 and C-27 resonances in the ^{1}H- and ^{13}C-NMR spectra of the R and S model steroids were used to determine the 24R configuration of gomophioside B (*cfr.* section 9.4).

123 Pisasteroside F

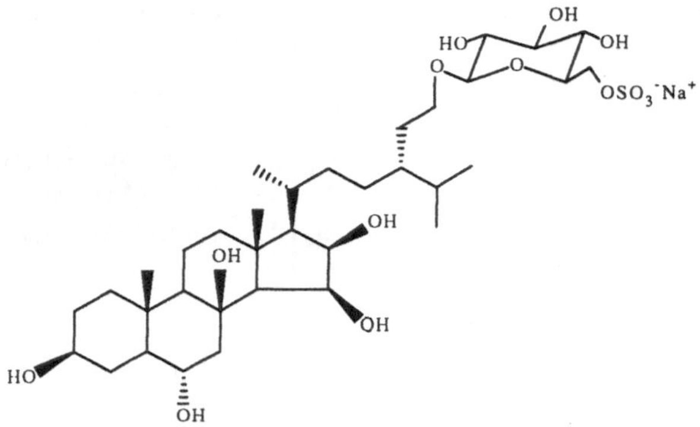

123

OCCURRENCE: *Pisaster giganteus* (73).
PHYSICAL DATA: $[\alpha]_D = +10.0°$ (MeOH); FAB MS (− ve ion), m/z 737 $[M_{SO_3^-}]$; selected ^{1}H-NMR signals: δ 0.88 (d, $J = 7.0$ Hz, 26- or 27-H$_3$), 0.91 (d, $J = 7.0$ Hz, 27- or 26-H$_3$); selected ^{13}C-NMR signals: δ19.3, 19.9 (C-26, -27).

Pisasteroside F shares the same aglycone with the previously isolated halityloside B (**125**) and the β-glucopyranosyl 6′-sulphate with pisasteroside A (**116**).

124 Pisasteroside C

OCCURRENCE: *Pisaster brevispinus* (72).
PHYSICAL DATA: $[\alpha]_D = +12.0°$ (MeOH); FAB MS (− ve ion), m/z 705 $[M_{SO_3^-}]$; ^{1}H-NMR signals of the xylosyl unit: δ 4.29 (d, $J = 7.5$ Hz, 1′-H), 3.28 (dd, $J = 7.0, 9.0$ Hz, 2′-H), 3.51 (t, $J = 9.0$ Hz, 3′-H), 4.19 (ddd, $J = 5.0, 9.0, 10.0$ Hz, 4′-H), 3.30 (t, $J = 10.0$ Hz, 5′-H), 4.15 (dd, $J = 10.0, 5.0$ Hz, 5′-H).

The aglycone is the 24(28)-dehydroderivative of the aglycone in pisasteroside F (**123**). The signal of the C-25 proton appeared at δ 2.88 (m) and based on relative shielding arguments between the E and Z isomers, the C-28 proton was assigned Z stereochemistry. The upfield

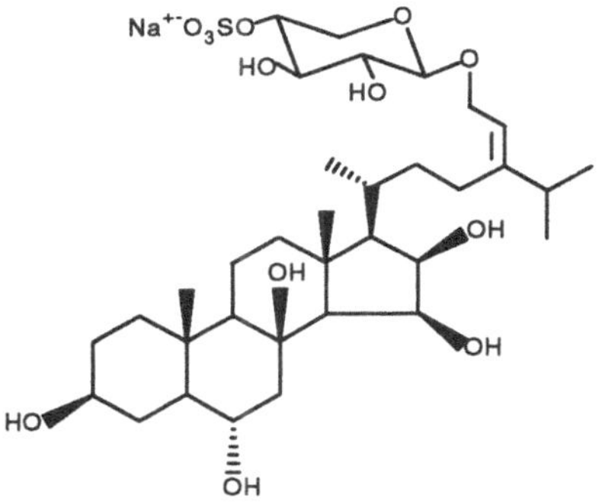

124

shift of the C-4′ proton of the xylopyranosyl moiety from δ 4.19 in **124** to δ 3.50 in the desulphated derivative showed that C-4 bears the sulphate.

125 Halityloside B

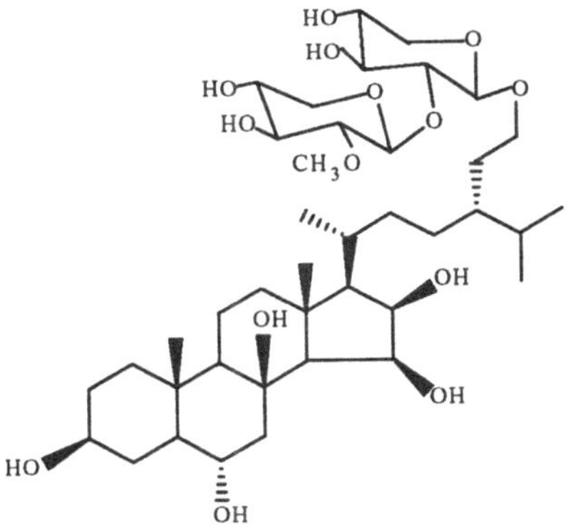

125

OCCURRENCE: *Halityle regularis* (*112*), *Nardoa novaecaledoniae* and *N. gomophia* (*70*), *Sphaerodiscus placenta* (*134*) and *Culcita novaeguineae* (*44*).

PHYSICAL DATA: $[\alpha]_D = -5.0°$ (MeOH); FAB MS (+ ve ion), m/z 797 $[M + Na]^+$;

ACTIVITY: Inhibits weakly cell division of fertilized sea urchin eggs at 10^{-5} M (0.25% inhibition) (*35*).

Halityloside B (**125**) is related to the major constituent halityloside A (**126**), but lacks the 4β-hydroxyl group. The structure was derived by spectral comparison with **126**. On treatment with $Me_2CO/TsOH$ it gave a monoacetonide instead of the bis-acetonide formed by **126**.

126 Halityloside A

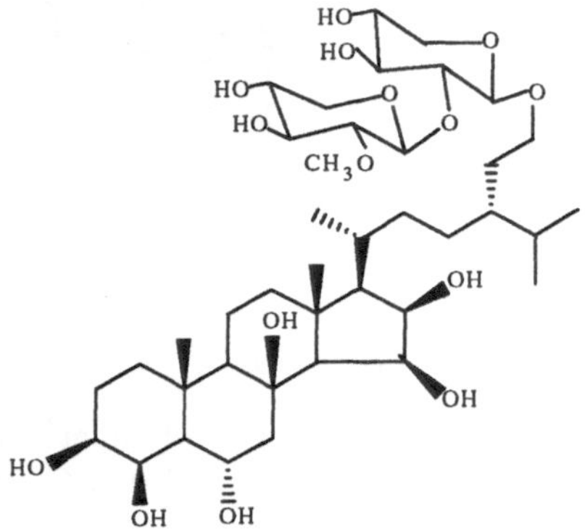

126

Occurrence: *Halityle regularis* (*112*), *Nardoa novaecaledoniae* and *N. gomophia* (*70*), and *Sphaerodiscus placenta* (*134*), and *Culcita novaeguineae* (*44*).

Activity: Cytotoxic against turbinate cells (10 μg/ml), causes inhibition of cell division of fertilized sea urchin eggs at 10^{-5} M concentration, and possess weak antiviral activity (*34, 35*).

Physical data: $[\alpha]_D = -3.1°$ (MeOH); FAB MS (+ ve ion), m/z 813 [M + Na]⁺; selected ¹H-NMR signals: δ 0.88 (d, $J = 7.0$ Hz, 26- or 27-H₃), 0.91 (d, $J = 7.0$ Hz, 27- or 26-H₃), 1.19 (s, 19-H₃), 1.27 (s, 18-H₃), 2.50 (dd, $J = 12.5, 4.0$ Hz, 7β-H), 4.22 (dt, $J = 4.0, 10.5$ Hz, 6β-H), 4.29 (br s, 4α-H), 4.25 (dd, $J = 6.6, 7.0$ Hz, 16α-H), 4.42 (dd, $J = 5.6, 6.6$ Hz, 15α-H); selected ¹³C-NMR signal: 69.7 (C-29) ppm.

¹³C-NMR assignments of the disaccharide moiety of halityloside A (**126**)

Position	xylose	2-O-methylxylose
1	103.6	104.4
2	81.5	84.7
3	76.9	77.7
4	71.3	71.3
5	66.5	66.6
OCH₃		60.7

References, pp. 297–308

Halityloside A is the major glycoside component of the extracts from *H. regularis*. The structure of the disaccharide moiety, later found in many other glycosides, was elucidated by acid methanolysis followed by *p*-bromobenzoylation which afforded methyl 2,3,4-tri-O-(*p*-bromobenzoyl)-α-D-xylopyranoside and methyl 2-O-methyl-3,4-di-O-(*p*-bromobenzoyl)-α-D-xylopyranoside (CD, 236/253 A = − 53.3; calc. A = − 62.0), and by ^{13}C-NMR spectrometry of the intact glycoside, which shows low field signals at 84.1 ppm due to C-2 of 2-O-methylxylose and 81.5 ppm due to the glycosidated C-2 of the second xylosyl unit. The FAB MS had peaks due to the sequential loss of 146 (2-O-methylxylose) and 132 (xylose) mass units from $[M + H]^+$, supporting the sequence 2-O-methylxylose → xylose → aglycone. The 15β,16β-dihydroxy steroidal structural feature first encountered in halityloside A, was determined by detailed ^{13}C-NMR analysis, which included calculation of shifts for compounds with 15β,16α-di-OH and 15β,16β-di-OH stereochemistry taking as point of departure halityloside D (**102**) which contains a 3β,4β,6α,8,15β-pentahydroxy cholestane nucleus, and was supported by formation of a bis-acetonide on treatment of **126** with Me$_2$CO/TsOH. The chemical shift of the isopropyl methyls indicated the 24R configuration (*cfr*. section 9.4).

127 Halityloside H

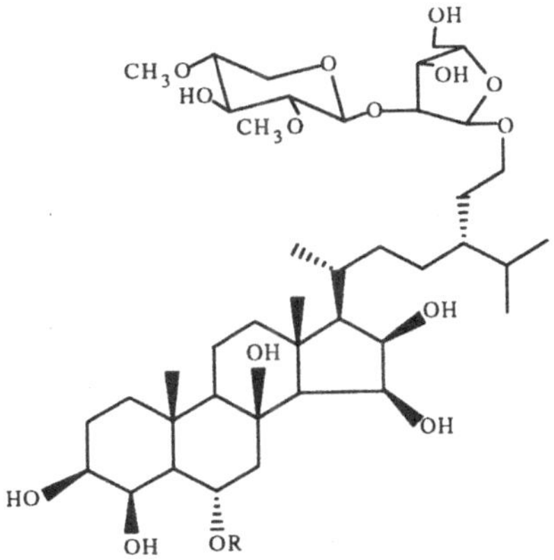

127, R=H

128, R=SO$_3^-$Na$^+$

OCCURRENCE: *Halityle regularis* (*112*), *Nardoa gomophia* (*70*).
PHYSICAL DATA: $[\alpha]_D$ = − 5.0° (MeOH) FAB MS (+ ve ion), m/z 827 $[M + Na]^+$; selected ^{13}C-NMR signals: 68.1 (C-29) ppm.

128 Halityloside H, 6-O-sulphate

Occurrence: *Halityle regularis (112), Nardoa gomophia (70)*.
Physical data: FAB MS (+ ve ion), m/z 929 $[M_{SO_3Na} + Na]^+$; selected
^{13}C-NMR signals: 68.1 (C-29) ppm.

Halityloside H (**127**) shares the same aglycone as halityloside A (**126**)
and the same disaccharide unit as halityloside D (**102**). A small change in
the chemical shift of C-29 on passing from **127** to **126** is observed. The
chemical shift of the 6-H signal at δ 4.95, down-field from that in **127**
(4.22) placed the sulphate at C-6 in **128**. Significant shifts were also
observed for the resonances of the 19-methyl (δ 1.32 *vs.* 1.19 in **127**) and
H-7β (δ 2.73 *vs.* 2.50 in **127**)

129 Moniloside E

Occurrence: *Fromia monilis (109)*.
Physical data: $[\alpha]_D = \pm 0°$ (MeOH); FAB-MS (− ve ion), m/z 789
$[M-H]^-$.

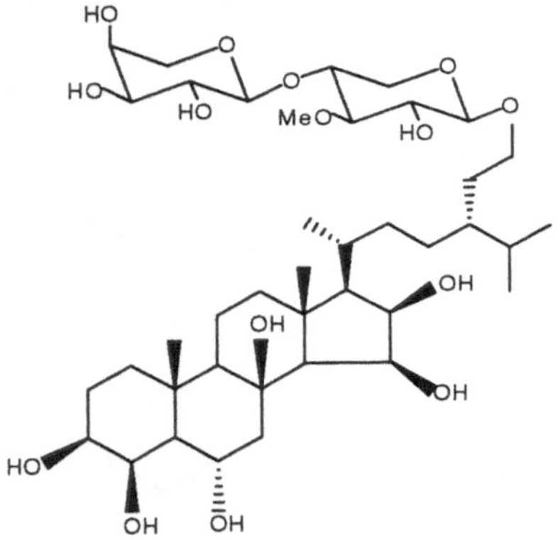

129

130, Δ^{22E}

130 Moniloside F (Δ^{22E}-moniloside E)

Occurrence: *Fromia monilis (109)*.
Physical data: FAB MS (− ve ion), m/z 787 $[M - H]^-$.
Activity: Both **129** and **130** are toxic at a concentration of 10 μg/5·10^4
cells against VERO cells.

Assignments of NMR signals of the disaccharide moiety of **129** and **130**.

Position	α-L-arabinopyranosyl		3-O-methyl-β-D-xylopyranosyl		
	^{1}H	^{13}C	^{1}H	^{13}C	Lit.[a]
1	4.32 (d, $J = 7.5$ Hz)	104.9	4.28 (d, $J = 7.5$ Hz)	103.3	104.9
2	3.54–3.60[b]	72.4	3.22–3.77[c]	74.3	75.0
3	3.54–3.60[b]	74.2	3.22–3.77[c]	85.2	87.6
4	3.84 (m)	69.5	3.78 (m)	76.3	71.0
5	3.93 (dd, $J = 12.5, 3.0$ Hz) 3.54–3.60[b]	66.8	4.05 (dd, $J = 11.5, 5.0$ Hz) 3.22–3.77[c]	64.0	66.5
OCH$_3$			3.64 (s)	60.5	60.8

[a] from borealoside C(**71**) (*93*).
[b] overlapping signals
[c] overlapping signals

Moniloside E, isolated in 1.7 mg amount from *F. monilis* (4.5 kg fresh material) contains the same steroidal aglycone as halityloside A (**126**) and many other glycosides. The disaccharide unit is new and contains 3-O-methylxylopyranoside and arabinopyranosyl units. Arabinose is a common sugar among the glycosides of polyhydroxysteroids but has never been found previously as a pyranose in this group of glycosides. The FAB MS shows intense peaks at m/z 657 and 511 corresponding to the sequential loss of arabinose (132 mass units) and O-methylxylose (146 mass units) from [M–H]$^-$, indicating the sequence. ^{13}C-NMR data and comparison with those of methyl 3-O-methyl-β-D-xylopyranoside (**73**) established the nature of interglycosidic linkage and the location of the methoxyl group at C-3′.

131 Moniloside G

OCCURRENCE: *Fromia monilis* (*109*).
PHYSICAL DATA: $[\alpha]_D = -17.5°$ (MeOH); FAB MS ($-$ ve ion), m/z 935 [M–H]$^-$

132 Moniloside H (Δ^{22E}-moniloside G)

PHYSICAL DATA: $[\alpha]_D = +23.5°$ (MeOH); FAB MS (-ve ion), m/z 933 [M–H]$^-$
OCCURRENCE: *Fromia monilis* (*109*).
ACTIVITY: Both **131** and **132** are toxic at a concentration of 10 μg/5·10^4 cells against VERO cells.

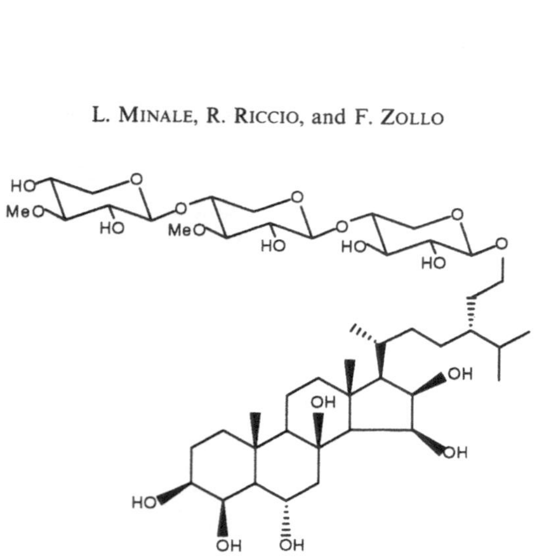

131

132, Δ^{22E}

Assignments of ^{13}C-NMR signals of the carbohydrate moiety in **131** and **132**.

Position	xylose	3-O-methylxylose	(T)-3-O-methylxylose
1	105.0	103.6	103.8
2	74.8	74.3	73.8
3	75.9	85.1	87.2
4	78.2	76.6	70.8
5	64.4	64.4	66.9
OCH$_3$		60.6	60.6

Monilosides G (**131**) and H (**132**) are the first triglycosides to be found among the group of "glycosides of polyhydroxysteroids" from starfishes. The FAB MS of **131** shows peaks at m/z 789 (loss of 146 mass units), 643 (loss of 146 mass units from 789) and 511 (loss of 132 mass units from 643) corresponding to the sequential loss of two O-methyl xylosyl units, and one xylosyl unit, thus indicating the sequence. The location of the methoxyl group at C-3 in both methoxylated xylose units and the nature of the interglycosidic linkages were derived from an accurate ^{13}C-NMR spectral analysis and comparison with references. Permethylation follow-ed by acid methanolysis and p-bromobenzoylation of the hydrolysis mixture afforded methyl 4-O-(p-bromobenzoyl)-2,3-di-O-methyl-α-D-xylopyranoside identified by the ^{1}H-NMR spectrum, thus confirming the 1-4 glycosidic linkages. The region of the ^{1}H-NMR spectrum containing the resonances of the sugar portions and the hydroxymethine and methylene protons of the steroid portion is rather crowded by overlapping signals. Thus only a few chemical shifts of the sugar protons could be determined: *i.e.* doublets with $J = 7.5$ Hz at δ 4.21, 4.35 and 4.39 for the anomeric protons, double doublets ($J = 5.0$, 11.0 Hz) at δ 3.88 for H-5eq of the terminal unit and at δ 4.03 and 4.09 for the 4-substituted internal units, and a one proton triplet ($J = 9.0$ Hz) shifted upfield to δ 3.08 for 3-H of the terminal 3-O-methylxylose unit.

133 Pisasteroside E

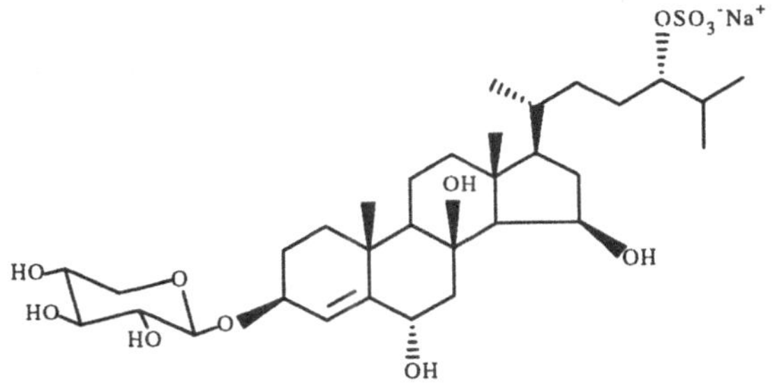

133

OCCURRENCE: *Pisaster giganteus (73)*
PHYSICAL DATA: FAB MS (− ve ion), m/z 661 $[M_{SO_3^-}]$; selected ^{1}H-NMR signals:δ 1.21 (s, 19-H$_3$), 1.31 (s, 18-H$_3$), 4.14 (q, $J = 6.5$ Hz, 24-H), 4.43 (d, $J = 7.5$ Hz, 1′-H), 4.45 (m, 15α-H), 4.60 (dd, $J = 12.5$, 5.0 Hz, 6β-H), 5.82 (br s, 4-H).

Pisasteroside E (**133**) is isomeric with pisasteroside D (**156**), differing in the stereochemistry at C-6 and C-15. Spectral comparison confirmed formulation **133** for pisasteroside E.

134 Coscinasteroside B

OCCURRENCE: *Coscinasterias tenuispina (42)*, *Pycnopodia helianthoides (74)*.
PHYSICAL DATA: FAB MS (− ve ion), m/z 663 $[M_{SO_3^-}]$; FAB MS (+ ve ion), m/z 709 $[M_{SO_3Na} + Na]^+$; selected ^{1}H-NMR signals: δ 1.01 (s, 18-H$_3$), 1.19 (s, 19-H$_3$), 1.43 (d, $J = 10.0$ Hz, 14-H), 3.90 (m, W$_{1/2} = 7.0$ Hz, 6α-H), 4.25 (d, $J = 7.5$ Hz, 1′-H), 4.90 (dt, $J = 3.0$, 10.0 Hz, 15β-H); ^{13}C-NMR spectrum of the steroid portion in Table 5.

Coscinasteroside B contains the common 5α-cholestane-3β,6β,8,15α,24-pentaol aglycone later found in several glycosides such as

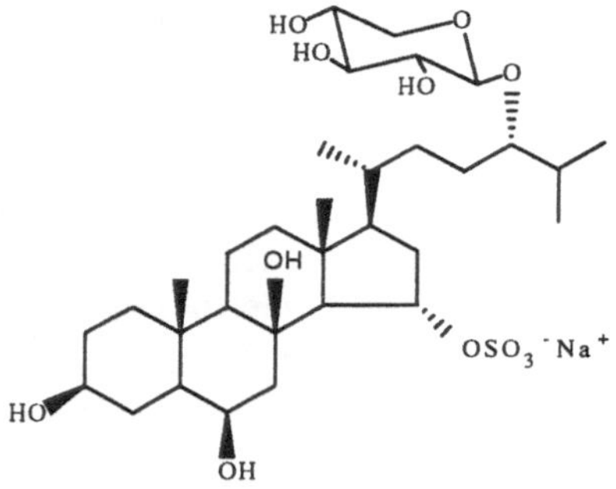

134

135, Δ^{22E}

136 and **137** which differ from each other only in the relative positions of the sulphate and the xylosyl residues, and **138** which is a diglycoside. The axial orientation of the hydroxyl group at C-6 was defined by the small width of the 6-H signal and supported by the downfield shift of the 19-methyl signals. The upfield shift of 15-H from δ 4.90 to 4.30 ppm in the desulphated derivative established the location of sulphate there. The ^{13}C-NMR spectrum confirmed the formulation for the aglycone and indicated C-24 as the site of glycosidation.

135 Scoparioside D (Δ^{22E} coscinasteroside B)

OCCURRENCE:*Astropecten scoparius (84)*.
PHYSICAL DATA: $[\alpha]_D = + 11.6°$ (MeOH); FAB MS (-ve ion), m/z 661 $[M_{SO_3^-}]$; selected ^{1}H NMR signals: δ 3.75 (t, $J = 7.5$ Hz, 24-H), 5.36 (dd, $J = 15.0$, 7.5 Hz, 23-H), 5.49 (dd, $J = 15.0$, 7.5 Hz. 22-H).
    Scoparioside D is the Δ^{22E} analog of **134**. Enzymatic hydrolysis with *Charonia lampas* glycosidase mixture removed xylose; subsequent solvolysis removed the sulphate to give the (22E)5α-cholest-22-ene-3β,6α,8,15α,24-pentaol aglycone. The chemical shift difference of the 24-H resonance in the R and S models (*i.e.* δ 3.71 and 3.68, respectively) was used to determine the C-24R configuration in the steroid aglycone (24-H:δ 3.71) and accordingly in **135**.

136 Pisasteroside B

OCCURRENCE: *Pisaster ochraceus (72)*.
PHYSICAL DATA: $[\alpha]_D = + 6.0°$ (MeOH); FAB MS (− ve ion), m/z 663 $[M_{SO_3^-}]$; selected ^{1}H-NMR signals: δ 4.14 (q, $J = 6.0$ Hz, 24-H), 4.30 (dt, $J = 3.0$, 10.0 Hz, 15β-H), 4.39 (d, $J = 7.5$ Hz, 1'-H).

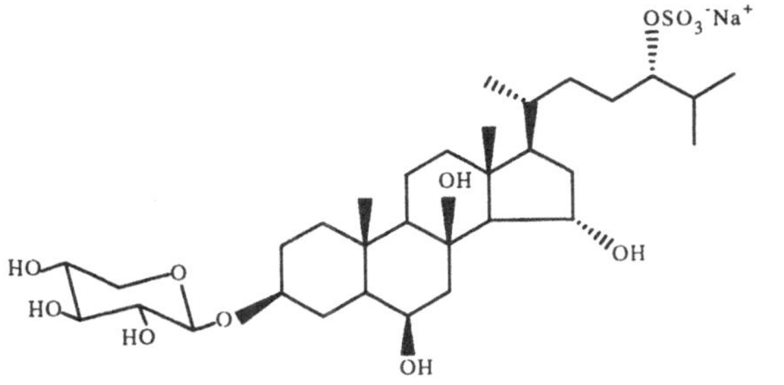

136

Pisasteroside B is isomeric with the previous coscinasteroside B (**134**) and with **137** described below. Comparison of their spectral data was of value for the structure assignment.

137 Aphelasteroside A

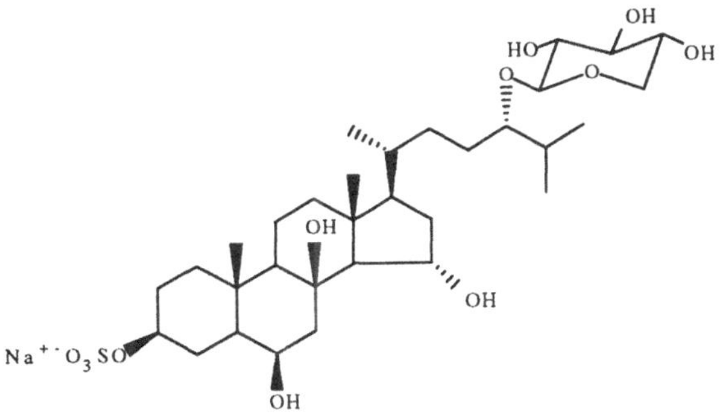

137

OCCURRENCE: *Aphelasterias japonica* (*143*).
PHYSICAL DATA: FAB MS (− ve ion), m/z 663 [$M_{SO_3^-}$]; selected ¹H-NMR signals: δ 3.30 (m, 24-H), 4.17 (d, *J* = 7.5 Hz, 1′H), 4.22 (m, 3α-H).

The ¹H- and ¹³C-NMR spectra and comparison with those of **136** established the location of sulphate at C-3 and xylose at C-24.

138 5-Deoxyisonodososide

OCCURRENCE: *Acanthaster planci* (*120*). *Choriaster granulatus* (*144*).
PHYSICAL DATA: [α]$_D$ = − 13.0° (MeOH); FAB MS (+ ve ion), m/z 753 [M + Na]⁺ 695 [M + H]⁺ ; selected ¹H-NMR signals: δ 1.20 (s, 19-H₃),

138

3.72 (m, 3α-H), 3.91 (q, J = 3.0 Hz, 6α-H), 4.46 (d, J = 7.5 Hz, 1′-H), 4.95 (br s, 1″-H); ^{13}C-NMR spectrum of the steroid portion in Table 2.

The FAB MS showed two groups of fragments corresponding to losses of 132 mass units (arabinose) and 146 mass units (2-O-methylxylose), respectively, from [M + H]$^+$ and stepwise water elimination. This indicated that the two monosaccharides were not linked to each other as in nodososide (**146**) and the many other 2-O-methyl-β-D-xylopyranosyl (1- > 2)-α-L-arabinosides isolated from starfishes. The nature of the sugars and of the steroid portions were confirmed by 2D-COSY and ^{1}H-^{13}C hetero-correlation spectroscopy (HETCOR), which provided unambiguous assignments for all protons in this system. The ^{13}C-NMR indicated C-3 and C-24 as sites of glycosidation. A larger downfield shift has invariably been observed for C-24 in 24-O-xylopyranosides (δ_C *ca* 86.5), than in 24-O-arabinofuranosyl steroids (δ_C 85.0). Thus, the relative positions of the monosaccharides in this group of glycosides can be determined by ^{13}C-NMR spectrometer. In **138** the relative positions of the attachment of the two monosaccharides were confirmed by mild acid hydrolysis to remove selectively the arabinofuranosyl unit and subsequent acetylation. The ^{1}H-NMR spectrum of the acetate derivative defined the substitution pattern; the multiplet assigned to 3-H remained unchanged at δ 3.70, whereas the isolated multiplet assigned to 24-H was shifted downfield to δ 4.68 compared with δ 3.40 prior to acetylation.

139 Granulatoside B

Occurrence: *Choriaster granulatus* (*144*).
Physical Data: [α$_D$] = − 12.0° (MeOH); FAB MS (+ ve ion), m/z 737 [M + Na]$^+$; selected ^{1}H-NMR signals: δ 0.78 (s, 18-H$_3$), 1.07 (s, 19-H$_3$),

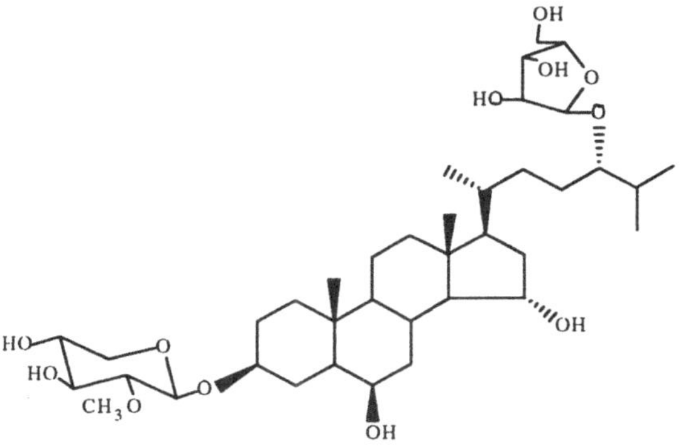

139

3.81 (br s, 6α-H), 3.92 (dt, $J = 3.0, 10.0$ Hz, 15β-H), 4.44 (d, $J = 7.5$ Hz, 1′-H), 4.95 (br s, 1″-H); selected ^{13}C-NMR signal: 84.9 (C-24); ^{13}C-NMR spectrum of the steroid portion in Table 5.

Granulatoside B is the 8-deoxy analog of 5-deoxyisonodososide (**138**). The lack of the hydroxyl group at C-8 was indicated by the ^{13}C-NMR spectrum which indicated the absence of tertiary hydroxyl groups.

140 Laeviuscoloside I (= forbeside I)

OCCURRENCE: *Henricia laeviuscula* (*94*).
PHYSICAL DATA: FAB MS (− ve ion), m/z 627 [M-H]$^{-}$; selected ^{1}H-NMR signals: δ 1.47 (s, 19-H$_3$), 4.31 (m, 4α-H and 6α-H), 4.50 (d, J = 7.5 Hz, 1′-H).

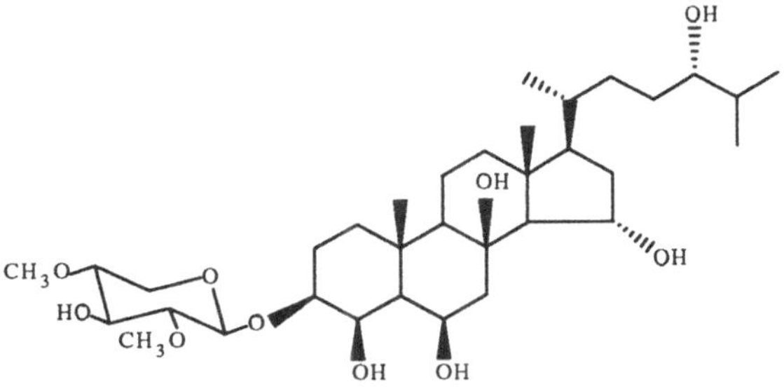

140

141, Δ^{22E}

141 Laeviuscoloside H (Δ^{22E}-laeviuscoloside I)

OCCURRENCE: *Henricia laeviuscula (94).*
PHYSICAL DATA: FAB MS (− ve ion), m/z 625 [M–H]$^-$; selected ^{1}H-NMR signals: δ 1.47 (s, 19-H$_3$), 4.31 (m, 4α-H and 6α-H), 4.50 (d, J = 7.5 Hz, 1'-H).

These compounds were isolated from *H. laeviuscola* as a mixture resistant to attempts of separation. Laeviuscoloside I (**140**) was subsequently isolated from *A. forbesi.* The introduction of a hydroxyl group at 4β position in the 3β,6β,8,15α-tetrahydroxycholestane skeleton causes the 6-H signal to move to δ 4.31, thus overlapping with the 4-H signal, and the 19-methyl signal to shift to δ 1.47 ppm. The ^{13}C NMR spectrum of laeviuscoloside I gave support to the structural assignment *(145).* The quartet at δ 3.71 for 24-H of **141** observed in the spectrum of the mixture was used to argue for the 24R-configuration *(cfr.* **135**). The 24S configuration is tentatively assigned to the saturated **140** by analogy with nodososide (**146**) and the many 24-hydroxysteroids isolated from starfishes.

142 Granulatoside A

OCCURRENCE: *Choriaster granulatus (144), Thromidia catalai (71).*
PHYSICAL DATA: [α]$_D$ = − 16.0° (MeOH); FAB MS (+ ve ion), m/z 769 [M + Na]$^+$; selected ^{1}H-NMR signals : δ 4.50 (d, J = 7.5 Hz, 1'-H), 4.95 (d, J = 1.4 Hz, 1''-H); ^{13}C-NMR spectrum of the steroid portion in Table 5.

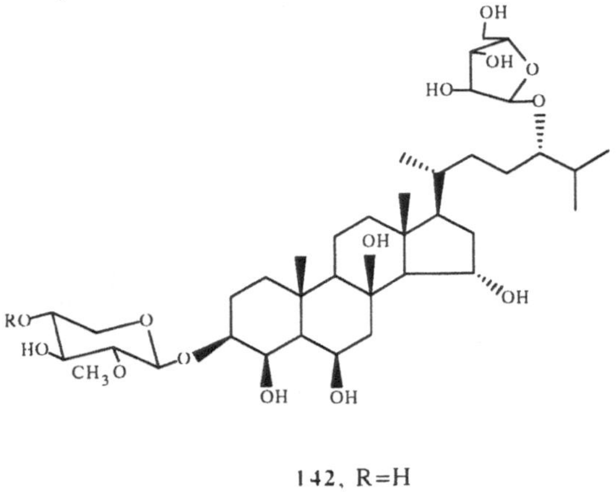

142, R=H

143, R=CH$_3$

Granulatoside A is related to 5-deoxynodososide (**138**) by having an additional hydroxyl group at the 4β position. The nature of the sugars and of the steroid portion were confirmed by ^{1}H-^{13}C heteronuclear correlation spectroscopy (HETCOR). The small shifts observed for C-1 and C-2 of the xylose unit (C-1:102.4; C-2:84.6) relative to **138** (C-1:103.3; C-2:85.2) were used to assign the hydroxyl group to C-4β, thus supporting the location of xylose at C-3 and, accordingly, arabinose at C-24, as suggested by the chemical shift of C-24 at 85.0 ppm. Mild acid hydrolysis, which selectively removed arabinose, confirmed the location of arabinose at C-24 (δH_{24} after acetylation, δ 4.68 ppm).

143 Laeviuscoloside G (= forbeside J)

OCCURRENCE: *Henricia laeviuscola (94). Asterias forbesi (145).*
PHYSICAL DATA: $[\alpha]_D = -0.8°$ (MeOH); FAB MS (− ve ion), m/z 759 [M−H]$^-$; selected ^{13}C-NMR signal: δ 85.1 (C-24).

This compound is closely related to granulatoside A (**142**) by methylation of the C-4 hydroxyl of the xylosyl residue.

144 Laeviuscoloside F

OCCURRENCE: *Henricia laeviuscola (94).*
PHYSICAL DATA: $[\alpha]_D = 0°$ (MeOH); FAB MS (− ve ion), m/z 743 [M−H]$^-$; selected ^{13}C-NMR signal: δ 85.1 (C-24).

Laeviuscoloside F (**144**) is related to granulatoside B (**139**) by having a hydroxyl group at position 4β of the steroid portion and a methoxyl group at position 4 of the xylosyl residue. The structure was derived by

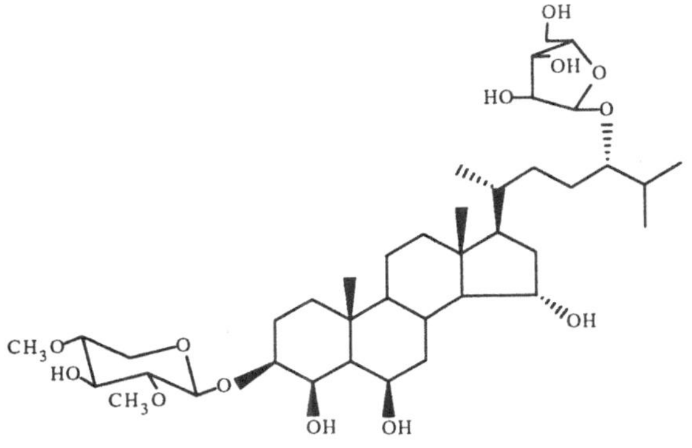

144

spectral comparison. Mild hydrolysis confirmed the relative positions of the two monosaccharides. The 24S configuration is suggested because the shifts of the isopropyl methyls of the 24 (+) MTPA ester match those found in the spectrum of the (+) MTPA ester of a (24S) 24-hydroxy model steroid and are quite different from those of the 24R-isomer (*cfr.* **61**).

145 Isonodososide

145

Occurrence: *Acanthaster planci (120).*
Physical Data: $[\alpha]_D = -15.0°$ (MeOH); FAB MS (+ ve ion), m/z 769 $[M + Na]^+$; selected ^{1}H-NMR signals: δ 0.99 (s, 18-H$_3$), 1.34 (s, 19-H$_3$), 3.63 (t, $J = 3.0$ Hz, 6α-H), 4.18 (m, 3α-H), 4.30 (dt, $J = 3.0$, 9.6 Hz, 15β-H), 4.46 (d, $J = 7.5$ Hz, 1'-H), 4.95 (d, $J = 1.4$ Hz, 1''-H); ^{13}C-NMR spectrum of the steroid portion in Table 5.

Isonodososide is isomeric with nodososide (**146**). Introduction of an hydroxyl group at the 5α position of the 3β,6β,8,15α-tetrahydroxycholestane nucleus is evidenced in the ^{1}H-NMR spectrum by downfield shifts of 3-H to δ 4.18 and of the 19-H$_3$ signal to δ 1.34 s (*cfr.*5-deoxyisonodososide, **138**). The relative positions of xylose and arabinose, indicated by both the ^{13}C-NMR and ^{1}H-NMR spectra were confirmed by mild acid hydrolysis.

146 Nodososide

Occurrence: *Protoreaster nodosus (110), Pentaceraster alveolatus (68), Acanthaster planci (146), Linckia laevigata (146).*

146

ACTIVITY: Cytotoxic against turbinate cells at a dose of 10 µg/ml and possesses stimulation effect on human lymphoma cells and mouse T-cell lymphoma cells (**36**).

PHYSICAL DATA: $[\alpha]_D = -21.0$ (MeOH); FAB MS (+ ve ion), m/z 769 $[M + Na]^+$; for assignments of signals of the disaccharide moiety see **91**; ^{13}C-NMR spectrum of the steroid portion in Table 5.

Nodososide (**146**) was the first representative of the group of poly-hydroxylated steroids to be found in starfishes. It was initially isolated from *P. nodosus* (*110*). Analysis of the 500 MHz ^{1}H- and ^{13}C-NMR spectra established the structure of the steroid which was confirmed by chemical transformations. Acetylation of **146** gave the 3,15,3′,5′,3″,4″-hexaacetate which was converted by oxidation to the 6-keto derivative. The acetate formed a phenylboronate which involves the 6β-OH because there was no reaction of the 6-keto derivative with phenylboronic anhydride, thus requiring that one *tert*-hydroxyl group be situated at the 8β-position. The structure of the disaccharide moiety, later found in many glycosides from starfishes, was elucidated by a) the formation of methyl   2-O-methyl-3,4-di-O-(*p*-bromobenzoyl)-α-D-xylopyranoside [CD: 236/253, Δε + 12/-38, A-50; D-series (*123*)] and methyl 2,3,4-tri-O-(*p*-bromobenzoyl)-α-L-arabinopyranoside [CD: 236/253, Δε-30/ + 95, A + 125; L-series (*123*)] on acid methanolysis followed by *p*-bromoben-zoylation; b) the ^{1}H- and ^{13}C-NMR spectral analysis of **146** which established that arabinose was in the furanose form and also the nature of the interglycosidic linkage, and c) the ^{1}H NMR spectrum of the hexaacetate in which the signals of H-2 of the arabinosyl residue were essentially unchanged.

The 24S configuration was assigned by application of the gas-chromatographic modification of Horeau's method to the (24S) 3β,5,6β,15α-tetramethoxy-5α-cholest-8(9)-en-24-ol obtained by methylation of nodososide and subsequent acid methanolysis (*147*).

147 Echinasteroside B$_2$

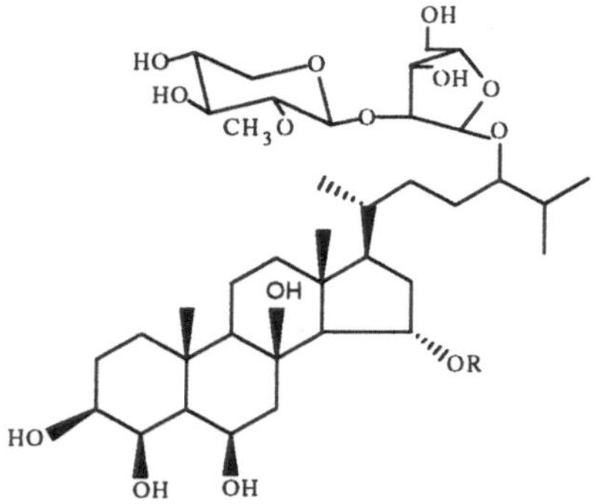

147, R=H

148, R=Ac

148 Echinasteroside B$_1$

OCCURRENCE : *Echinaster sepositus (148)*.

149 Laeviuscoloside B

OCCURRENCE: *Henricia laeviuscola (94)*.
PHYSICAL ATA: $[\alpha]_D = +1.5°$ (MeOH); FAB MS ($-$ve ion), m/z 681 $[M_{SO_3^-}]$; selected ^{1}H-NMR signals: δ 1.19 (s, 18-H$_3$), 1.20 (s, 19-H$_3$), 3.71

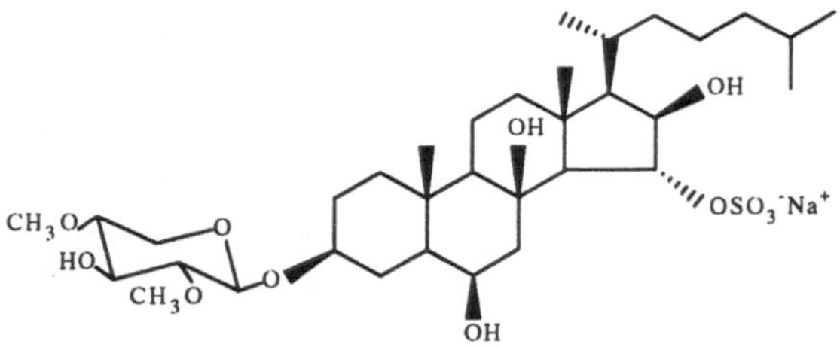

149

(m, 3α-H), 3.85 (q, J = 3.0 Hz, 6α-H), 4.35 (dd, J = 7.5, 2.5 Hz, 16α-H), 4.79 (dd, J = 11.0, 2.5 Hz, 15β-H).

Laeviuscoloside B, a minor glycoside from *H. laeviuscola*, is a member of the series characterized by the steroidal aglycone 3β,68β,8,16β-tetrahydroxy-15α-sulphoxy cholestane nucleus. The structure of **149** was derived by spectral comparison with attenuatoside S-I (**157**), the first isolated compound of this series.

150 Laeviuscoloside C [4(5)-dihydroloeviusoloside D]

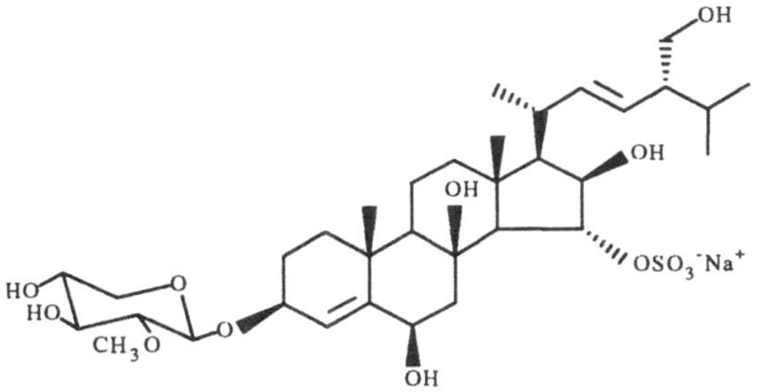

150, 4(5)-dihydro

151

OCCURRENCE: *Henricia laeviuscola (94)*.
PHYSICAL DATA: $[\alpha]_D$ = + 12.0° (MeOH); FAB MS (− ve ion), m/z 705 $[M_{SO_3^-}]$.

151 Laeviuscoloside D

OCCURRENCE: *Henricia laeviuscola (94)*.
PHYSICAL DATA: FAB MS (− ve ion), m/z 703 $[M_{SO_3^-}]$; selected ^{1}H-NMR signals: δ 1.24 (s, 18-H$_3$), 1.40 (s, 19-H$_3$), 3.48 (1H, dd, J = 10.0, 9.0 Hz, 28-H$_2$), 3.63 (1H, dd, J = 10.0, 5.0 Hz, 28-H$_2$), 4.23 (m, 3α-H), 4.34 (t, J = 3.0 Hz, 6α-H), 4.35 (dd, J = 7.5, 2.5 Hz, 16α-H), 4.80 (dd, J = 11.0, 2.5 Hz, 15β-H), 5.29 (dd, J = 15.0, 8.0 Hz, 23-H), 5.63 (dd, J = 15.0, 7.5 Hz, 22-H), 5.66 (br s, 4-H).

The structure of the side was assigned by ^{1}H-NMR analysis and confirmed by comparison with model Δ^{22E}, 24-hydroxymethylsteroids and Δ^{22E}, 24-methyl-26-hydroxysteroids, synthesized in our laboratory (*140, 172*). The 24R configuration was determined by ^{1}H-NMR spectral comparison of the 28(+) MTPA derivative of *151* with those of the

(+) MTPA ester of (24R)- and (24S)-model compounds (172; *cfr.* section 9.3).

152 Echinasteroside A

OCCURRENCE: *Echinaster sepositus* (*128*), *Henricia laeviuscola* (*94*).
PHYSICAL DATA: FAB MS (+ ve ion), m/z 749 $[M_{SO_3Na} + Na]^+$; selected
^{1}H-NMR signals: δ 0.91 (d, $J = 7.0$ Hz, 27-H$_3$), 0.97 (d, $J = 7.0$ Hz, 28-H$_3$), 1.04 (d, $J = 7.0$ Hz, 21-H$_3$), 1.53 (dd, $J = 15.0, 3.0$ Hz, 7-H), 2.69 (dd, $J = 15.0, 3.0$ Hz, 7-H), 3.46 (1H, dd, $J = 10.0, 9.0$ Hz, 26-H$_2$), 3.58 (1H, dd, $J = 10.0, 5.0$ Hz, 26-H$_2$), 5.39 (dd, $J = 15.0, 8.0$ Hz, 23-H), 5.46 (dd, $J = 15.0, 7.5$ Hz, 22-H); ^{13}C-NMR spectrum of the steroid portion in Table 5.

152

The structure was elucidated by analysis of the ^{1}H- and ^{13}C-NMR spectra. A NOEDS experiment showed the spatial proximity of the olefinic proton at C-4 and the proton at C-6 and also indicated the 6β-OH stereochemistry. The presence of the hydroxyl group at C-8 was confirmed by formation of a phenylboronate when the desulphated echinasteroside A was treated with phenylboronic anhydride. The structure of the steroid side chain also followed from the ^{1}H-NMR spectrum and sequential decoupling. The 24R, 25S absolute configuration was assigned by comparison with the stereoselectively synthesized stereo-isomers of Δ^{22E}, 24-methyl-26-hydroxysteroids (*140*). ^{1}H- and ^{13}C-NMR spectra differentiate between the *threo* and *erythro* isomers, whereas differentiation between individual steroisomers of *threo* and *erythro* pair is achieved by ^{1}H-NMR analysis of the MTPA derivatives (*cfr.* section 9.5).

153 Echinasteroside ·B

OCCURRENCE: *Echinaster sepositus* (*128*), *Henricia laeviuscola* (*94*).

PHYSICAL DATA: FAB MS (+ ve ion), m/z 765 $[M_{SO_3Na} + Na]^+$; FAB MS (− ve ion), m/z 719 $[M_{SO_3^-}]$; selected ^{1}H-NMR signals : $\delta\,0.87$ (d, J = 7.0 Hz, 26- or 27-H$_3$), 0.90 (d, J = 7.0 Hz, 27- or 26-H$_3$), 3.65 (ABq, 29-H$_2$); selected ^{13}C NMR data: δ 19.1, 20.1 (C-26 or C-27), 62.3 (C-29) ppm.

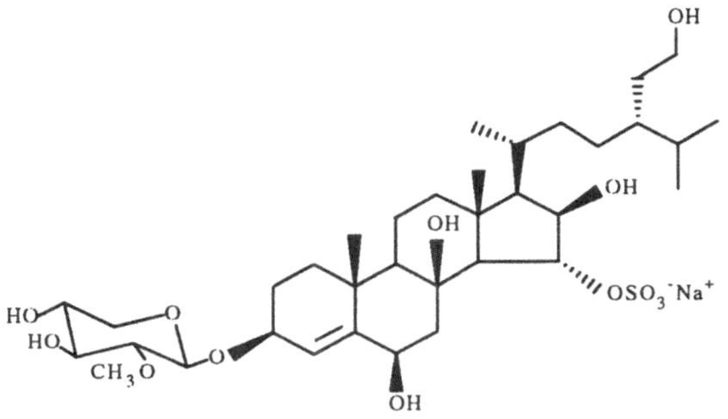

153

154, 4(5)-dihydro

154 Laeviuscoloside E [4(5)-dihydroechinasteroside B]

OCCURRENCE: *Henricia laeviuscola* (*94*).
PHYSICAL DATA: FAB MS (− ve ion), m/z 721 $[M_{SO_3^-}]$.

The 24R configuration in both **153** and **154** is based on the chemical shifts of the isopropyl methyls and comparison with 29-hydroxyclionasterol (24R) and 29- hydroxysitosterol (24S) models.

155 Aphelasteroside B

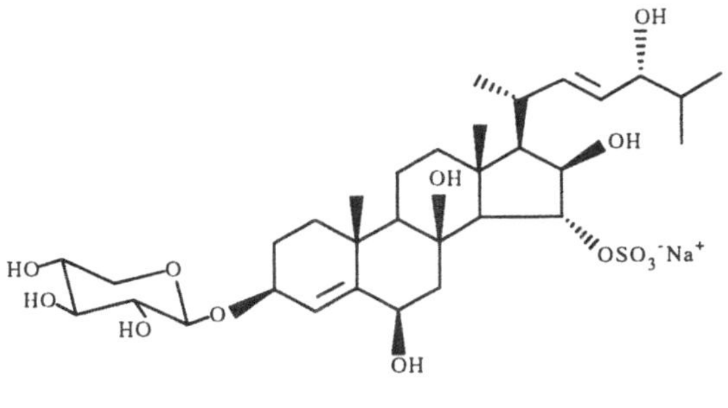

155

Occurrence: *Aphelasterias japonica* (*143*).
Physical Data: $[\alpha]_D = -16.8°$ (MeOH); FAB MS (− ve ion), m/z 675 $[M_{SO_3^-}]$.

The 24R configuration is proposed based on the chemical shift of 24-H at δ 3.70 and analogy with laeviuscoloside H (**141**).

156 Pisasteroside D

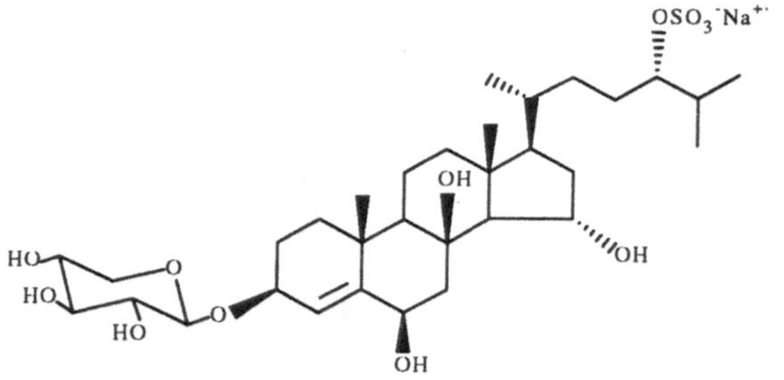

156

Occurrence: *Pisaster giganteus* (*73*).
Physical Data: $[\alpha]_D = +10.0°$ (MeOH); FAB MS (− ve ion), m/z 661 $[M_{SO_3^-}]$; selected ^{1}H-NMR signals: δ 4.14 (q, $J = 6.5$ Hz, 24-H), 4.33 (m, 15β-H and 6α-H).

The ^{1}H-NMR signals at δ 5.70 (br s, 4-H), 4.25 (m, 3α-H), 4.33 (m, 6α-H) and 1.38 (s, 19-H$_3$) indicated Δ^4, 3β,6β,8-hydroxylation in the steroid nucleus. The signals at δ 4.33 and 4.14 were assigned to 15-H and 24-H. Analysis of the ^{13}C-NMR spectra of **156** and its desulphated derivative established location and nature of the sugar at C-3 and the sulphate at C-24.

157 Attenuatoside S-I

Occurrence: *Hacelia attenuata* (*127*).
Physical Data: $[\alpha]_D = +10.0°$ (MeOH); FAB MS (+ ve ion), m/z 769 $[M_{SO_3Na} + K]^+$, 753 $[M_{SO_3Na}]^+$, 731 $[M_{SO_3Na} + H]^+$; selected ^{13}C-NMR data: 69.6 (C-29) ppm.

158 Attenuatoside S-II [Δ^{22E}-attenuatoside S-I]

Occurrence: *Hacelia attenuata* (*127*).
Physical Data: $[\alpha]_D = +7.4°$ (MeOH); FAB MS (+ ve ion), m/z 767 $[M_{SO_3Na} + K]^+$, 751 $[M_{SO_3Na} + Na]^+$, 729 $[M_{SO_3Na} + H]^+$.

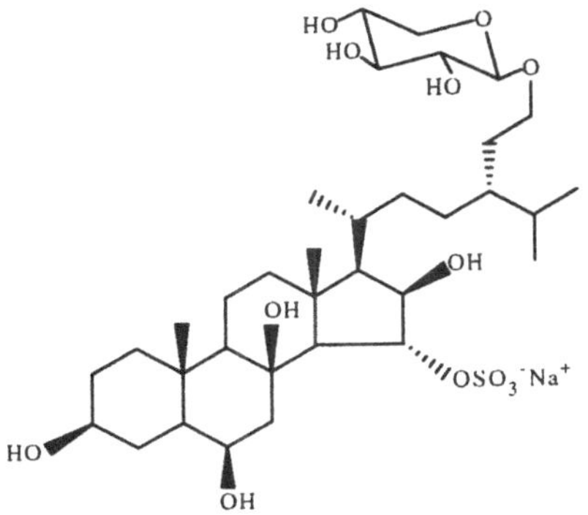

157

158, Δ^{22E}

The structure of attenuatoside S-I was derived by accurate ^{1}H- and ^{13}C-NMR analysis (^{1}H-NMR data *cfr.* **149**). Assignments of the ^{13}C resonances of the side chain carbons were made by comparison with 29-hydroxyclionasterol and 29-hydroxysitosterol, which also indicated the location of xylose at C-29. The 24R configuration in **157** is based on the chemical shifts of the isopropyl methyls (δ_H 0.88 t; δ_C 19.1 and 19.7 ppm, (*cfr.* Sect. 9.4).

159 Attenuatoside S-III

OCCURRENCE: *Hacelia attenuata* (*127*).
PHYSICAL DATA: $[\alpha]_D = +3.0°$ (MeOH); FAB MS (+ ve ion), m/z 769 $[M_{SO_3Na} + K]^+$, 753 $[M_{SO_3Na} + Na]^+$, 731 $[M_{SO_3Na} + H]^+$.

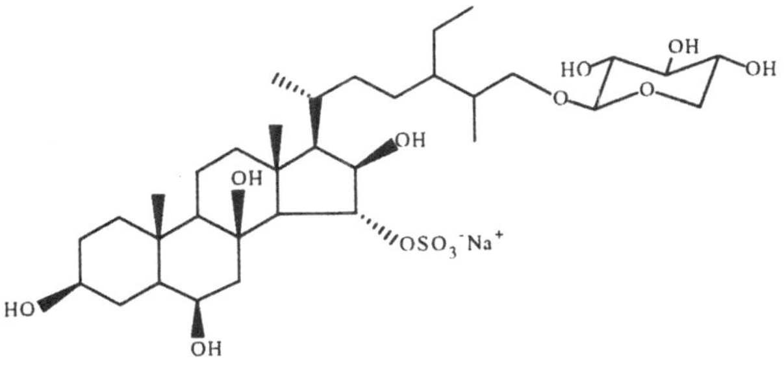

159

The structure of attenuatoside S-III isolated in very small amounts is proposed on the basis of ^{1}H-NMR spectrum which is very similar to that of the isomeric attenuatoside S-I (**157**) except in the methyl region, where a methyl doublet is replaced by a methyl triplet, thus requiring the presence of an ethyl group.

160 Forbeside L

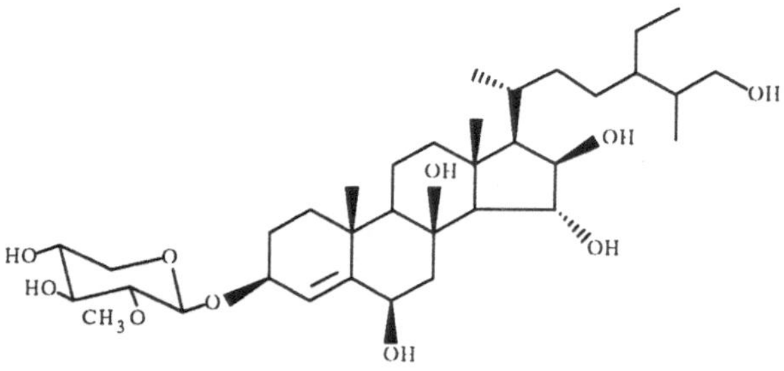

160

Occurrence: *Asterias forbesi* (*145*).
Physical Data: FAB MS (− ve ion), m/z 639 [M–H]$^-$.

The structure of the steroid nucleus was derived by analysis of the ^{1}H- and ^{13}C-NMR spectra and comparison with echinasteroside A (**152**). The 2D-COSY spectrum revealed the presence of an ethyl group and a CH (CH$_3$)CH$_2$OH; the hydroxymethylene carbon signal was observed at 66.8 ppm [in 24-(β-hydroxyethyl) steroids, δ$_C$ 62.3]. The proton chemical shifts for the side chain of forbeside L are consistent with those of desulphated attenuatoside S-III (**159**).

161 Forbeside K

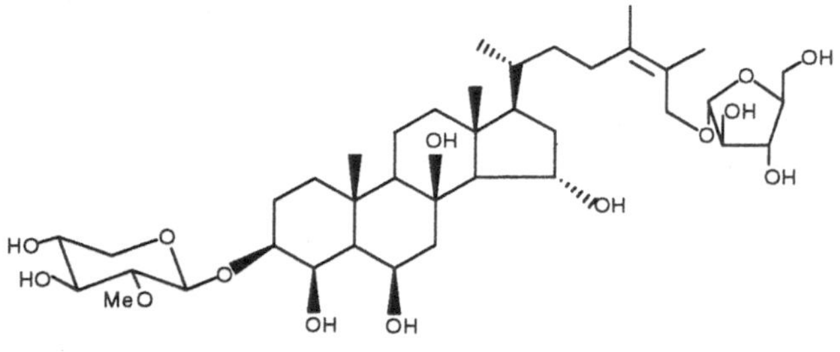

161

OCCURRENCE: *Asterias forbesi* (*145*).

PHYSICAL DATA: $[\alpha]_D = -9.2°$ (H_2O); FAB MS ($-$ ve ion), m/z 757 [M-H]$^-$; selected ^{1}H-NMR signals: δ 1.69 (s, 27- and 28-H$_3$), 4.09 (s, 26-H$_2$), 4.46 (d, $J = 7.5$ Hz, 1'-H), 4.83 (d, $J = 1.4$ Hz, 1''-H).

The ^{1}H- and ^{13}C-NMR data for the tetracyclic steroidal nucleus are coincident with those of forbeside I (*140*). After subtracting the ^{13}C signals for the sugar moieties and those due to the steroid from C-1 to C-21, the remaining seven carbon signals were assigned to C-22 to C-28 of the side chain with the aid of a selective INEPT experiment. The locations of the xylosyl unit at C-3 and the arabinosyl unit at C-26 of the aglycone were inferred from the coincident ^{13}C-NMR chemical shifts for the steroid ring system with those of similar glycosides (*e.g.* granulatoside A, *142*).

162

OCCURRENCE: *Poraster superbus* (*149*).

PHYSICAL DATA: $[\alpha]_D = -67.0°$ (MeOH); FAB MS ($+$ ve ion), m/z 649 [M + Na]$^+$, 627 [M + H]$^+$; selected ^{1}H NMR signals: δ 3.58 (1H, dd, $J = 12.0, 6.5$ Hz, 26-H$_2$), 3.65 (1H, dd, $J = 12.0, 5.0$ Hz, 26-H$_2$), 4.02 (dd, $J = 8.0, 2.5$ Hz, 16α-H), 4.17 (dd, $J = 11.0, 2.5$ Hz, 15β-H), 4.78 (1H, br s, 28-H$_2$), 4.87 (1H, *br s*, 28-H$_2$)

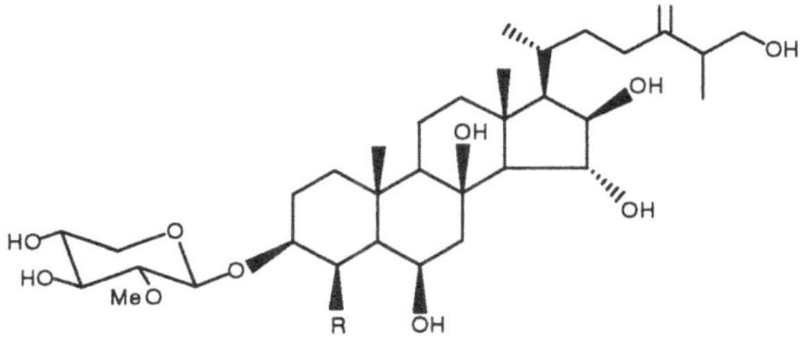

162, R=H

163, R=OH

163 Thromidioside

OCCURRENCE: *Thromidia catalai* (*71*).

PHYSICAL DATA: $[\alpha]_D = +10.0°$ (MeOH); FAB MS ($-$ ve ion), m/z 641 [M-H]$^-$.

164 Crossasteroside P$_1$

Occurrence: *Crossaster papposus (150).*
Physical Data: $[\alpha]_D = -12.9°$ (MeOH).

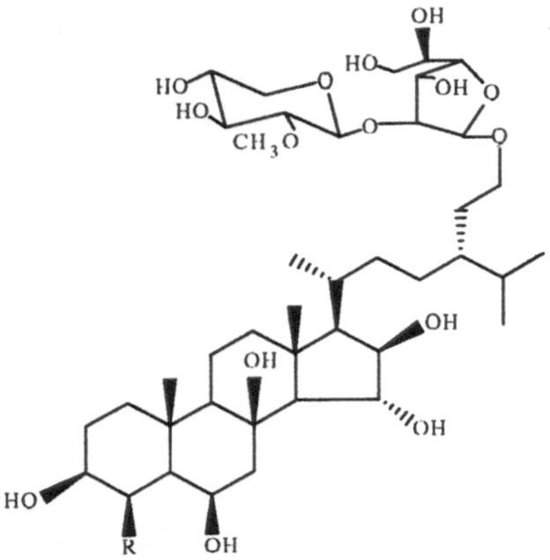

164, R=H

165, R=OH

165 Crossasteroside P$_2$
Occurrence: *Crossaster pappopus (150).*
Physical Data: $[\alpha]_D = -14.0°$ (MeOH).

Crossasterosides P$_1$ and P$_2$ are the second example of galactofuranosides found in starfishes after indicoside A (**121**). Acid hydrolysis gave D-galactose and 2-O-methyl-D- xylose. ^{1}H- and ^{13}C-NMR data established the dimensions of the rings of the monosaccharides. The sequence of the sugars was determined by acetylation and ^{1}H-NMR analysis of the acetate, in which the 2′-H signal of the galactofuranose was essentially unchanged. The 24R configuration was assigned on the basis of the chemical shifts of the isopropyl methyls (δ_H 0.80 d, 0.82 d; δ_C 18.6, 19.8 ppm in d$_5$-pyridine) and comparison with (24R) and (24S)-24-ethyl-5α-cholest-7-ene-3β,29-diols (*cfr.* section 9.4).

166 Halityloside F

Occurrence *Halityle regularis (112)*, *Gomophia watsoni (131)*, *Nardoa gomophia (70)*, *Culcita novaeguineae (44)*.
Physical Data: $[\alpha]_D = -14.1°$ (MeOH); FAB MS (+ ve ion), m/z 767 [M + Na]$^+$; selected ^{1}H-NMR signals: δ 1.21 (s, 19-H$_3$), 1.30 (s, 18-H$_3$),

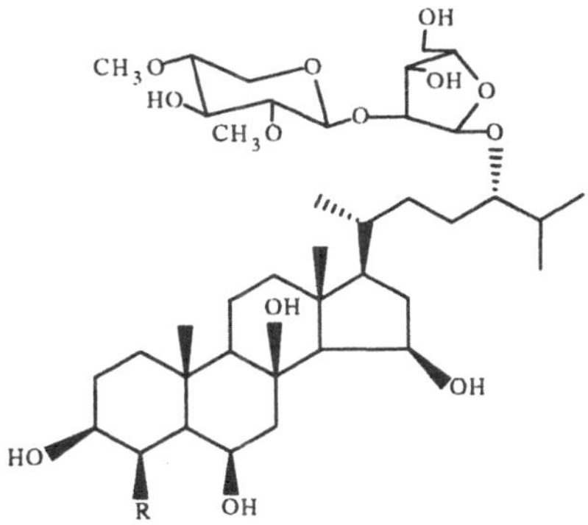

166, R=H

167, R=OH

3.88 (q, $J = 3.0$ Hz, 6α-H), 4.45 (m, 15α-H); ^{13}C-NMR spectrum of the steroid portion in Table 5.

Halityloside F is epimeric with halityloside E (**93**) and differs from **93** in the stereochemistry at C-6, which in **166** is 6β-OH. In the ^{1}H-NMR spectrum the CH$_3$-19 singlet was shifted downfield to δ 1.21 and there was an apparent quartet at δ 3.88 with $J = 3$ Hz, characteristic for a 6β-hydroxyl group. The ^{13}C-NMR spectrum supported the structural assignment.

167 Gomophioside A

OCCURRENCE: *Gomophia watsoni* (*131*), *Culcita novaeguineae* (*44*).
PHYSICAL DATA: $[\alpha]_D = -31.3°$ (MeOH); FAB MS (+ ve ion), m/z 783 [M + Na]$^+$; selected ^{1}H-NMR signals: δ 1.30 (s, 18-H$_3$), 1.47 (s, 19-H$_3$), 4.10 (m, W$_{1/2} = 7.0$ Hz, 4α-H), 4.29 (m, W$_{1/2} = 7.0$ Hz, 6α-H), 4.45 (m, 15α-H); ^{13} C-NMR of the steroid in Table 5.

Gomophioside A is the 4β-hydroxy derivative of **166**, also found in *G. watsoni*, its structure was derived from spectral data and comparison with **166**.

168 Moniloside D

OCCURRENCE: *Fromia monilis* (*109*).
ACTIVITY: Toxic at a concentration of 10 μg/5 · 10^4 cells (vero cells).
PHYSICAL DATA: $[\alpha]_D = -18.4°$ (MeOH); FAB MS (− ve ion), m/z 613

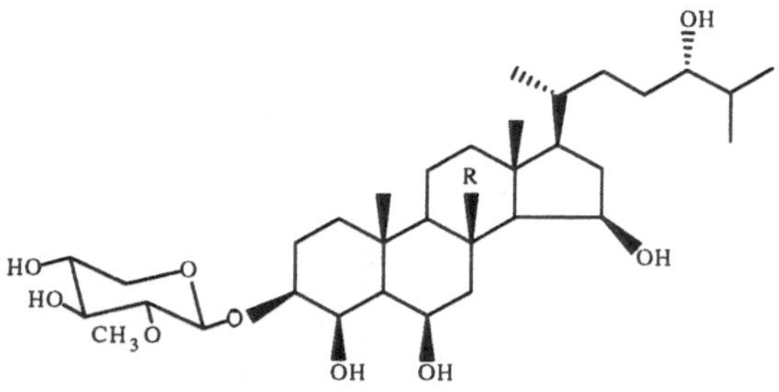

168, R=OH

169, R=H

[M–H]⁻; selected ¹H-NMR signals: δ 4.30 (2H, m, 4α-H and 6α-H), 4.50 (d, $J = 7.5$ Hz, 1'-H).

The structure of moniloside D is close to that of laeviuscoloside I (**140**) and differs from **140** in the stereochemistry at C-15, which in **168** is 15β-OH, and by the lack of a methoxyl group at 4'-OH of the xylose. The location of the 2-O-methyl xylose at C-3 follows from the chemical shift of the anomeric proton at δ 4.50 (*cfr.* **140**). The xylosyl residue at C-3 also moves the signal of 4-H to δ 4.30 which overlaps the signal of 6-H. Confirmation comes from the ¹³C-NMR spectrum.

169 Moniloside C

Occurrence: *Fromia monilis* (*109*).
Activity: Toxic at a concentration of 10 μg/5 · 10⁴ cells (vero cells).
Physical Data: [α]_D = − 33.0° FAB MS (− ve ion), m/z 597 [M–H]⁻; selected ¹H-NMR signals: δ 1.01 (s, 18-H₃), 1.40 (s, 19-H₃), 4.19 (2H, m, 6α, 15α-H), 4. 29 (br s, 4α-H), 4.50 (d, $J = 7.5$ Hz, 1'-H).

Moniloside C is related to the moniloside D by lack of the hydroxyl group at C-8, as indicated by a comparative spectral analysis.

5.1 Spectroscopy

Structure determination of the glycosides of polyhydroxysteroids has mainly involved the use of ¹H- and ¹³C-NMR spectroscopy. Selected ¹H-NMR chemical shifts and coupling data for some typical steroidal skeletons were reported in the preceeding sections which also gave ¹H- and ¹³C-NMR data of a number of representative saccharide moieties. The greatest difficulty associated with the assignment of ^{1}H-NMR signals of an oligosaccharide is that most of the resonances are confined within a narrow spectral window; even so, the analysis of some resolved resonances such as the anomeric protons, the 2-H and 5-H_2 protons of xylopyranose and 5-H_2 protons of arabinofuranose units, is very useful for structure assignment. Comparison of the chemical shifts of the saccharide signals with reference NMR data of known compounds has been used to ascertain not only the position of the substituents (*e.g.* methyl and sulphate groups) but also to provide complete structures of the oligosaccharide moiety subsequently confirmed by ^{13}C-NMR data.

We wish to make only a few remarks upon this aspect. The chemical shift of the signal of the anomeric proton of the xylopyranose unit is sensitive to location of the sugar at C-3 or C-24 of the aglycone. In the ^{1}H-NMR spectra of 3-O-xylopyranosides the resonance of the anomeric proton appears at δ 4.40–4.44 ppm and is shifted to δ 4.48–4.50 ppm when a hydroxyl group is located at the 4β-position of the aglycone, whereas in the spectra of the 24-O-xylopyranosides the same signal is seen at δ 4.24–4.26 ppm. The presence of an upfield double doublet at δ 2.84–2.95 ppm (J = 9.5, 7.5 Hz, 2-H) is diagnostic of the presence of a 2-O-methylxylopyranose unit. The chemical shift of this signal, which does not overlap with anything else, is very sensitive to structural changes in its neighbours. In the spectra of steroidal 3-O-(2′-O-methyl) xylopyranosides the signal is at δ 2.84–2.85 ppm and is shifted to δ 2.93–2.95 ppm by the presence of an additional hydroxyl group at the 4β-position of the aglycone; in 2,4-di-O-methylxylopyranosides the resonance is observed at δ 2.90 ppm. Substitution of the 4-OH of the xylopyranose moiety is also accompanied in the ^{1}H-NMR spectrum by a downfield shift of the 5-H equatorial signal from the usual δ 3.85 (dd, J = 10.5, 4.5 Hz) ppm position to δ 4.00–4.02 ppm. The anomeric proton of the α-arabinofuranosyl moieties in steroidal 24-O-arabinosides gives rise to a narrow doublet (J = 1.5 Hz) at δ 4.95–4.97 ppm which is shifted to 5.07–5.11 ppm in β-D-xylopyranosyl (1 → 2) α-arabinofuranosyl moieties. The presence of a sulphate group at C-2 or at C-3 of the arabinofuranosyl moiety causes a larger downfield shift of the anomeric proton signal to *ca.* δ 5.18 ppm.

Although ^{1}H-NMR spectroscopy has been an important source of information for the structure determination of these mono- and disaccharides, the potential of ^{13}C-NMR spectroscopy has long been recognized. Assignments of the ^{13}C-NMR resonances have been made mostly by comparison with reported data of the constituent monosaccharides, taking into account the expected shifts caused by glycosidation and by substitution with methyl and sulphate groups. Carbon-13 NMR spectroscopy has also provided an excellent tool for substantiating the structure of the steroidal aglycones and for determining the location of the saccharide moiety. Assignments of ^{13}C-NMR resonances of steroidal aglycones have been made by information obtained through DEPT pulse sequences, comparison of the spectra with those of reference steroids, making use of additive substituent parameters published for monohydroxy (113) and dihydroxysteroids (114), and comparison within this series of aglycones, taking into account structural variations and known chemical shift rules. Chemical shifts of the aglycone carbons in representative glycosides of polyhydroxysteroids are compiled in Table 5.

Large shielding γ-gauche effects are observed for the methine carbons C-5 (− 8.3 ppm), C-9 (− 8.8 ppm) and C-14 (− 5.5 ppm) in amurensoside B (62) upon introduction of a 7α-hydroxyl group (axial) in the structural framework of amurensoside A (61). We also note two large deviations from the additivity experienced by the vicinal hydroxyl groups at C-6 and C-7 in the axial-equatorial configuration (*ca.* − 6 ppm). From a comparison of the spectrum of amurensoside A (61) with that of amurensoside C (63) it was *e.g.* quite obvious that 61 had a sulphate group at C-15. In 63 the C-15 resonance is shifted greatly upfield (+ 7.7 ppm), whereas resonances assigned to C-14 and C-16 are shifted downfield (− 2.4 and − 4.2 ppm, respectively). Introduction of an axial hydroxyl group at C-8 (steroid 64) into the 3β,6α,15α-trihydroxycholestane skeleton gives rise to decreased shielding γ-gauche effects [C-6 (− 4.3 ppm), C-11 (− 3.1 ppm) and C-15 (− 4.9 ppm)], compared with those reported, *i.e.* − 6.5 ppm for methylene carbons and − 7.8 ppm for methine carbons (*113*). In the 8β-hydroxysteroid (64) the 8β-hydroxyl group suffers from two 1,3-*syn*-diaxial interactions with the C-18 and C-19 methyl groups and it has been assumed that the reduced γ-gauche effects observed are associated with skeletal deformations taking place to relieve the steric hindrance. It is also interesting to note that the two δ interactions of the 8β-hydroxyl group with the angular methyl groups results in different δ shifts on the methyl resonances. Introduction of the 7α-hydroxyl group in the 3β,6α,8,15α-tetrahydroxy skeleton (steroid 72) induces a substitution effect similar to that observed in 62 *vs* 61, with smaller shielding γ-gauche effects on methine carbons C-9 (− 5.4 ppm) and C-14 (− 4.4 ppm). Steroid 75 is the first example in Table 5 of a compound bearing a 4β-hydroxyl group and comparison of spectral data with those of 64 evidences the expected substitution effects caused by the 4β-hydroxyl group on adjacent carbon signals. Shielding γ-gauche effects are exerted on C-2 (− 6.1 ppm) and C-6 (− 1.9 ppm) while the C-19 methyl signal is significantly deshielded

(+ 2.8 ppm) due to a δ effect (*113*). This comparison also substantiates the assignments made for C-18 and C-19 signals in **64**. It is worth of note that the resonance of C-24 is affected significantly by the structure of the sugar glycosidically linked to it: in 24-O-xylopyranosides the C-24 resonance is observed at *ca.* 86.5 ppm (see **61, 62, 72**), whereas in 24-O-arabinofuranosides it is observed in the range 83.5–85.0 ppm (see for example **64** and **76**). In the 5α-hydroxysteroid **76**, C-1 and C-7 methylene carbons experience large upfield shifts (− 6.4 and − 7.0 ppm, respectively) when compared with corresponding signals in **64**, whereas the C-3 hydroxymethine carbon undergoes a smaller shielding effect (− 3.1 ppm) with a deviation from the additivity of about 3 ppm. In 5α-androstane-3β,5α-diol, VAN ANTWERP *et al.* report a deviation of 2 ppm (*114*). The vicinal hydroxyl groups at C-5 and C-6 in the axial-equatorial arrangement as in **76** produce deviations from the additivity for the carbinol carbon atoms smaller than the *ca.* − 6.0 ppm reported for other axial-equatorial 1,2-diols (*113, 114*) (*e.g.* in **76** C-6: exp 69.5, calc 72.0 ppm; deviation − 2.5 ppm). In miniatoside A (**80**) the vicinal hydroxyl groups at C-15 and C-16 on the five-membered ring are in a *trans* arrangement and this is reflected in smaller deviations of the corresponding chemical shifts from the predicted positions; in indicoside C (**107**), isomeric with **80** but with the *cis*-15β,16β-dihydroxy stereochemistry and with the two vicinal hydroxyl groups on the five-membered ring almost eclipsed, the deviations from additivity are the largest among all compounds studied [− 11.8 ppm for C-15 using the 15β-OH steroid **85** as starting structure and the reported substituents effects in hydroxylated steroidal nuclei (*113*)]. The presence of the C-17 side chain *cis* to the 16β-hydroxyl group is also cause of a large deviation from the additivity for the chemical shift of the hydroxyl bearing carbon-16 (− 9.8 ppm in **80** using **64** as starting structure and − 21.3 ppm in **107** using **85** as starting structure). The 15β-hydroxysteroid aglycone of **85** is isomeric with the 15α-hydroxysteroid aglycone of **64** and accordingly the major differences in their spectral data are observed for the resonances of C-14 (62.6 *vs.* 66.6), C-17 (58.1 *vs.* 55.4) and C-18 (16.6 *vs.* 15.5). Smaller but significant differences are also observed for the resonances at C-6 and C-11 which experience a downfield shift in the 15β-hydroxysteroid (C-6: 67.8 *vs.* 66.6; C-11: 19.8 *vs.* 19.1). These decreased γ-gauche effects of the 8β-hydroxyl group in 15β-hydroxysteroids can probably be associated with further skeletal deformations beyond that caused by the 1,3-diaxial interactions among the 8β-hydroxyl group and the angular methyls, in order to relieve the additional 1,3-*syn* interaction between the 8β and 15β hydroxyl groups. This *syn* arrangement of the hydroxyl groups also results in a downfield shift of the C-8 signal (77.6 ppm *vs.* 75.5 ppm). The spectrum of indicoside

A (121), which has a 3β,6α,7α,8,15β-pentahydroxysteroidal aglycone isomeric with that of crossasteroside A (72) (15α-hydroxy-stereochemistry), has shown shifts vs. 85 already observed in 72 vs. 64. We note largest γ-effects due to the hydroxyl group at C-7α exhibited by C-9 (− 7.3 ppm) and C-14 (− 6.0 ppm) in 121 relative to those observed in the 15α-hydroxy isomer 72. Imbricatoside A (106) is representative of steroidal structures with both 4β- and 7α-hydroxyl groups in the common 3β,6α,8,15β-tetrahydroxylated skeleton. Introduction of an hydroxyl group at C-16β in the 3β,6α,8,15β-tetrahydroxylated skeleton, as in indicoside C (107), results in the same sensitive changes of spectral characteristics already discussed above. Coscinasteroside B (134) is the first example in Table 5 of a steroidal aglycone with an axial 6β-hydroxyl group and a number of changes are observed for the resonances of the carbons in rings A and B relative to the 6α-hydroxy isomer 64. Changes in chemical shifts are more prominent for C-5, C-6 and C-7; small but significant changes are also observed for the resonances of the remote δ -carbon C-1, found downfield at 41.4 ppm (vs. 39.1 ppm in 64), and of the quaternary γ-carbon C-8 at 77.1 ppm (vs. 75.5 ppm in 64). The spectrum of 5-deoxynodososide (138) illustrates the effects of glycosidation at C-3; the resonance of C-3 is seen at 80.3 ppm, 7.8 ppm downfield compared with 134, whereas the resonances of the γ-carbons C-2 and C-4 move upfield by ca. 1.5 and 3.3 ppm, respectively. ^{13}C-NMR spectroscopy also provides information for the location of the arabinopyranosyl moiety at C-24 (δ C-24 at 84.8 ppm; in 24-O-xylopyranosides δ C-24 is at 86.0–86.5 ppm) and accordingly of the xylopyranosyl moiety at C-3.

Granulatoside B (139), 8-deoxy analog of the previous 5-deoxyisono-dososide (138), is one of the rare cases of polyhydroxysteroids from starfishes which lack the 8-hydroxyl group. Comparison within 138 and 139 shows that the shielding γ-gauche effects at C-6 (− 1.6 ppm), C-11 (− 2.3 ppm) and C-15 (− 4.2 ppm) due to the presence of the 8-hydroxyl group in 138 are further decreased relative to those observed in 64 vs. 63, probably because of the additional 1,3-syn-diaxial interaction between the 6β and 8β hydroxyl groups. Comparison between 5-deoxyisonodoso-side (138) and granulatoside A (142) provides a further example of the chemical shift variations upon introduction of a 4β-hydroxyl group, whereas the ^{13}C-NMR spectral data of isonodososide (145) and nodoso-side (146) illustrate once again the substitution effects caused by introduction of a 5α-hydroxyl group. Echinasteroside A (152) here exemplifies the group of 3-O-β-xylopyranosides having a 15-sulphated Δ^4-3β,6β,8,15α,16β-pentahydroxysteroidal aglycone with different side chains. Finally halityloside F (166) and gomophioside A (167), related by the presence of an additional hydroxyl group at C-4β in 167, are

Table 5. Carbon Chemical shifts[a] of Steroidal Aglycones of Selected Glycosides of Polyhydroxy Steroids.

C	61 (43)	62 (43)	63 (43)	64 (116)	72 (121)	75 (93)	76 (124)	80 (117)	85 (132)	121 (142)	106 (139)
1	38.6	38.3	38.8	39.1	39.5	39.9	32.7	39.6	39.5	39.4	39.7
2	31.9	32.0	32.0	32.3	31.5	26.2	30.8	31.5	31.5	31.5	26.4
3	72.0	72.0	72.0	71.2	72.2	73.6	68.1	72.2	72.2	72.3	73.6
4	33.0	32.9	33.1	32.9	32.3	69.1	41.8	32.4	32.4	32.1	69.7
5	52.7	44.4	52.9	53.5	44.6	57.0	78.4	53.7	53.9	44.4	51.3
6	70.2	72.0	70.3	66.6	68.8	64.7	69.5	67.7	67.8	69.5	66.7
7	41.4	71.3	41.3	50.6	76.4	50.6	43.6	49.0	49.0	76.5	76.7
8	35.1	39.9	35.4	75.5	77.9	77.4	75.7	76.0	77.6	79.5	79.4
9	55.3	46.5	55.5	56.6	51.2	58.4	49.0	57.4	57.5	50.2	56.7
10	37.2	37.2	37.4	37.1	37.8	38.1	38.1	37.9	38.0	37.5	37.7
11	22.1	22.2	22.2	19.1	19.4	19.1	19.4	19.4	19.8	19.5	19.0
12	42.0	41.6	42.0	42.2	42.7	42.8	43.0	43.2	43.4	43.1	43.1
13	43.8	43.0	44.8	44.7	45.7	44.7	45.5	45.3	44.4	44.3	44.3
14	61.5	56.0	63.9	66.6	62.2	67.3	66.9	64.5	62.6	56.6	60.7
15	81.7	82.0	74.0	69.1	68.9	69.9	69.9	80.8	71.3	71.2	71.3
16	38.7	38.6	42.9	41.4	41.3	41.7	40.5	83.0	42.6	42.7	42.8
17	55.1	55.3	54.9	55.4	56.6	55.9	55.9	60.6	58.1	58.0	58.1
18	13.7	13.5	13.9	15.5	15.4	15.4	15.5	17.0	16.6	16.5	16.5
19	13.9	12.8	13.9	14.3	13.9	17.1	17.3	14.2	14.1	13.9	16.8
20	36.6	36.8	40.9	35.5	36.2	36.2	36.3	31.2	36.4	36.8	36.4
21	19.0	19.0	21.3	18.7	18.9	18.9	19.0	18.5	19.1	19.0	19.1
22	32.7	32.7	140.8	31.8	32.9	32.8	32.9	35.1	33.0	30.0	33.1
23	29.0	28.9	128.4	28.5	28.9	28.7	28.5	28.8	28.8	31.9	28.9
24	86.3	86.3	89.0	83.5	86.3	86.2	84.4	42.6	85.5	76.5	85.0
25	32.0	31.9	33.9	31.3	31.9	31.9	31.5	31.0	31.6	34.5	31.9
26	18.2	18.3	19.2	18.2	16.1	18.3	18.1	19.3	18.3	17.2	18.3
27	18.5	18.5	18.4	18.0	18.4	18.3	18.4	19.9	18.3	17.4	18.3
28								31.9		72.5	
29								67.9			

C	107 (132)	134 (42)	138 (120)	139 (144)	142 (144)	145 (120)	146[b] (110)	152 (128)	166 (112)	167 (131)
1	39.5	41.4	41.4	39.9	41.0	34.4	34.3	39.7	41.4	41.0
2	31.5	31.7	30.2	30.6	25.2	28.8	31.8	27.9	31.7	26.6
3	72.2	72.5	80.3	80.5	80.5	77.4	67.3	77.3	72.4	73.2
4	32.4	36.5	33.2	33.3	74.6	38.3	42.4	126.7	36.4	77.4
5	53.8	48.8	49.0	49.1	50.6	76.3	75.7	148.6	49.8	51.1
6	67.8	74.3	74.2	72.6	76.2	78.0	77.9	76.0	74.8	76.3
7	49.0	44.8	45.6	41.6	45.3	40.4	41.8	44.8	44.3	44.1
8	77.4	76.9	77.1	31.4	76.6	77.3	76.7	76.4	79.4	79.1
9	57.3	57.1	57.4	55.1	57.8	49.0	48.7	57.7	57.1	57.4
10	38.0	36.7	36.8	36.8	36.9	39.3	39.1	37.8	36.8	36.9
11	19.4	19.8	19.8	22.1	19.3	19.7	19.4	19.5	19.9	19.3
12	43.4	42.7	42.9	41.8	42.7	42.9	42.4	42.9	43.3	43.1
13	44.4	44.9	45.6	45.0	45.5	45.5	44.8	44.2	44.4	44.3
14	61.0	64.6	66.6	64.0	66.7	66.5	66.3	62.2	61.9	61.8
15	71.3	78.0	70.2	74.4	70.2	70.1	69.2	87.7	71.4	71.4
16	72.7	38.6	41.6	40.9	41.6	41.7	40.9	80.3	42.4	42.1
17	62.5	56.0	56.1	55.9	56.0	55.9	55.0	61.0	58.1	58.0
18	17.9	15.3	15.4	13.8	15.4	15.4	15.6	17.0	16.6	16.6
19	14.0	15.8	15.7	16.1	18.6	18.0	18.2	22.6	15.7	18.6
20	31.0	36.1	36.2	36.8	36.2	36.3	35.4	34.0	36.4	36.5
21	18.5	18.9	19.1	19.2	19.1	19.0	18.9	20.1	19.1	19.1
22	32.5	32.5	32.9	33.1	32.9	32.8	31.9	136.2	33.0	32.9
23	28.7	29.0	29.2	29.1	29.0	29.5	27.8	134.2	28.9	28.6
24	85.5	86.3	84.8	84.9	85.0	84.8	83.5	40.4	84.8	84.5
25	31.7	32.0	32.0	31.9	32.0	31.8	30.6	42.2	31.7	31.6
26	18.5	18.3	18.2	18.3	18.2	18.3	18.2	66.9	18.2	18.3
27	18.4	18.6	18.5	18.5	18.5	18.4	18.2	17.5	18.3	18.2
28								14.5		
29										

[a] Data (in ppm) were mostly obtained at 62.9 MHz from solution in d_4-methanol and are referred to the central line of the solvent signal (49.0 ppm), [b] data from solution in d_5-pyridine. Amurensoside A (61), amurensoside B (62), amurensoside C (63), asterosaponin P-1 (64), crossasteroside A (72), borealoside D (75), 6-epinodososide (76), miniatoside A (80), indicoside B (85), indicoside A (121), imbricatoside A (106), indicoside C (107), coscinasteroside B (134), 5-deoxynodososide (138), granulatoside B (139), granulatoside A (142), isonodososide (145), nodososide (146), echinasteroside A (152), halityloside F (166), gomophioside A (167).

representative of the smallest group of glycosides of polyhydroxysteroids with the 3β,6β,8,15β-hydroxylation pattern.

6. Glycosides of Steroid Phosphates

170 Tremasterol A
OCCURRENCE: *Tremaster novaecaledoniae (40).*
PHYSICAL DATA: $[\alpha]_D = +40.0°$ (MeOH); FAB-MS (− ve ion), m/z 803 $[M_{Na}]^-$.

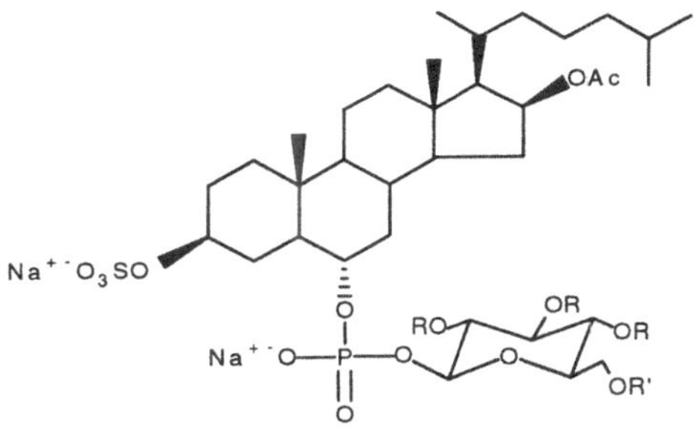

170, R=R'=H

171, R=R'=Ac

172, R=H, R'=Ac

171 Tremasterol B

OCCURRENCE: *Tremaster novaecaledoniae (40).*
PHYSICAL DATA: FAB-MS (− ve ion), m/z 971 $[M_{Na}]^-$.

172 Tremasterol C

OCCURRENCE: *Tremaster novaecaledoniae (40).*
PHYSICAL DATA: FAB-MS (− ve ion), m/z 845 $[M_{Na}]^-$.

We recently had the opportunity to examine a deep water starfish species, *Tremaster novaecaledoniae*, considered to be a living fossil and discovered at 530 m depth during exploration of the bathial zone off New

Caledonia. Analysis of the polar extracts from this organism resulted in the isolation of a new group of glycosides of polyhydroxysteroids with phosphate and sulphate conjugation, tremasterols A–C (**170–172**) (*40*). To the best of our knowledge this is the first reported isolation of steroids with phosphate conjugation. Their structures were essentially derived from spectral data. The FAB mass spectrum of tremasterol A (**170**) displayed molecular ion species at m/z 803 [M_{Na}]⁻ and 781 [M_{Ha}]⁻ together with fragments at m/z 641 (100%) and 619 (90%), correspond- ing to the loss of the glucosyl residue, and at m/z 539 and 521, interpreted as losses of SO_3 from m/z 619 and $NaHSO_4$ from m/z 641. In addition to the β-glucosyl moiety, the low-field region of the ^1H nmr spectrum revealed one proton signals at δ 4.06 (6-H), 4.24 (3-H) and 5.27 (15-H) ppm. The multiplet at 4.06 ppm, assigned to 6β-H, appeared as a dddd (*J* = 9.5, 9.5, 7.5 and 4.5 Hz) with one more coupling constant than what was expected for a 6α-hydroxycholestane in which the 6β proton is usually seen as an apparent double triplet with *J* = 4.5 and 9.5 Hz. This was the first indication of the presence of a phosphate ($J_{H-C-O-P} = ca.$ 8 Hz) also linked to C-1 of the β-glucosyl residue (1′-H δ 4.89, t, *J* = 7.5 Hz). Confirming evidence came from the proton noise decoupled ^{13}C-NMR spectrum, in which the signals due to C-5, C-6, C-1′ and C-2′ appeared as doublets because of ^{31}P-^{13}C couplings through two and three bonds and from the ^{31}P NMR spectrum which showed a triplet (*J* = 7.5 Hz) at δ 3.54 ppm downfield from the external standard (H_3PO_4 85% in D_2O), converted into a doublet by irradiation at either δ 4.06 (6-H) and 4.89 (1′-H). The location of acetoxy group at C-16β was also derived from ^1H and ^{13}C-NMR data. Definitive structural information was obtained through removal of glucose by very mild acid treatment, followed by solvolysis in pyridine-dioxane to remove sulphate.

7. Polyhydroxysteroids

The glycosides of polyhydroxysteroids are often accompanied by various polyhydroxysteroids. Polyhydroxsteroids are not uncommon in marine species, and have been isolated from soft corals, gorgonians, nudibranchs, sponges and ophiuroids; however starfishes appear to be the richest source of new polyhydroxysteroids. They have been found in almost all species examined, quite constantly as complex mixtures; so far more than eighty polyhydroxysteroids from starfishes have been re- ported. The 3β,6α(orβ),8,15α(orβ),16β-pentahydroxycholestane structure is a common feature. The major structural subgroup possesses a 26- hydroxyl function and usually has the 25S-configuration, whereas in a less common subgroup the side chain is hydroxylated at C-24 with the 24S-configuration. Additional hydroxyl groups are found at positions 4β,5α,7α(orβ) and occasionally at 14α, all disposed on one side of the steroidal nucleus, thus bestowing an amphiphilic character on the molecules with an hydrophilic and an hydrophobic region. In part they occur in sulphated form, with the sulphate group located at position 3β,6α,15α or 24.

All such compounds are listed below in the standardized form arranged according to their hydroxylation pattern.

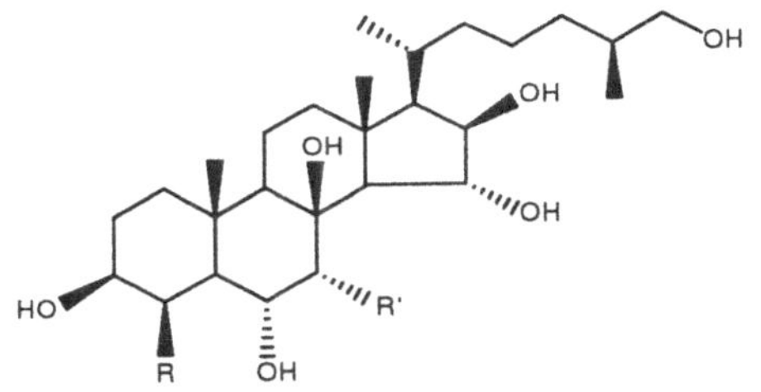

173, R=R'=H

174, R=H; R'=OH

175, R=R'=OH

173   176, R=OH, R'=H

OCCURRENCE: *Protoreaster nodosus (125), Poraster superbus (149), Pentaceraster alveolatus (124), Asterina pectinifera (63,119), Patiria pectinifera (151), Patiria miniata (117)* and *Rosaster sp. (152).*

ACTIVITY: Moderate cytotoxic and antitumoral activities *(36,119)*, no activity in the sea urchin eggs test below 10^{-5} M concentration.

PHYSICAL DATA: m.p. 285–287 °C; $[\alpha]_D = +13.8°$ (MeOH); EI-MS, m/z 450 $[M-H_2O]^+$; selected ^{1}H-NMR signals: δ 0.93 (d,J = 7.0 Hz, 21- or 27-H$_3$), 0.95 (d,J = 7.0 Hz, 27- or 21-H$_3$), 1.03 (s, 19-H$_3$), 1.15 (s, 18-H$_3$), 3.30 (m)-3.44 (dd,J = 10.5, 6.0 Hz, 26-H$_2$, 3.56 (m, 3α-H), 3.64 (dd, J = 9.5, 3.0 Hz, 6β-H), 4.00 (dd, J = 8.5, 2.5 Hz, 16α-H), 4.06 (dd, J = 11.5, 2.5 Hz, 15β-H); ^{13}C-NMR spectrum in Table 6.

174

OCCURRENCE: *Protoreaster nodosus (125), Poraster superbus (149), Pentaceraster alveolatus (124), Asterina pectinifera (63,119), Patiria pectinifera*

(*151*), *Patiria miniata* (*117*), *Rosaster sp.* (*152*), *Pycnopodia helianthoides* (*74*), *Solaster borealis* (*93*).
ACTIVITY: Moderate cytotoxic and antitumoral activities (*36, 119*), no activity in the sea urchin eggs test below 10^{-5} M concentration.
PHYSICAL DATA: m.p. 255–258 °C; $[\alpha]_D$ = + 33.8° (MeOH); EI-MS, m/z 484 [M]$^+$; selected ^{1}H-NMR signals: δ 3.78 (2H, br m, 6β-H and 7β-H), 4.00 (dd, J = 8.5, 2.5 Hz, 16α-H), 4.13 (dd, J = 11.5, 2.5 Hz, 15β-H); ^{13}C-NMR spectrum in Table 6.

175

OCCURRENCE: *Protoreaster nodosus* (*125*), *Poraster superbus* (*149*), *Pentaceraster alveolatus* (*124*), *Asterina pectinifera* (*63, 119*), *Patiria pectinifera* (*151*), *Patiria miniata* (*117*), *Rosaster sp.* (*152*), *Pycnopodia helianthoides* (*74*), *Solaster borealis* (*93*); as 6-O-sulphated in *Oreaster reticulatus* (*116*).
ACTIVITY: Moderate cytotoxic and antitumoral activities (*36,119*), no activity in the sea urchin egg test below 10^{-5} M concentration.
PHYSICAL DATA: m.p. 263–266 °C; $[\alpha]_D$ = + 10.0° (MeOH); FD-MS, m/z 501 [M + H]$^+$, 523 [M + Na]$^+$; selected ^{1}H-NMR signals: δ 1.19 (s, 19-H5), 3.50 (m, 3α-H), 3.86 (d, J = 3.0 Hz, 7β-H), 4.21 (br m, 4α-H and 6β-H); ^{13}C-NMR spectrum in Table 6.

176

OCCURRENCE: *Protoreaster nodosus* (*153*), *Pentaceraster alveolatus* (*124*), *Patiria miniata* (*117*), *Solaster borealis* (*93*).
PHYSICAL DATA: m.p. 241–243 °C; $[\alpha]_D$ = + 28.4° (MeOH); selected ^{1}H-NMR signals: δ 4.01 (dd, J = 8.5, 2.5 Hz, 16α-H), 4.10 (m, 6β-H and 15β-H), 4.28 (br s, 4α-H); ^{13}C-NMR spectrum in Table 6.

177

OCCURRENCE: *Protoreaster nodosus* (*153*).
PHYSICAL DATA: m.p. 253–256 °C; $[\alpha]_D$ = + 27.8° (MeOH); EI-MS, m/z 464 [M–H$_2$O]$^+$; selected ^{1}H-NMR signals: δ 0.82 (d, J = 7.0 Hz, 27- or 28-H$_3$), 0.84 (d, J = 7.0 Hz, 28- or 27-H$_3$), 0.95 (d, J = 7.0 Hz, 21-H$_3$).

178

OCCURRENCE: *Protoreaster nodosus* (*153*).
PHYSICAL DATA: m.p. 185–188 °C; $[\alpha]_D$ = + 20.0° (MeOH); EI-MS, m/z 478 [M–H$_2$O]$^+$; selected ^{1}H-NMR signals: δ 0.91 (d, J = 7.0 Hz, 27-H$_3$),

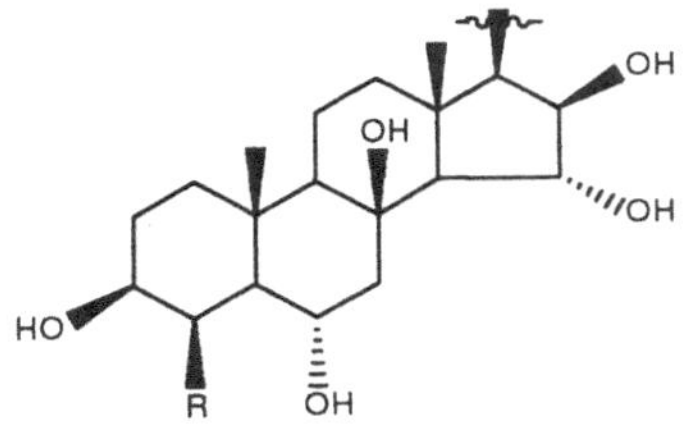

177 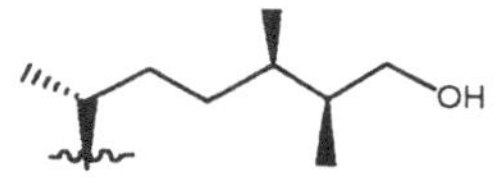 ; R=H

178 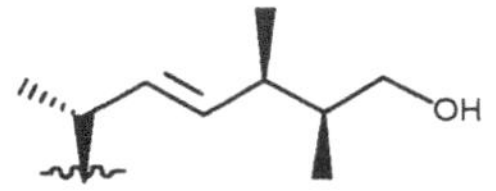 ; R=OH

179 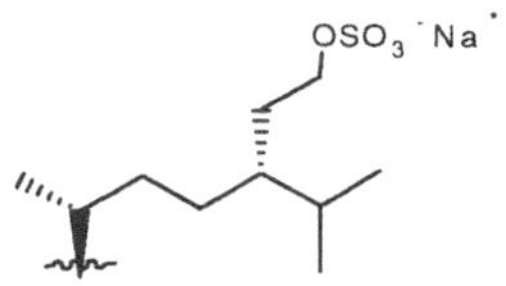 ; R=H

0.99 (d, J = 7.0 Hz, 28-H$_3$), 1.04 (d, J = 7.0 Hz, 21-H$_3$), 1.17 (s, 18-H$_3$), 3.41 (dd, J = 10.5, 6.0 Hz) − 3.56 (dd, J = 10.5, 6.0 Hz, 26-H$_2$), 3.93 (dd, J = 8.0, 2.5 Hz, 16α-H), 4.10 (dd, J = 11.0, 2.5 Hz, 15β-H), 5.45 (m, 22- and 23-H).

179

OCCURRENCE: *Poraster superbus (149)*.
PHYSICAL DATA: $[\alpha]_D$ = + 4.9° (MeOH); FAB-MS (+ ve ion), m/z 621 [M + Na]$^+$, 599 [M + H]$^+$; selected ^{1}H-NMR signals: δ 0.89 (d, J = 7.0 Hz, 26- or 27-H$_3$), 0.92 (d, J = 7.0 Hz, 27- or 26-H$_3$), 0.96 (d, J = 7.0 Hz, 21-H$_3$), 4.06 (m, 29-H$_2$; after desulphation shifted to δ 3.60)

Compounds **173–175** were the first polyhydroxysteroids to be isolated from a starfish, the Pacific Ocean species *Protoreaster nodosus*. They were later found in several other species. Structures were based on the results of NMR and mass spectral analyses, chemical transformations and related spectroscopic data. Acetylation of **173** followed by oxidation with Jones reagent afforded 3β,6α,15α,26-tetra (acetyloxy)-8-hydroxy-5α-cholestan-16-one. The elimination of the side chain with migration of one hydrogen (m/z 464, Mc Lafferty rearrangement) and 18-methyl fission (m/z 449) in the mass spectrum is diagnostic of 16-ketosteroids. Acetylation of **174** and **175** afforded the corresponding tetraacetates (3β,6α,15α,26-acetyloxy) which, on oxidation with Jones reagent, gave the 3β,6α,15α,26-tetra(acetyloxy)-7α,8-dihydroxy-5α-cholestan-16-one, and 3β,6α,15α,26-tetra(acetyloxy)-7α,8-dihydroxy-5α-cholestane-4,16-dione, respectively. Treatment of **173, 174** and **175** with *p*-bromobenzoyl chloride in pyridine led to the formation of the corresponding 3,6,15,26-

tetra (p-bromobenzoates) whose CD curves displayed a strong positive first ($\Delta\varepsilon_{252}$ + 37.2) and negative ($\Delta\varepsilon_{235}$ − 20.0) second Cotton effect in agreement with the clockwise twist (positive chirality) of the three interactions, 3β/6α,3β/15α and 6α/15α-dibenzoates, in a cholestane skeleton with the absolute 5α-H configuration.

Steroids **176–178** were isolated as very minor components from the same species *P. nodosus*.

The 25S configuration is based on the pattern of the 26-methylene proton signals in the 26(+)MTPA and 26(−)MTPA derivatives (*cfr.* section 9.2). The threo configuration at C-24 and C-25 in **177** and **178** is now proposed after stereoselective synthesis of the four stereoisomeric 24-methyl-26-hydroxy steroid models and comparison with their NMR spectra (*140, cfr.* section 9.5). The absolute configuration shown in **177** and **178** is preferred only because of the analogy with echinasteroside A (**152**). The 24R configuration in **179** is suggested on the basis of the reported chemical shifts of the isopropyl methyls (*cfr.* section 9.4).

180

OCCURRENCE: *Halityle regularis* (*112*), *Nardoa gomophia* (*70*), *Pycnopodia helianthoides* (*74*), *Dermasterias imbricata* (*139*), *Astropecten scoparius* (*84*).
ACTIVITY: Inhibits cell division of fertilized sea urchin eggs at 10^{-5} M concentration (25% inhibition) (*35*).
PHYSICAL DATA: $[\alpha]_D$ = 0° (MeOH); EI-MS, m/z 450 $[M-H_2O]^+$; selected ^{1}H-NMR signals: δ 0.95 (d, J = 7.0 Hz, 21- or 27-H_3), 0.98 (d, J

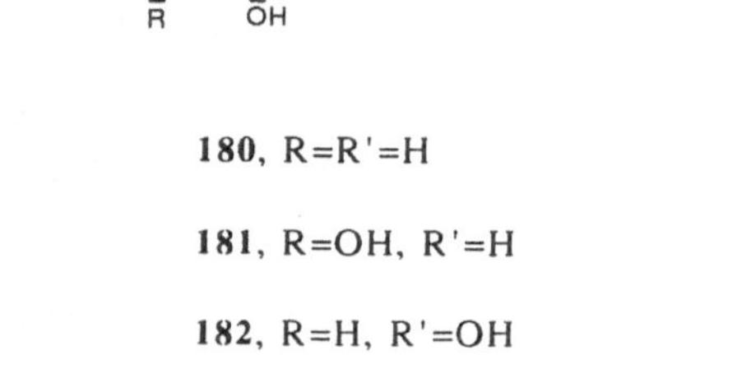

180, R=R'=H

181, R=OH, R'=H

182, R=H, R'=OH

183, R=R'=OH

$= 7.0$ Hz, 27- or 21-H$_3$), 1.02 (s, 19-H$_3$), 1.27 (s, 18-H$_3$), 3.62 (m, 3α-H), 3.74 (td, $J = 10.5$, 4.0 Hz, 6β-H), 4.25 (t, $J = 6.8$ Hz, 16α-H), 4.40 (dd, $J = 6.8$, 5.6 Hz, 15α-H).

181

OCCURRENCE: *Halityle regularis* (*112*), and as the 3-sulphated derivative from *Coscinasterias tenuispina* (*42*).

PHYSICAL DATA: $[\alpha]_D = + 6.3°$ (MeOH); EI-MS, m/z 484 [M]$^+$; selected ^{1}H-NMR signals: δ 1.19 (s, 19-H$_3$), 1.27 (s, 18-H$_3$), 3.50 (m, 3α-H), 4.22 (td, $J = 10.5$, 4.0 Hz, 6β-H), 4.29 (br s, 4α-H), in 3-sulphated δ 4.16–4.31 (3H, br m, 3α-H, 6β-H, 16α-H), 4.63 (br s, 4α-H); ^{13}C-NMR spectrum in Table 6.

182

OCCURRENCE: *Pycnopodia helianthoides* (*74*), *Asterina pectinifera* (*63*), *Astropecten scoparius* (*84*).

ACTIVITY: Inhibits cell division of fertilized sea urchin eggs at 10^{-5} M concentration (25% inhibition) (35).

PHYSICAL DATA: $[\alpha]_D = + 16.0°$ (MeOH); FAB-MS (+ ve ion), m/z 485 [M + H]$^+$; selected ^{1}H-NMR signals: δ 1.42 (d, $J = 5.5$ Hz, 14-H), 3.87 (dd, $J = 12.5$, 2.5 Hz, 6β-H), 3.90 (d, $J = 2.5$ Hz, 7α-H), 4.25 (t, $J = 6.5$ Hz, 16α-H), 4.50 (dd, $J = 6.5$, 5.5 Hz, 15α-H); ^{13}C-NMR spectrum in Table 6.

183

OCCURRENCE: *Asterina pectinifera* (*63, 119*), *Patiria miniata* (*117*), *Solaster borealis* (*93*), *Astropecten scoparius* (*84*); also isolated as the 6-O-sulphate from *Asterina pectinifera* (*63*).

ACTIVITY: Inhibits cell division of fertilized sea urchin eggs at 10^{-5} M concentration (50% inhibition) and also has antifungal activity (minimum inhibitory amount against Cladosporium cucumerinum, 10 μg) (35).

PHYSICAL DATA: $[\alpha]_D = + 0.4°$ (MeOH); FAB-MS (+ ve ion), m/z 523 [M + Na]$^+$, 501 [M + H]$^+$; selected ^{1}H-NMR signals: δ 1.19 (s, 19-H$_3$), 1.28 (s, 18-H$_3$), 3.47 (m, 3α-H), 3.98 (d, $J = 2.5$ Hz, 7β-H), 4.22 (br s, 4α-H), 4.32 (td, $J = 11.0$, 2.5 Hz, 6β-H); ^{13}C-NMR spectrum in Table 6.

Compounds 180–183 differ from 173–176 in being 15β,16β-dihydroxy instead of 15α,16β-dihydroxy derivatives. The *cis*-15,16-dihydroxy stereochemistry was supported by formation of a monoacetonide on treatment of the hexaol 180 with Me$_2$CO and TsOH. On similar treatment the heptaol 181 afforded a bis-acetonide.

184

OCCURRENCE: *Dermasterias imbricata* (*139*).
ACTIVITY: Antifungal activity (minimum inhibitory amount against *Cladosporium cucumerinum*, 5 μg) (35).
PHYSICAL DATA: $[\alpha]_D = +7.1°$ (MeOH); FAB-MS (− ve ion), m/z 501 [M − H]⁻; selected ¹H-NMR signals: δ 1.00 (d, $J = 7.0$ Hz, 27-H₃), 1.02 (s, 19-H₃), 1.04 (d, $J = 6.5$ Hz, 21-H₃), 1.05 (d, $J = 6.0$ Hz, 14-H), 1.31 (s, 18-H₃), 4.40 (br t, $J = 6.0$ Hz, 15α-H), 5.31 (m, 22-H and 23-H).

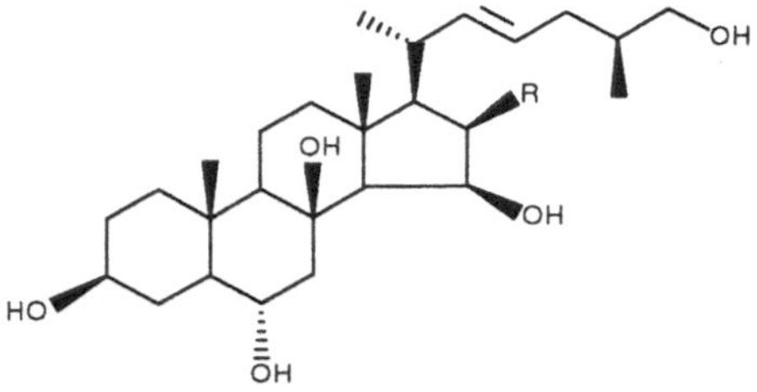

184, R=H

185, R=OH

185

OCCURRENCE: *Dermasterias imbricata* (*139*).
PHYSICAL DATA: $[\alpha]_D = +12.0°$ (MeOH); FAB-MS (− ve ion), m/z 465 [M–H]⁻; selected ¹H-NMR signals: δ 1.30 (s, 18-H₃), 4.15 (t, $J = 6.7$ Hz, 16α-H), 4.38 (dd, $J = 6.7, 7.0$ Hz, 15α-H).

Steroid **184** has the same nuclear hydroxylation pattern as 24-hydroxysteroid **227** from *Gomophia watsoni*, this latter also being the aglycone of many glycosides. Steroid **185** is the Δ²²-analog of **180**. The E configuration was deduced from the ¹³C-NMR data.

186

OCCURRENCE: *Dermasterias imbricata* (*139*).
PHYSICAL DATA: $[\alpha]_D = +3.5°$ (MeOH); FAB-MS (− ve ion), m/z 479 [M–H]⁻; selected ¹H-NMR signals: δ 1.00 (d, $J = 7.5$ Hz, 21-H₃), 3.41 (dd, $J = 11.0, 7.5$ Hz) – 3.63 (dd, $J = 11.0, 6.0$ Hz, 26-H₂), 4.78 (br s) – 4.85 (br s) (28-H₂); ¹³C-NMR spectrum in Table 6.

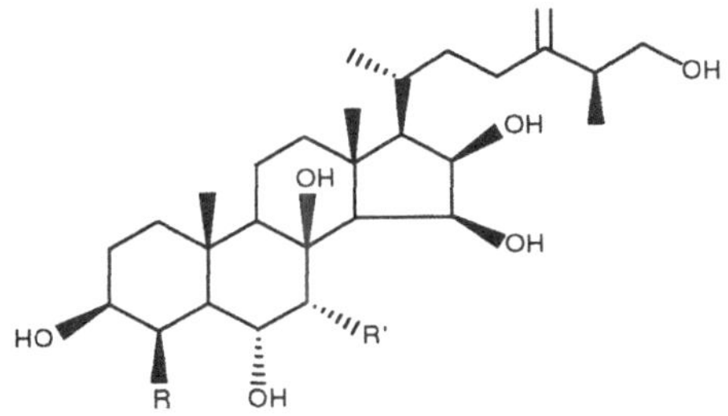

186, R=R'=H

187, R=OH, R'=H

188, R=R'=OH

187

OCCURRENCE: *Culcita novaeguineae* (*44*).
PHYSICAL DATA: $[\alpha]_D = -10.0°$ (MeOH); FAB-MS (+ ve ion), m/z 627
[M + Na + thioglycerol]$^+$, 605 [M + H + thioglycerol]$^+$.

188

OCCURRENCE: *Patiria miniata* (*117*), *Astropecten scoparius* (*84*), *Solaster borealis* (*93*).
PHYSICAL DATA: $[\alpha]_D = +8.5°$ (MeOH); FAB-MS (+ ve ion), m/z 643
[M + Na + thioglycerol]$^+$, 621 [M + H + thioglycerol]$^+$.

Steroids **186–188** are 24-methylene analogs of **180, 181** and **183**. The presence of the exomethylene function gives rise to very intense pseudomolecular ions associated with thioglycerol in the positive ions FAB mass spectra (glycerol/thioglycerol matrix), whereas the spectra run in the negative ion mode are usually quite poor. The chemical shift differences of the 26-methylene protons in the 26(+)MTPA (δ 4.26 dd – 4.31 dd) and 26(–) MTPA (δ 4.20 dd – 4.41 dd) esters were used to determine the 25S configuration of steroid **188** (*117*, cfr. section 9.2).

189

OCCURRENCE: *Dermasterias imbricata* (*139*).
PHYSICAL DATA: $[\alpha]_D = +41.0°$ (MeOH); FAB-MS (– ve ion), m/z 483
[M–H]$^-$; selected ^{1}H-NMR signals: δ 1.04 (s, 19-H$_3$), 1.35 (s, 18-H$_3$), 4.00

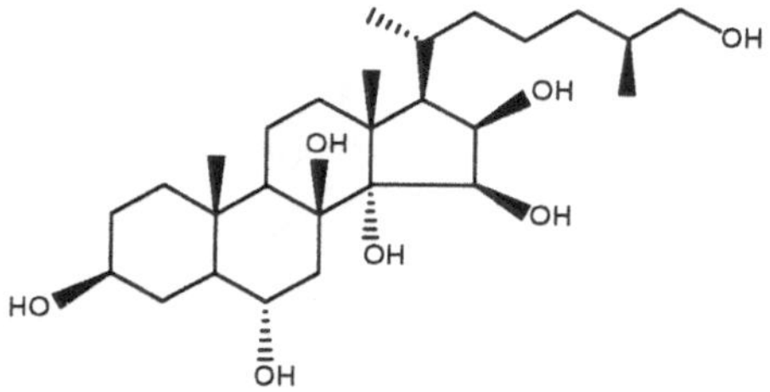

189

190, 24-methylene

(d, $J = 6.5$ Hz, 15α-H), 4.37 (t, $J = 6.5$ Hz, 16α-H); ^{13}C-NMR spectrum in Table 6.

190

OCCURRENCE: *Dermasterias imbricata (139)*.
ACTIVITY: Inhibits cell division of fertilized sea urchin eggs at 10^{-5} M concentration (50% inhibition).
PHYSICAL DATA: $[\alpha]_D = 30.5°$ (MeOH); FAB-MS (− ve ion), m/z 495 [M–H]$^-$.

Large upfield shifts were found for C-7 (− 4.6 ppm), C-9 (− 8.9 ppm), C-12 (− 5.4 ppm) and C-17 (− 10.0 ppm) in the 14-hydroxysteroids **189** and **190** compared with their 14-deoxy counterparts **180** and **186** (Table 6).

191

OCCURRENCE: *Patiria miniata (117)*.

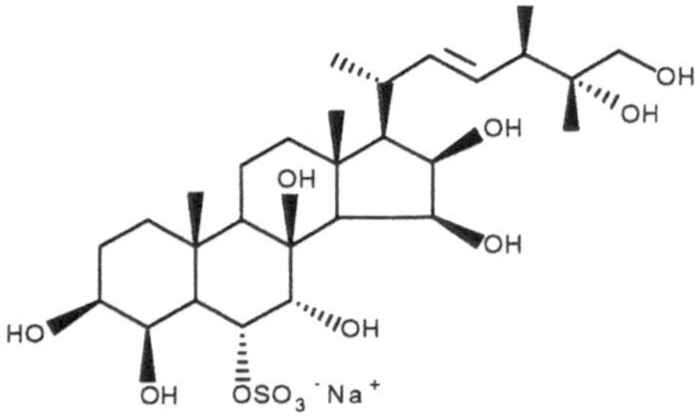

191

PHYSICAL DATA: $[\alpha]_D = +22.9°$ (MeOH); FAB-MS (− ve ion), m/z 607 $[M_{SO_3^-}]$; selected ^{1}H-NMR signals: δ 1.01 (d, $J = 7.0$ Hz, 28-H$_3$), 1.07 (d, $J = 7.0$ Hz, 21-H$_3$), 1.10 (s, 27-H$_3$), 3.53 (m, 3α-H), 4.14 (t, $J = 7.0$ Hz, 16α-H), 4.26 (br s, 4α-H), 4.30 (d, $J = 3.0$ Hz, 7β-H), 4.50 (dd, $J = 7.0$, 5.5 Hz, 15α-H), 5.09 (dd, $J = 11.5$, 2.5 Hz, 6β-H), 5.54 (m, 22-H and 23-H).

The configuration of the stereogenic carbons in the side chain is proposed by comparing the ^{1}H and ^{13}C-NMR data with those of **220** from *Archaster tipicus* (*cfr.* section 9.7).

192

OCCURRENCE: *Hacelia attenuata* (*154*).
PHYSICAL DATA: m.p. 197–199°; $[\alpha]_D = 0°$ (MeOH); EI-MS, m/z 452 $[M]^+$; selected ^{1}H-NMR signals: δ 0.93 (d, $J = 7.0$ Hz, 27-H$_3$), 0.95 (s, 18-H$_3$), 0.99 (d, $J = 7.0$ Hz, 21-H$_3$), 1.07 (s, 19-H$_3$), 3.30 (m) − 3.46 (dd, $J = 12.0$, 6.0 Hz, 26-H$_2$), 3.54 (m, 3α-H), 3.76 (br s, 6α-H), 3.80 (dd, $J = 10.5$, 3.0 Hz, 15β-H), 4.00 (dd, $J = 8.5$, 3.0 Hz, 16α-H); ^{13}C-NMR spectra in Table 6.

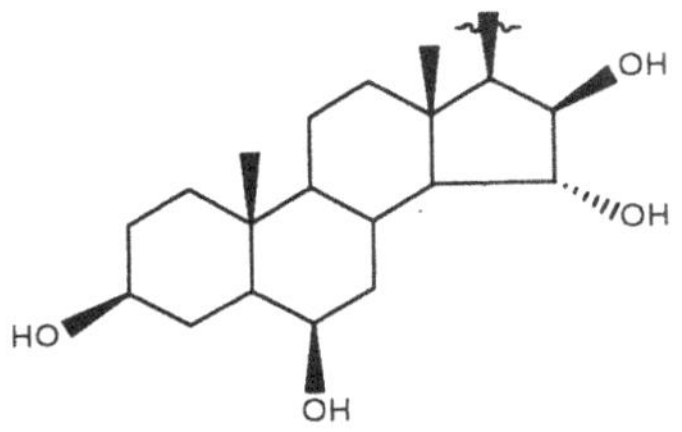

192

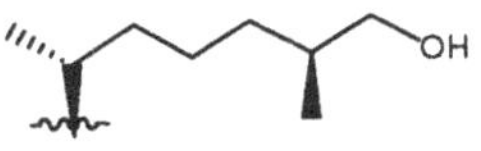

193

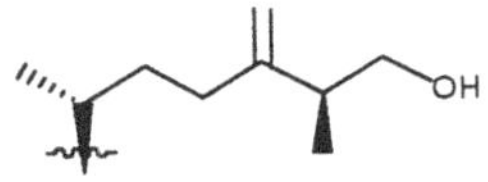

193

OCCURRENCE: *Sphaerodiscus placenta* (*134*).
PHYSICAL DATA: $[\alpha]_D = +28.1°$ (MeOH); EI-MS, m/z 449 $[M^+-CH_3]$.

Acetylation of 192 followed by oxidation with Jones reagent led to the formation of 3β,6β,15α,26-tetra(acetyloxy)-5α-cholestan-16-one. This transformation was accompanied by downfield shifts in the resonances of 15β-H to δ 4.92 as a doublet with $J = 14$ Hz, and of the 18-methyl to δ 1.01 (δ 0.93 in the tetraacetate), consistent with a β-oriented hydroxyl group on C-16.

194

OCCURRENCE: *Luidia maculata* (*155*), *Solaster borealis* (*93*), *Rosaster sp.* in the 15-sulphated form (*152*) and *Myxoderma platyacanthum* (*85*) in both 15-sulphated and non-sulphated form.

PHYSICAL DATA: $[\alpha]_D = +12.1°$ (MeOH); EI-MS, m/z 468 $[M]^+$ (21%), 450 $[M^+-H_2O]$ (50%); 432 $[M^+-2H_2O]$ (60%), 414 $[M^+-3H_2O]$ (100%); selected ^{1}H-NMR signals: δ 0.94 (18-H$_3$), 1.21 (s, 19-H$_3$), 3.50 (br s, 6α-H), 3.76 (dd, $J = 10.5, 3.0$ Hz, 15β-H), 4.00 (dd, $J = 8.5, 3.0$ Hz, 16α-H), 4.04 (m, 3α-H); ^{13}C-NMR spectrum in Table 6.

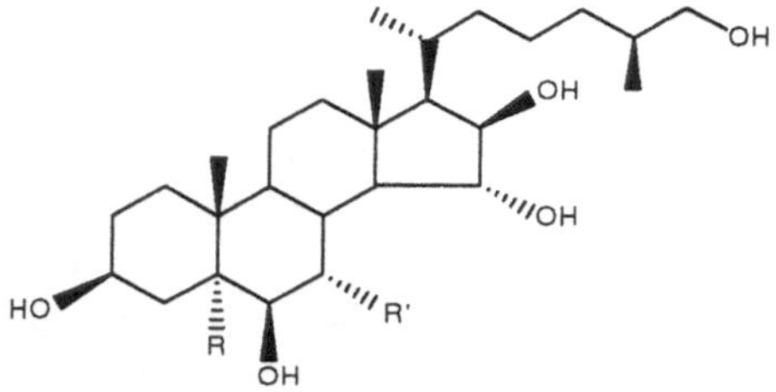

194, R=OH, R'=H

195, R=H, R'=OH

196, R=R'=OH

195

OCCURRENCE: *Luidia maculata* (*155*).

PHYSICAL DATA: m.p. 238–241°C; $[\alpha]_D = +3.8°$ (MeOH); EI-MS, m/z 450 $[M^+-H_2O]$ (50%); 432 $[M^+-2H_2O]$ (60%), $[M^+-3H_2O]$ (100%); selected ^{1}H-NMR signals: δ 0.96 (18-H$_3$), 1.04 (s, 19-H$_3$), 3.60 (m, 3α-H), 3.62 (t, $J = 3.5$ Hz, 6α-H), 3.91 (t, $J = 3.5$ Hz, 7β-H), 3.89 (dd, $J = 10.5, 3.0$ Hz, 15β-H), 4.04 (dd, $J = 7.5, 3.0$ Hz, 16α-H); ^{13}C-NMR spectrum in Table 6.

196

OCCURRENCE: *Luidia maculata* (*155*).

PHYSICAL DATA: m.p. 243–246°C; $[\alpha]_D = -4.7°$ (MeOH); EI-MS, m/z 484 $[M]^+$ (< 1); 448 $[M^+-2H_2O]$ (100%); selected ^{1}H-NMR signals: δ 1.17 (s, 19-H$_3$), 3.55 (d, $J = 3.0$ Hz, 6α-H), 4.02 (t, $J = 3.0$ Hz, 7β-H); ^{13}C-NMR spectrum in Table 6.

The 3β,5,6β-trihydroxy functionality of **194** is a common element in marine polyhydroxysteroids and is also encountered in many poly-hydroxysteroids from starfishes. Structures of **194–196** were deduced from ^{1}H and ^{13}C-NMR spectral data and comparison with reference steroids and within this series.

The 25S configuration is based on the ^{1}H-NMR pattern of the 26-methylene proton signals in the 26(+)MTPA and 26(−)MTPA deriv-atives (for more detailed discussion see (*85*) *cfr.* also section 9.2).

197

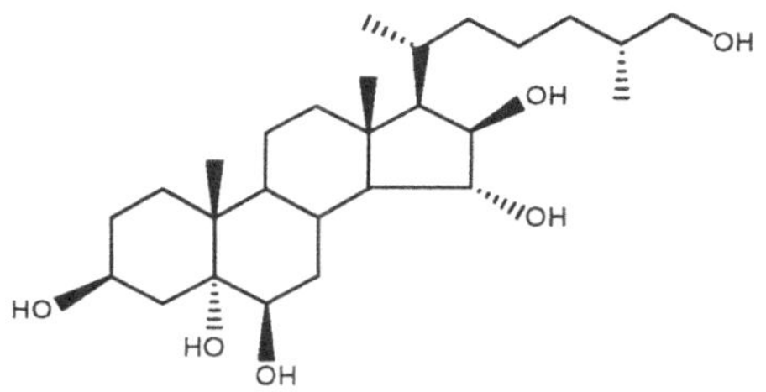

197

OCCURRENCE: *Tremaster novaecaledoniae* (*156*), also present as the 15-sulphated derivative.

197 differs from its 25S-isomer **194** only in the chemical shifts of the side chain carbons, especially C-24, C-26 and C-27. The Δδ values of the corresponding carbons are so small [δ_C 34.1, 68.4 and 17.3 for the (25S)-isomer and δ_C 34.7, 68.6 and 17.1 for the (25R)-isomer] that the differenti-ation can be accomplished only by measuring the spectrum of the mixture. Very small differences are also observed in the ^{1}H-NMR spectra of the two isomers; at 500 MHz the 27-methyl protons are seen at δ 0.934 ppm in the spectrum of the (25S)-isomer (**194**), whereas in that of the (25R)-isomer the same signal is observed at δ 0.925 ppm.

In the ^{1}H-NMR spectra of 26(+)- and 26(−)-MTPA derivatives of **197** the resonances of the 26-methylene protons are much closer [δ 4.18 dd − 4.23 dd] in the (−)MTPA ester than [δ 4.13 dd − 4.26 dd] in the (+)MTPA ester, thus reversing the behaviour of the MTPA esters of the (25S)-isomer **194**, and confirming the 25R configuration in steroid **197** from *T. novaecaledoniae*. The 25R configuration was also found in the steroid **198** from the same source. These compounds are the only examples with the 25R-configuration among the 26-hydroxysteroids from starfishes.

198

Occurrence: *Tremaster novaecaledoniae (156)*.
Physical data: $[\alpha]_D = +32.1°$ (MeOH); FAB-MS (− ve ion), m/z 451
$[M-H]^-$; selected ^{1}H-NMR signals: δ 0.945 (18-H$_3$), 1.15 (s, 19-H$_3$), 3.54
(m, 3β-H), 3.74 (br s, 6α-H).

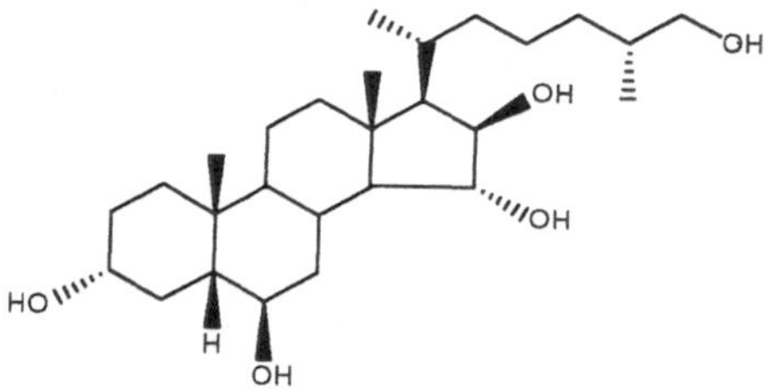

198

The appearance of the angular methyl carbon-19 at lower field 26.1
ppm, indicated that **198** was a 5β-H steroid. The multiplet at δ 3.54 ppm
had the shape characteristic 3α-hydroxy-5β-steroid and the narrowed
signal at δ 3.74 ppm was assigned to an equatorial proton (6β-OH). The
alternative 11β-position is ruled out because of the chemical shift of the
18-methyl signal at δ 0.945 and the upfield shift of a carbon triplet
methylene at 21.4 ppm, typical for C-11 in a steroid skeleton.

The 25R-configuration assigned to **198** is based on the same evidence
mentioned in the case of **197**.

199

Occurrence: *Sphaerodiscus placenta (134)*, *Crossaster papposus (157)*,
Culcita novaeguineae (44).
Physical data: EI-MS, m/z 450 $[M-H_2O]^+$ (80%); selected ^{1}H-NMR
signals: δ 1.15 (s, 18-H$_3$), 1.20 (s, 19-H$_3$), 3.89 (br s, 6α-H).

200

Occurrence: *Rosaster sp. (152)*, *Culcita novaeguineae (44)*.
Physical data: $[\alpha]_D = +5.5°$ (MeOH); FAB-MS (− ve ion), m/z 483
$[M-H]^-$; selected ^{1}H-NMR signals: δ 1.15 (s, 6H, 18- and 19-H$_3$), 3.72
(dd, $J = 3.1$, 2.9 Hz, 6α-H), 3.87 (d, $J = 3.1$ Hz, 7β-H), 4.21 (dd, $J = 10.0$,
2.5 Hz, 15β-H), 4.00 (dd, $J = 7.5$, 2.5 Hz, 16α-H); ^{13}C-NMR spectrum in
Table 6.

References, pp. 297–308

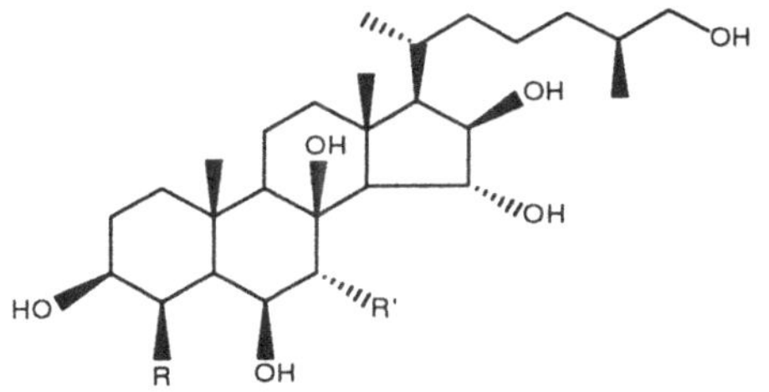

199, R=R'=H

200, R=H, R'=OH

201, R=R'=OH

201

OCCURRENCE: *Rosaster sp.* (*152*).

ACTIVITY: Possesses antifungal activity (minimum inhibitory amount against *Cladosporium cucumerinum*, 5μg) (*35*).

PHYSICAL DATA: $[\alpha]_D = +9.0$ (MeOH); FAB-MS (− ve ion), m/z 499 [M−H]⁻; selected ¹H-NMR signals: δ 1.15 (s, 18-H₃), 1.43 (s, 19-H₃), 3.50 (m, 3α-H), 3.85 (d, J = 3.0 Hz, 7β-H), 4.04 (dd, J = 3.1, 2.9 Hz, 6α-H), 4.10 (br s, 4α-H); ¹³C-NMR spectrum in Table 6.

Steroids **199–201** differ from the previous series **173–175** only in the stereochemistry at C-6, which in **199–201** is 6β-hydroxy.

202

OCCURRENCE: *Sphaerodiscus placenta* (*134*).
¹³C-NMR spectrum in Table 6.

203–205

OCCURRENCE: *Hacelia attenuata* (*130*).

Steroids **203–205** are very minor components of *Hacelia attenuata* (*130*), that they contain a shortened side chain was suggested mainly by analysis of their ¹H-NMR spectra. The 26-hydroxymethyl-27-nor-cholestane side chain of **203** and **204** received further support from the presence of a methylene carbon signal at 62.1 ppm in the ¹³C-NMR spectrum, in agreement with the predicted chemical shift (60.4 ppm) and with the chemical shift of 62.0 ppm assigned to the similarly placed C-29

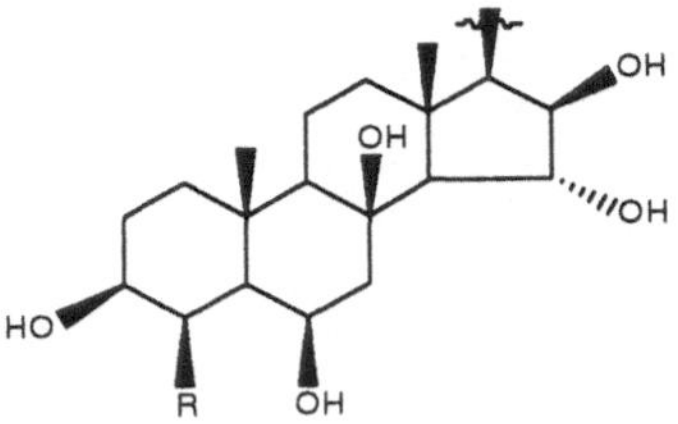

202 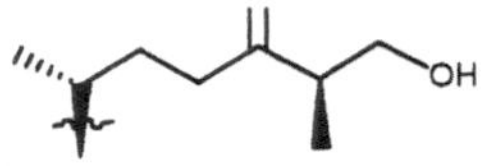   ; R=H

203   ; R=H

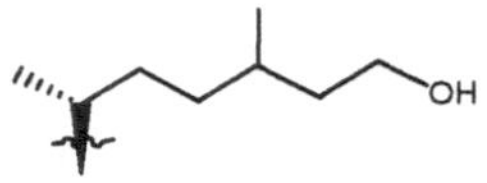

204   ; R=OH

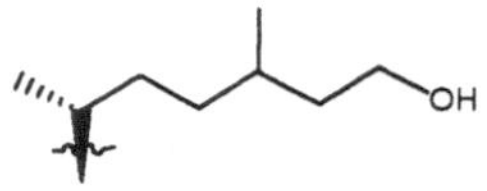

205   ; R=OH

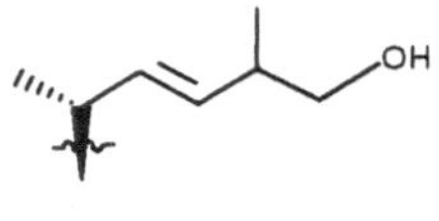

of 29-hydroxysteroids. Their occurrence may be of some interest as an indication of the capability of the starfish to oxidize dietary sterols.

206

Occurrence: *Solaster borealis* (*93*).
Physical data: $[\alpha]_D = +0°$ (MeOH); FAB-MS (− ve ion), m/z 483 [M–H]$^-$; selected ^{1}H-NMR signals: δ 1.18 (s, 18-H$_3$), 1.30 (s, 19-H$_3$), 3.62

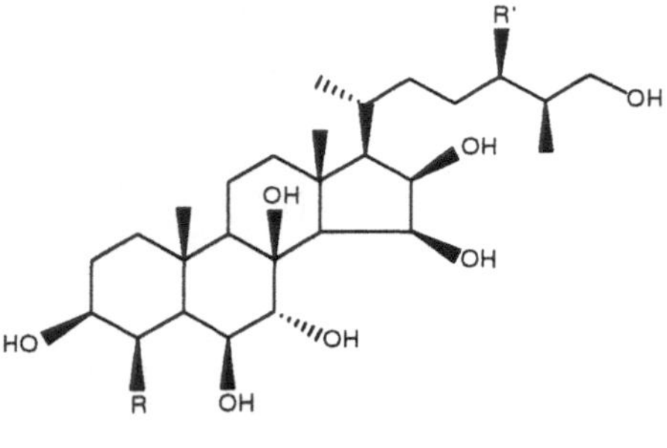

206, R=R'=H

207, R=OH, R'=H

208, R=OH, R'=H, Δ^{22E}

209, R=OH, R'=CH$_3$

(dd, $J = 3.0$, 2.9 Hz, 6α-H), 4.03 (d, $J = 3.0$ Hz, 7β-H), 4.25 (t, $J = 6.2$ Hz, 16α-H), 4.50 (dd, $J = 6.2$, 5.0 Hz, 15α-H).

207

OCCURRENCE: *Solaster borealis* (*93*).
PHYSICAL DATA: $[\alpha]_D = +10.0°$ (MeOH); FAB-MS (− ve ion), m/z 499 [M−H]⁻; selected ¹H-NMR signals: δ 1.30 (s, 18-H_3), 1.45 (s, 19-H_3), 4.00 (dd, $J = 3.0$, 2.9 Hz, 6α-H), 4.02 (d, $J = 3.0$ Hz, 7β-H), 4.08 (br s, 4α-H); ¹³C-NMR spectrum in Table 6.

208

OCCURRENCE: *Solaster borealis* (*93*).
PHYSICAL DATA: $[\alpha]_D = -4.7°$ (MeOH); FAB-MS (+ ve ion), m/z 499 [M + H]⁺; selected ¹H-NMR signals: δ 3.46 (1H, dd, $J = 10.5$, 6.2 Hz) − 3.55 (1H, dd, $J = 10.5$, 6.2 Hz, 26-H_2), 5.50 (dt, $J = 14.8$, 7.0 Hz, 23-H), 5.61 (dd, $J = 14.8$, 8.0 Hz, 22-H).

209

OCCURRENCE: *Solaster borealis* (*93*).
PHYSICAL DATA: $[\alpha]_D = +0°$ (MeOH); FAB-MS (− ve ion), m/z 513 [M−H]⁻

This group of steroids, only recently isolated from *Solaster borealis* (*93*), represents a further variation in the stereochemistry of the nuclear hydroxylation pattern. They are epimeric with the previous series 180–183, differing only in the stereochemistry at C-6, which in 206–209 is 6β-OH; they are equally epimeric with the previous series 199–201 differing only in the stereochemistry at C-15, which in 206–209 is 15β-OH. The 25S-configuration assigned to 206–208 is based on the MTPA method (*cfr.* section 9.2). The *threo*-configuration at C-24 and C-25 of 209 was assigned by spectral comparison with stereoselectively synthesized model compounds, while the (24R,25S)-configuration was assigned upon derivatization with chiral (+)MTPA chloride and ¹H-NMR spectral comparison with the 26(+)MTPA esters of the *threo* model compound pair (*i.e.* 24R,25S- and 24S,25R-isomers) (*cfr.* section 9.5).

210

OCCURRENCE: *Myxoderma platyacanthum* (*85*).
PHYSICAL DATA: $[\alpha]_D = +13.0°$ (MeOH); FAB-MS (− ve ion), m/z 451

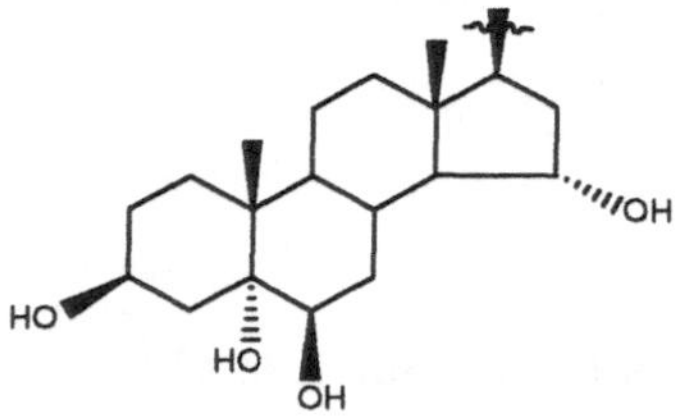

210

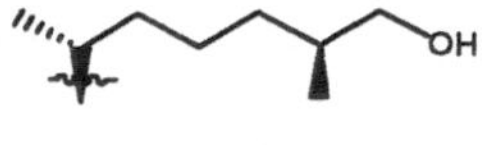

211, 15-O-sulphated

212

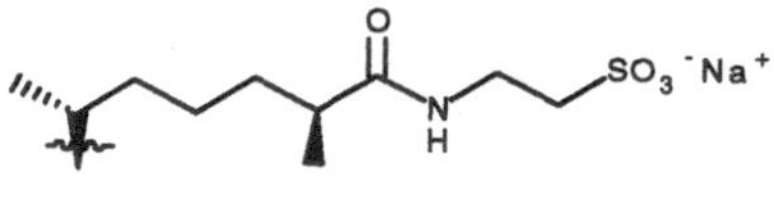

213

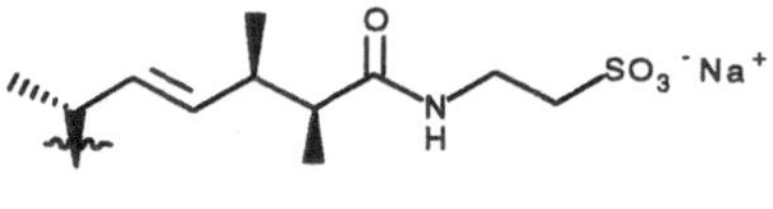

214

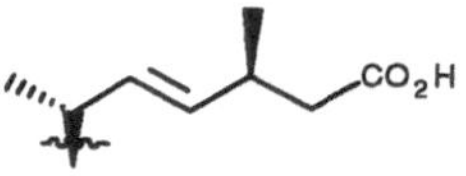

215

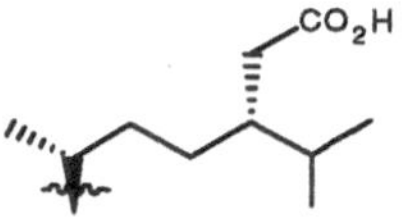

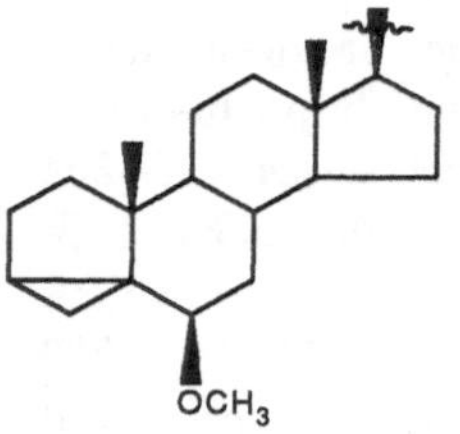

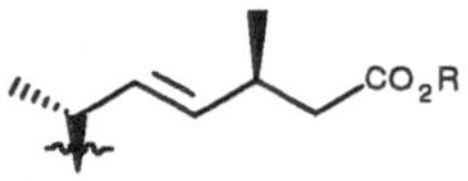

A, R=Et   B, R=Et

A$_1$, R=H   B$_1$,.R=H

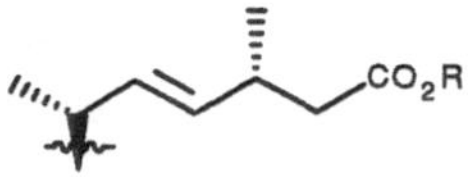

[M–H]$^-$; selected ^{1}H-NMR signals: δ 0.76 (s, 18-H$_3$), 1.20 (s, 19-H$_3$), 3.51 (br s, 6α-H), 3.89 (dt, J = 3.0, 9.0 Hz, 15β-H), 4.04 (m, 3α-H).

211

Occurrence: *Myxoderma platyacanthum* (85).
Physical data: [α]$_D$ = + 2.2° (MeOH); FAB-MS (− ve ion), m/z 531 [M$_{SO_3}$-]; selected ^{1}H-NMR signals: δ 4.51 (dt, J = 3.0, 9.0 Hz, 15β-H).

References, pp. 297–308

212

OCCURRENCE: *Myxoderma platyacanthum* (*85*).
PHYSICAL DATA: $[\alpha]_D = + 18.5°$ (MeOH); FAB-MS (− ve ion), m/z 572 $[M_{SO_3}\text{-}]$, FAB-MS (+ ve ion), m/z 618 $[M_{SO_3Na} + Na]^+$; selected ^{1}H-NMR signals: δ 1.12 (d, J = 7.0 Hz, 27-H$_3$), 2.99 (t, J = 6.0 Hz)–3.66 (t, J = 6.0 Hz) (taurine residue).

213

OCCURRENCE: *Myxoderma platyacanthum* (*85*).
PHYSICAL DATA: $[\alpha]_D = + 10.9°$ (MeOH); FAB-MS (− ve ion), m/z 584 $[M_{SO_3}\text{-}]$, FAB-MS (+ ve ion), m/z 630 $[M_{SO_3Na} + Na]^+$; selected ^{1}H-NMR signals: δ 0.99 (d, J = 7.0 Hz, 21-H$_3$), 1.01 (d, J = 6.5 Hz, 28-H$_3$), 1.10 (d, J = 7.0 Hz, 27-H$_3$), 5.26 (m, 2H, olefinic protons).

214

OCCURRENCE: *Myxoderma platyacanthum* (*85*).
PHYSICAL DATA: $[\alpha]_D = + 18.5°$ (MeOH); FAB-MS (− ve ion), m/z 463 $[M-H]^-$; selected ^{1}H-NMR signals: δ 2.11 (1H, dd, J = 15.0, 7.0 Hz)–2.22 (1H, dd, J = 15.0, 6.0 Hz, 25-H$_2$), 5.31 (2H, m, olefinic protons).

215

OCCURRENCE: *Myxoderma platyacanthum* (*85*).
PHYSICAL DATA: $[\alpha]_D = + 6.7°$ (MeOH); FAB-MS (− ve ion), m/z 493 $[M-H]^-$; selected ^{1}H-NMR signals: δ 0.87 (d, J = 7.0 Hz, 26- or 27-H$_3$), 0.93 (d, J = 7.0 Hz, 27- or 26-H$_3$), 2.32 (1H, dd, J = 15.0, 5.5 Hz, 28-H; the remaining 28-H signal is submerged in the 1.5–2.0 ppm region).

Steroids **210–215** isolated from *Myxoderma platyacanthum* (*85*) along with **194**, the latter in both 15-sulphated and non-sulphated form, are again an example of structural variety from the same organism. This is the first reported isolation from starfishes of sterols with a methyl group oxidized to carboxyl.

The configuration at C-25 of **210** and **211** was determined as usual from the NMR data of their 26(+) and (−)-MTPA esters (*cfr.* section 9.2). The configuration at C-24 and C-25 of **213** was suggested by spectral comparison with stereoisomeric 24-methyl-Δ^{22E}-26-oic acid steroidal models (*cfr.* section 9.5). The configuration at C-24 of **214** was suggested by spectral comparison with (24R)- and (24S)-24-methyl-27-nor-26-oic steroid models synthesized by exposing the *cis*-allylic alcohols **273** (22S) and **274** (22R) to ethyl orthoacetate which afforded the olefinic esters **A** and **B**, respectively, subsequently converted to the corresponding acids. Very small differences were observed in the ^{1}H-NMR spectra of the two synthetic epimers, mainly in the shifts of the olefinic protons, which appear as double doublets at δ 5.32, with the internal lines coincident in **A$_1$**, and separated by 1 Hz in **B$_1$**. In the spectrum of the natural **214** the olefinic pattern is superimposable on that observed in **A$_1$**. Likewise, the proposed configuration at C-24 of **215** is based on comparison with stereoisomeric model compounds (*cfr.* section 9.4).

216

Occurrence: *Archaster typicus* (*158*).
Physical data: $[\alpha]_D = +54.5°$ (MeOH); FAB-MS (+ ve ion), m/z 591 $[M_{SO_3Na} + Na]^+$, 561 $[M_{SO_3Na} + H]^+$; selected ^{1}H-NMR signals: δ 0.90 (d, $J = 7.0$ Hz, 21-H$_3$), 1.06 (s, 19-H$_3$), 1.19 (s, 18-H$_3$), 1.67 (br s, 27-H$_3$), 3.52 (m, 3α-H), 3.67 (dt, $J = 4.0, 11.0$ Hz, 6β-H), 3.94 (br s, 26-H$_2$), 4.97 (15β-H, partially under solvent), 5.40 (br t, $J = 6.5$ Hz, 24-H).

217

Occurrence: *Archaster typicus* (*158*).
Physical data: $[\alpha]_D = +43.6°$ (MeOH); FAB-MS (+ ve ion), m/z 607 $[M_{SO_3Na} + Na]^+$, 585 $[M_{SO_3Na} + H]^+$; selected ^{1}H-NMR signals: δ 1.18 (s, 18-H$_3$), 1.22 (s, 19-H$_3$), 3.48 (dt, $J = 11.0, 4.0$ Hz, 3α-H), 4.13 (dt, $J = 4.0, 11.0$ Hz, 6β-H), 4.28 (br s, 4α-H).

218

Occurrence: *Archaster typicus* (*158*).
Activity: Moderately cytotoxic, caused inhibition of growth of human lymphoma cells at a dose of 0.05 µg/ml (*36*).
Physical data: $[\alpha]_D = +56.2°$ (MeOH); FAB-MS (+ ve ion), m/z 509 $[M + Na]^+$, selected ^{1}H-NMR signals: δ 1.14 (s, 18-H$_3$), 1.33 (s, 19-H$_3$), 3.40 (m, 24-H), 3.98 (d, $J = 4.0$ Hz, 4α-H), 4.06 (m, 3α-H), 4.37 (dd, $J = 12.0, 5.0$ Hz, 6β-H), 4.66 (dd, $J = 4.2, 9.5$ Hz, 15β-H).

219

Occurrence: *Archaster typicus* (*158*).
Physical data: $[\alpha]_D = +40.0°$ (MeOH); selected ^{1}H-NMR signals: δ 1.04 (t, $J = 7.0$ Hz, 26-H$_3$), 2.50 (q, $J = 7.0$ Hz, 25-H$_2$).

216, R=H

217, R=OH

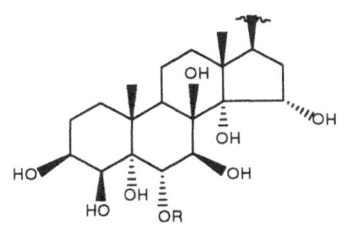

218

219

220

221

222, R=H

223, R=SO$_3^-$Na$^+$

224, R=SO$_3^-$Na$^+$

220

OCCURRENCE: *Archaster typicus* (*159*).
PHYSICAL DATA: m.p. 288–290°C; $[\alpha]_D = +33.3°$ (MeOH); EI-MS, m/z
492 [M–2H$_2$O]$^+$, 474, 456, 438; selected ^{1}H-NMR signals: δ 1.01 (d, J
= 7.0 Hz, 28-H$_3$), 1.12 (s, 27-H$_3$), 3.40 (1H, J = 11.0 Hz)–3.94 (1H, d, J
= 11.0 Hz, 26-H$_2$), 5.27 (dd, J = 15.0, 8.5 Hz, 23-H), 5.44 (dd, J = 15.0,
8.0 Hz, 22-H).

221

OCCURRENCE: *Archaster typicus* (*159*).
PHYSICAL DATA: $[\alpha]_D = + 18.4°$ (MeOH); EI-MS, m/z 492 $[M-2H_2O]^+$; selected ^{1}H-NMR signals: δ 1.18 (s, 26- or 27-H_3), 1.20 (s, 27- or 26-H_3), 3.62 (1H, dd, $J = 11.0$, 7.0 Hz)–3.86 (1H, dd, $J = 11.0$, 6.5 Hz, 28-H_2), 5.25 (dd, $J = 15.0$, 9.0 Hz, 23-H), 5.42 (dd, $J = 15.0$, 8.0 Hz, 22-H).

222

OCCURRENCE: *Archaster typicus* (*158*).
PHYSICAL DATA: m.p. 288–290°C; $[\alpha]_D = + 37.6°$ (MeOH); EI-MS, m/z 466 $[M-2H_2O]^+$, FAB-MS (+ ve ion), m/z 503 $[M + H]^+$, 525 $[M + Na]^+$; selected ^{1}H-NMR signals: δ 4.03 (m, 3α-H), 3.98 (d, $J = 4.0$ Hz, 4α-H), 4.05 (d, $J = 9.0$ Hz, 6β-H), 4.10 (d, $J = 9.0$ Hz, 7α-H), 4.39 (dd, $J = 5.0$, 9.5 Hz, 15β-H).

223

OCCURRENCE: *Archaster typicus* (*158*).
PHYSICAL DATA: $[\alpha]_D = + 56.2°$ (MeOH); FAB-MS (+ ve ion), m/z 643 $[M_{SO_3Na} + K]^+$, 627 $[M_{SO_3Na} + Na]^+$; selected ^{1}H-NMR signals: δ 4.08 (d, $J = 4.0$ Hz, 4α-H), 4.84 (d, $J = 10.0$ Hz, 6β-H), 4.40 (d, $J = 10.0$ Hz, 7α-H).

224

OCCURRENCE: *Archaster typicus* (*158*).
PHYSICAL DATA: $[\alpha]_D = + 47.4°$ (MeOH); FAB-MS (+ ve ion), m/z 671 $[M_{SO_3K} + K]^+$, 655 $[M_{SO_3Na} + K]^+$, 639 $[M_{SO_3Na} + Na]^+$.

Compounds $216 \div 224$ illustrate the variety of highly hydroxylated steroids encountered in *Archaster typicus* (158, 159); they were isolated in relatively large amounts as compared with the very limited fraction of steroidal glycosides. The nonaols **220–224** constitute, as far as we know, the most highly hydroxylated sterols isolated from a natural source whose structures were deduced by spectroscopic methods. ^{1}H–^{13}C-NMR cross correlation spectroscopy (one-bond and long range) and the NOEDS technique were used in the structure elucidation of **222**. In **222**, irradiation of the C-19 methyl signal caused a marked enhancement of the doublet at δ 4.05 (6-H), whereas a substantial enhancement at δ 4.10 (7-H) was observed in irradiation of the 9-H proton signal (δ 2.32 br d, $J = 11.0$ Hz). This showed that the C-6 hydroxyl group was on the α face and that at C-7 on the β face of the steroid skeleton. The 15-proton signal showed intense enhancement when the C-18 methyl signal was irradiated, thereby fixing $18\text{-}H_3$ and the 15-H protons on the same β face of the steroid. On treatment with acetone and TsOH, compound **222** formed a 3β,4β,14α,15α-bis-acetonide, thus confirming the 14α-hydroxy stereochemistry. Application of Horeau's method of kinetic resolution to the bis-acetonide allowed the configuration of C-24 to be assigned as 24R. Structure **222**, which has the remarkable feature of eight sequential hydroxyl groups protruding from the same side of the molecule, was confirmed by X-ray crystallography (*160*). The crystal packing is a consequence of the amphiphilic character of the molecule; the hydroxyl groups form an intricate and extensive network of hydrogen bonds linking the molecules in double layers which interact through their hydrophobic surfaces. Assignment of the stereochemistry at C-24 and C-25 in **220** and at C-24 in **221** required comparison with synthetic model compounds (*159*, *cfr*. section. 9.7).

225

OCCURRENCE: *Asterina pectinifera* (*119*).
ACTIVITY: Weak cytotoxicity, IC_{50} 14 and 23 μg/ml on the growth of L 1210 and KB cells in vitro, respectively.
PHYSICAL DATA: m.p. 236–239°C; $[\alpha]_D = + 53.0°$ (MeOH); FAB-MS (− ve ion), m/z 451 $[M - H]^-$.

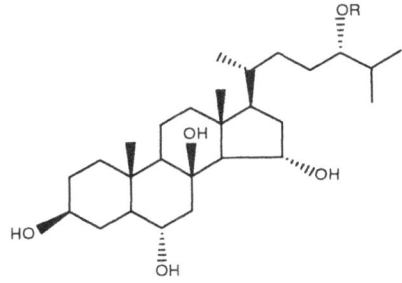

225, R=H

226, R=$SO_3^-Na^+$

226

OCCURRENCE: *Aphelasterias japonica* (*143*).
PHYSICAL DATA: $[\alpha]_D = + 23.5°$ (MeOH); FAB-MS (− ve ion), m/z 531 $[M_{SO_3}-]$; selected ^1H-NMR signals: δ 4.14 (q, $J = 6.5$, 24-H).

The pentaol 225 is the aglycone of the known glycoside astero-saponin P-1 (64).

227

Occurrence: *Gomophia watsoni (131)*, *Nardoa gomophia (70)*, *Dermasterias imbricata (139)*, *Culcita novaeguineae (44)*.
Activity: No activity in the fertilized sea urchin eggs test below 10^{-5} M concentration; antifungal activity (minimum inhibitory amount against *Cladosporium cucumerinum*, 3 µg) (35).
Physical data: $[\alpha]_D = + 12.0°$ (MeOH); EI-MS, m/z 434 $[M^+-2H_2O]$; ^{1}H-NMR, *cfr.* attenuatoside A-II (82).

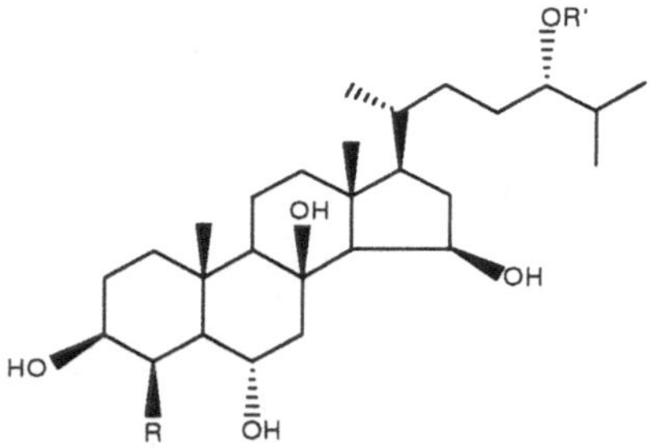

227, R=R'=H

228, R=H, R'=SO$_3^-$Na$^+$

229, R=OH, R'=H

228

Occurrence: *Astropecten scoparius (84)*.
Physical data: $[\alpha]_D = + 8.6°$ (MeOH); FAB-MS (− ve ion), m/z 531 $[M_{SO_3}^-]$.

229

Occurrence: *Gomophia watsoni (131)*, *Culcita novaeguineae (44)*.
Activity: Inhibits cell division of fertilized sea urchin eggs at 10^{-5} M concentration (25% inhibition) (35).
Physical data: $[\alpha]_D = + 7.2°$ (MeOH); EI-MS, m/z 450 $[M^+-H_2O]$; ^{1}H-NMR, *cfr.* attenuatoside B-2 (99).

Steroid **227** is isomeric with **225**, the only difference being the stereochemistry at C-15 which in **227** is 15β-hydroxy as confirmed by formation of a 8,15-phenylboronate. The 24S-configuration in **227** was suggested by the ^{13}C-NMR data in comparison with those of (24S)- and (24R)-hydroxycholestanol. The differences observed between the spectra of the two models are very small (161) and in the desulphated **228**, the 24S configuration was confirmed by using the MTPA method (cfr, section 9.1). Steroids **227** and **228** are the aglycones of many glycosides.

230

OCCURRENCE: *Dermasterias imbricata* (139).
ACTIVITY: Antifungal activity (minimum inhibitory amount against *Cladosporium cucumerinum*, < 1 μg).
PHYSICAL DATA: FAB-MS (− ve ion), m/z 449 [M–H]$^-$; selected ^{1}H-NMR signals: δ 1.33 (s, 18-H$_3$), 3.73 (t, J = 7.0 Hz, 24β-H), 5.39 (dd, J = 16.0, 7.0 Hz, 23-H), 5.46 (dd, J = 16.0, 7.5 Hz, 22-H).

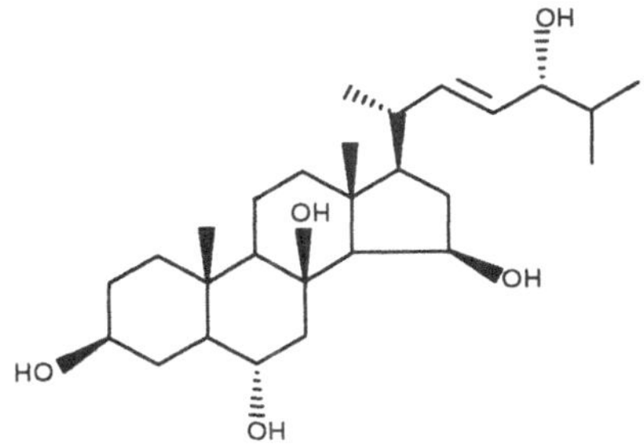

230

Steroid **230** was isolated in admixture with its 22,23-dihydroderivative **227**, a mixture which resisted attempts at separation. The significant ^{1}H-NMR shifts indicative of the presence of the Δ^{22E},24-hydroxy side chain are listed above. Assignment of the 24R configuration is based on the chemical shift of the 24-proton at δ 3.73 and comparison with Δ^{22E},(24R)- and Δ^{22E},(24S)-24-hydroxy model steroids.

231

OCCURRENCE: *Coscinasterias tenuispina* (42).
PHYSICAL DATA: $[\alpha]_D$ = + 8.2° (MeOH); FAB-MS (− ve ion), m/z 561 $[M_{SO_3}-]^-$, FAB-MS (+ ve ion), m/z 607 $[M_{SO_3Na} + Na]^+$, 585 $[M_{SO_3Na}$

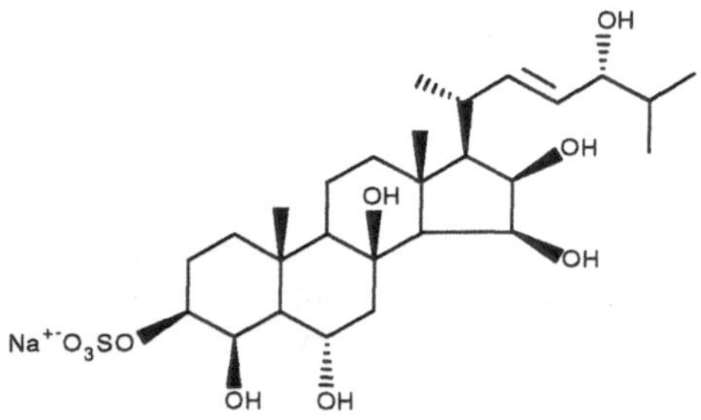

231

+ H]$^+$; selected ^{1}H-NMR signals: δ 3.72 (t, J = 8.0 Hz, 24-H), 4.23 (2H, m, 3α- and 6β-H), 4.16 (t, J = 7.0 Hz, 16α-H), 4.40 (dd, J = 7.0, 5.0 Hz, 15α-H), 4.63 (br s, 4α-H), 5.48 (dd, J = 15.0, 8.0 Hz, 23-H), 5.74 (dd, J = 15.0, 7.5 Hz, 22-H).

The hydroxylation pattern of the tetracyclic nucleus of **231** is common among polyhydroxysteroids and glycosides from starfishes. The 24R configuration is assigned on the same arguments used for **230**.

232

Occurrence: *Coscinasterias tenuispina (42)*.
Physical data: $[\alpha]_D$ = + 10.0° (MeOH); FAB-MS (− ve ion), m/z 529 [M_{SO_3}-]; ^{1}H-NMR, *cfr.* coscinasteroside B (**134**) and scopariosidce D (**135**).

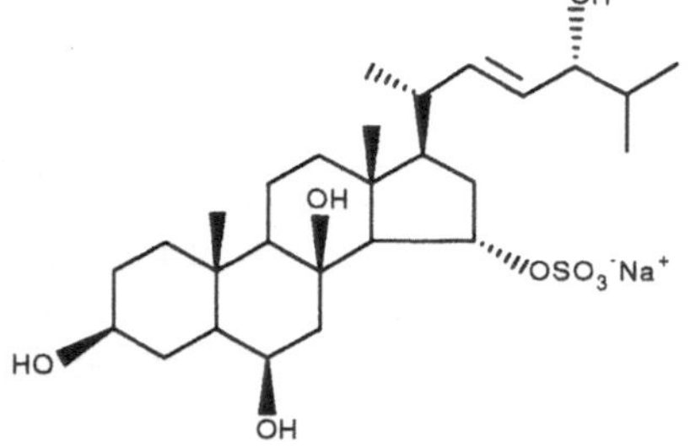

232

The sulphated steroid **232** was later found as the aglycone of scoparioside D (**135**).

233

OCCURRENCE: *Tremaster novaecaledoniae* (*156*).
PHYSICAL DATA: $[\alpha]_D = +25.0°$ (MeOH); FAB-MS (− ve ion), m/z 599 $[M_{(SO_3Na)SO_3}{}^-]$, 577 $[M_{(SO_3Na)SO_3}{}^-]$; selected ^{1}H-NMR signals: δ 0.69 (s, 18-H$_3$), 0.94 (d, $J = 7.0$ Hz, 21-H$_3$), 1.07 (s, 19-H$_3$), 3.60 (br d, $J = 8.5$ Hz, 22-H), 4.23 (m, 3α-H), 4.35 (td, $J = 4.0$, 10.5 Hz, 6β-H), 5.40 (br d, 11-H).

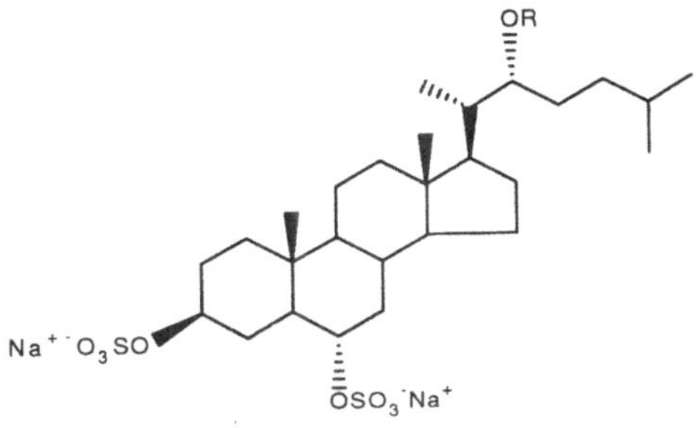

233, R=H, $\Delta^{9(11)}$

234, R=Ac, $\Delta^{9(11)}$

235, R=H

236, R=Ac

234

OCCURRENCE: *Tremaster novaecaledoniae* (*156*).
PHYSICAL DATA: $[\alpha]_D = +27.2°$ (MeOH); FAB-MS (− ve ion), m/z 641 $[M_{(SO_3Na)SO_3}{}^-]$, 619 $[M_{(SO_3H)SO_3}{}^-]$; selected ^{1}H-NMR signals: δ 0.66 (s, 18-H$_3$), 0.98 (d, $J = 7.0$ Hz, 21-H$_3$), 4.90 (br d, $J = 8.5$ Hz, 22-H).

235

OCCURRENCE: *Tremaster novaecaledoniae* (*156*).
PHYSICAL DATA: $[\alpha]_D = +36.4°$ (MeOH); FAB-MS (− ve ion), m/z 601 $[M_{(SO_3H)SO_3}{}^-]$, 579 $[M_{(SO_3H)SO_3}{}^-]$; selected ^{1}H-NMR signals: δ 0.75 (s, 18-H$_3$), 0.94 (s, 19-H$_3$).

236

OCCURRENCE: *Tremaster novaecaledoniae* (*156*).
PHYSICAL DATA: $[\alpha]_D = +21.2°$ (MeOH); FAB-MS (− ve ion), m/z 643 $[M_{(SO_3Na)SO_3}{}^-]$, 621 $[M_{(SO_3H)SO_3}{}^-]$.

This group of disulphated triols was isolated from the "living fossil" species *Tremaster novaecaledoniae* collected at a depth of 530 m off New Caledonia. The chemical shifts of carbons 20 and 23 in the spectra of **233** and **235** were used to determine the 22R configuration. A larger down-field shift (6.8–6.7 ppm) for the C-20 signal and a smaller one (3.6–3.7 ppm) for the C-23 signal relative to the reference cholesterol this indi-cated the 22R configuration. In the spectrum of the (22S)-isomer a larger effect is observed for the C-23 signal and a smaller one for the C-20 signal (*162*). The chemical shift of the 21-methyl signal at δ 0.92 (CDCl$_3$) in the spectrum of 5α-cholest-9(11)-ene-3β,6α,22-triol 22-acetate obtained from **233** by acetylation followed by solvolysis confirmed the 22R configura-tion. In the 22S isomer the C-21 methyl signal is shifted slightly downfield (δ 0.97 ppm) (*163*).

237

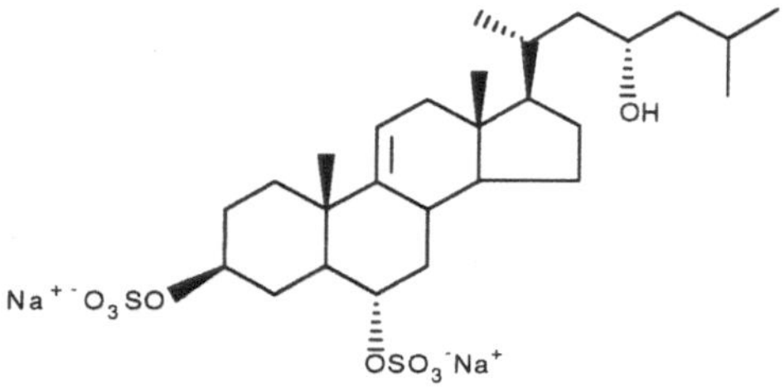

237

OCCURRENCE: *Aphelasterias japonica* (*143*).
PHYSICAL DATA: $[\alpha]_D = +24.2°$ (MeOH); FAB-MS (− ve ion), m/z 599 [M$_{(SO_3Na)SO_3}$−], 577 [M$_{(SO_3H)SO_3}$−]; selected ^{1}H-NMR signals: δ 3.75 (m, 23-H).

The presence of a hydroxyl group at C-23 of the side chain was supported by a signal at 67.0 ppm in the ^{13}C-NMR spectrum. The 23S configuration is suggested by the signal of the C-18 methyl in the spectrum of the benzoate, which is shifted upfield to δ 0.56 (0.70 in **237**), (*164*).

238

OCCURRENCE: *Tremaster novaecaledoniae* (*156*).
PHYSICAL DATA: FAB-MS (− ve ion), m/z 671 [M$_{(SO_3Na)(PO_3H^-)}$], 649 [M$_{(SO_3H)(PO_3H^-)}$]; selected ^{1}H-NMR signals: δ 0.81 (d, $J = 7.0$ Hz, 26- or 27-H$_3$), 0.84 (d, $J = 7.0$ Hz, 27- or 26-H$_3$), 4.06 (dq, $J = 9.5, 4.5$ Hz, 6β-H).

The presence of the phosphate at C-6 initially indicated by the shape of the 6-proton signal, was confirmed by the noise decoupled ^{13}C-NMR

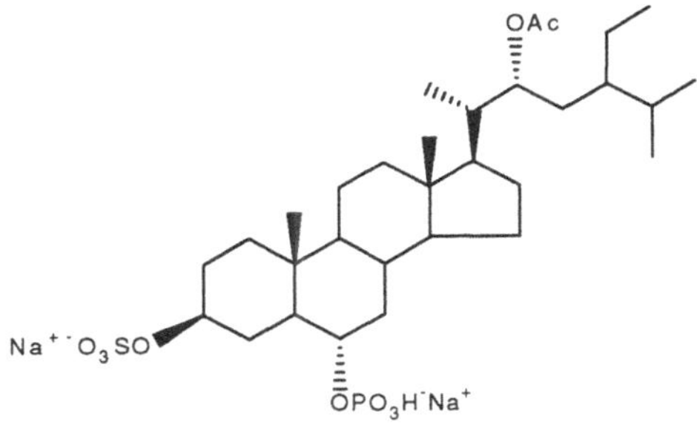

238

spectrum in which the C-5 and C-6 signals appeared as doublets because of $^{31}P-^{13}C$ couplings through two and three bonds.

239–244

OCCURRENCE: *Euretaster insignis (49)*.

This unique group of 3β,21-dihydroxysteroids occurs as a mixture of disulphates, in *Euretaster insignis* which was resistant to attempts at

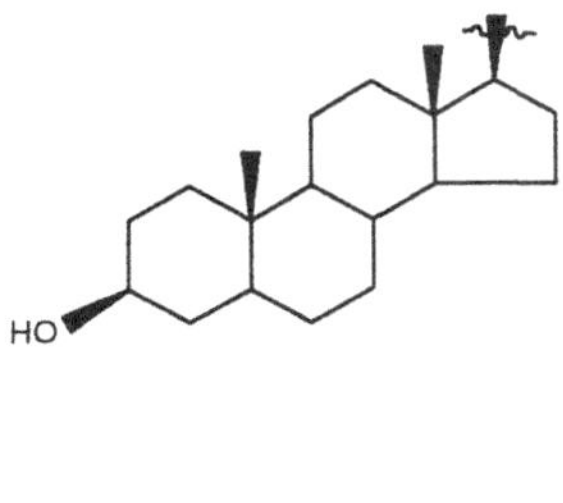

239

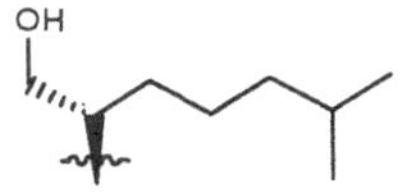

240

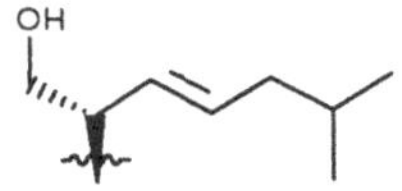

241

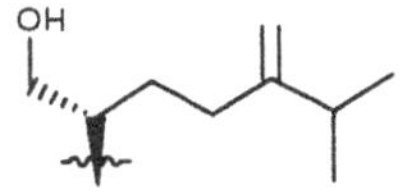

242

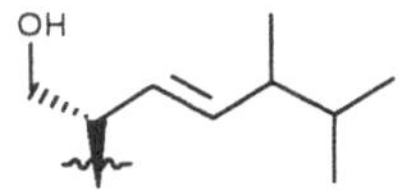

243 Δ⁵,

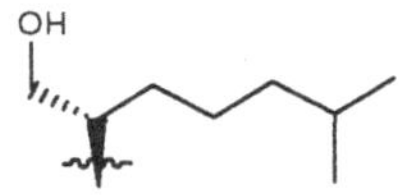

244 Δ⁵,

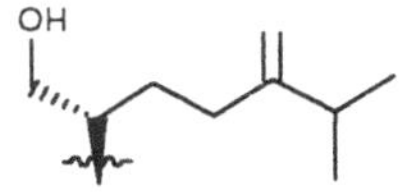

Table 6. *Carbon Chemical Shifts*[a] *of Selected Polyhydroxsteroids*

3β,6β,15α,16β-OH's series   3β,6α,8,15α,16β-OH's series

C	192 (154)	5α-OH 194 (155)	7α-OH 195 (155)	5α,7α-OH 196 (155)	173 (125)	7α-OH 174 (125)	4β,7α-OH 175 (125)	4β-OH 176 (153)
1	39.8	31.7	39.9	31.7	39.6	39.6	39.7	39.9
2	32.2	33.5	32.3	33.6	31.5	31.5	26.1	26.3
3	72.5	68.4	72.5	67.7	72.2	72.3	73.6	73.6
4	36.4	41.6	35.9	41.5	32.4	32.3	69.5	69.2
5	49.0	76.7	42.9	78.2	53.7	44.5	47.9	57.2
6	72.5	76.6	76.4	77.3	67.6	68.9	66.1	64.7
7	40.6	35.4	73.1	74.8	49.0	76.5	76.6	50.5
8	31.2	31.2	35.7	36.1	75.9	77.7	77.6	76.0
9	55.8	46.7	49.0	41.3	57.4	51.2	52.1	58.4
10	36.6	39.5	36.6	39.9	37.8	37.8	37.9	38.0
11	21.9	22.0	21.9	22.0	19.4	19.3	18.6	18.9
12	41.9	42.1	41.9	42.1	43.2	43.2	42.9	43.2
13	44.7	44.9	44.7	44.8	45.3	45.5	45.4	45.4
14	61.1	61.2	56.8	56.7	64.5	59.6	59.5	64.6
15	85.0	85.0	84.2	84.2	80.7	79.3	79.7	80.8
16	82.9	83.2	82.6	82.6	83.0	82.7	82.6	83.1
17	59.9	60.1	60.9	60.9	60.6	61.4	61.3	60.7
18	15.0	15.2	15.0	15.0	16.9	16.9	16.8	16.9
19	16.3	17.2	16.0	17.8	14.2	13.9	16.8	17.0
20	30.9	31.0	30.9	31.0	30.6	30.6	30.5	30.6
21	18.6	18.6	18.6	18.5	18.4	18.4	18.3	18.3
22	37.4	37.5	37.4	37.4	37.2	37.1	36.9	37.2
23	24.8	24.8	24.8	24.8	24.9	24.9	24.8	24.8
24	34.9	35.0	34.9	35.0	35.0	35.0	34.9	35.0
25	37.0	37.0	37.0	37.0	37.1	37.1	37.0	37.0
26	68.4	68.6	68.6	68.6	68.5	68.5	68.5	68.6
27	17.3	17.3	17.3	17.3	17.4	17.4	17.3	17.2
28								

[a]Data (in ppm) were mostly obtained at 62.9 MHz from solution in d_4-methanol and are referred to the central line of the solvent signal (49.0 ppm).

separation. After solvolysis to remove the sulphate groups, the mixture of dihydroxysteroids was fractionated by HPLC to afford pure **243** and two additional fractions still containing a mixture of compounds. Acetylation followed by column chromatography over silica gel impregnated with silver ion afforded the invidual compounds.

Interestingly this starfish is apparently devoid of asterosaponins.

3β,6α,8,15β,16β-OH's series   3β,6β,8,15α,16β-OH's series 3β,6β,8,15β,16β-OH's series

4β-OH	7α-OH	4β,7α-OH		14α-OH		7α-OH	4β,7α-OH	4β,7α-OH
186	**181**	**182**	**183**	**189**	**202**	**200**	**201**	**207**
(139)	(112)	(74)	(63)	(139)	(134)	(152)	(152)	(93)
39.5	39.7	39.4	39.6	39.8	41.4	41.4	41.1	41.0
31.5	26.2	31.5	26.3	31.6	31.7	31.7	26.6	26.7
72.1	73.7	72.3	73.7	72.3	72.5	72.5	73.2	73.0
32.2	69.2	32.2	69.6	32.5	36.4	35.8	77.7	77.7
53.8	57.4	44.5	47.8	53.7	49.0	42.8	45.3	45.2
68.1	64.8	69.5	66.7	68.1	74.2	74.1	80.1	81.2
50.3	50.0	76.8	77.1	45.7	45.5	78.4	73.4	73.4
77.2	77.2	79.2	79.1	80.9	76.9	77.8	78.2	78.0
57.4	58.6	50.3	50.0	48.6	57.3	51.4	52.1	50.8
38.0	38.3	37.6	37.0	38.0	36.7	36.4	36.6	36.6
19.6	18.9	19.2	18.6	18.4	19.7	18.3	19.0	18.6
43.1	43.6	43.3	43.1	38.0	43.3	43.2	43.1	43.0
44.4	44.6	44.5	44.4	47.6	45.4	45.6	45.5	44.9
61.2	61.4	55.4	55.3	82.4	64.0	59.5	59.4	54.8
71.1	71.3	71.3	71.2	76.3	81.0	80.1	80.0	71.2
72.8	72.8	72.8	72,8	72.4	83.1	82.8	82.7	72.6
63.1	63.2	63.2	63.1	53.3	60.6	61.6	61.6	63.2
17.8	17.9	17.8	17.8	17.2	16.8	16.8	16.8	17.8
14.1	16.9	13.9	16.8	14.2	15.8	16.4	19.2	19.1
30.7	31.0	31.0	31.0	30.7	30.6	30.6	30.6	31.0
18.3	18.4	18.4	18.5	18.9	18.4	18.3	18.3	18.5
35.6	37.1	37.1	37.1	37.6	35.6	37.0	37.0	37.1
33.0	24.8	24.8	24.8	24.9	32.9	24.8	24.8	24.8
154.0	35.0	35.0	34.9	35.0	154.0	35.0	35.0	35.0
43.6	37.0	37.0	37.0	37.0	43.5	37.1	37.1	37.0
67.6	68.6	68.6	68.5	68.6	67.6	68.6	68.6	68.5
17.9	17.3	17.3	17.3	17.9	17.2	17.2	17.2	17.3
109.2					109.2			

7.1 Spectroscopy

Selected ^{1}H-NMR data for the polyhydroxylated sterols are given in the previous sections. ^{13}C-NMR chemical shifts for representative series of polyhydroxylated steroidal nuclei are compiled in Table 6. The basis for the assignments was discussed in section 5.1.

8. Steroidal Glycosides and Polyhydroxysteroids from Ophiuroidea (brittle stars).

The study of natural products from starfishes and sea cucumbers has received considerable attention recently especially because of their content of toxic saponins. On the other hand brittle stars (ophiuroids) have received only moderate attention in comparison with the two above mentioned classes. Papers dealing with their sterol content have appeared only sporadically. As a direct consequence of our efforts to isolate biologically active compounds from starfishes, we had occasion to investigate some ophiuroid species from which we were able to isolate a number of sulphated polyhydroxysteroids and two steroidal glycosides. In contrast with the commonly encountered hydroxylation at C-26 among the polyhydroxysteroids from starfishes, the polar steroids from ophiuroids are characterized by hydroxylation at C-21 found only, among starfish metabolites, in the steroids from *Euretaster insignis* (*49*). Hydroxylation at C-11 and C-12, and the *cis*-A/B ring fusion found in many polar steroids isolated from ophiuroids, are further distinctive features.

All compounds are listed below in the usual standardized form.

245 Longicaudoside A

OCCURRENCE: *Ophioderma longicaudum* (*14*).
PHYSICAL DATA: $[\alpha]_D = 0°$ (MeOH); FAB-MS (+ ve ion), m/z 709 $[M_{SO_3Na} + K]^+$, 693$[M_{SO_3Na} + Na]^+$, 671 $[M_{SO_3Na} + H]^+$; selected ^{1}H-NMR signals: δ 0.85 (s, 18-H$_3$), 1.08 (s, 19-H$_3$), 3.55 (dd, J = 11.0, 4.1 Hz, 12α-H), 3.72 (q, J = 3.0 Hz, 6α-H), 3.78 (1H, dd, J = 11.0, 3.2 Hz, 21-H, the other 21-H signal under the methanol signal), 4.72 (br s, 3β-H).

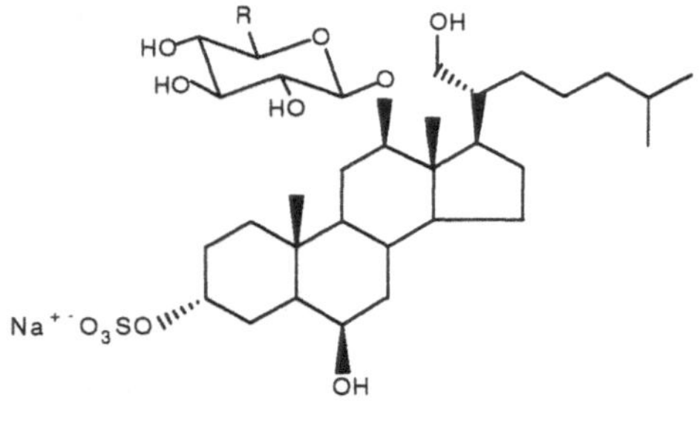

245, R=H

246, R=CH$_2$OH

246 Longicaudoside B

OCCURRENCE: *Ophioderma longicaudum* (*14*).
PHYSICAL DATA: $[\alpha]_D = +3.7°$ (MeOH); FAB-MS (+ ve ion), m/z 723
$[M_{SO_3Na} + Na]^+$, 701 $[M_{SO_3Na} + H]^+$.
The ^{1}H-NMR spectra of both compounds exhibited a narrow low-field signal at $\delta 4.72$, shifted to $\delta 4.10$ in the spectra of the desulphated derivatives indicative of a 3α-OSO$_3$-5α-stanol structure. The chemical shift of carbon-19 at 15.3 ppm eliminated the alternative 3β-OSO$_3$-5β-stanol structure. ^{13}C-NMR signals at 63.4 ppm (CH$_2$-OH) and at 23.0 and 23.1 ppm (26- and 27-H$_3$) established the presence of the 21-hydroxylated side chain. A C-26 hydroxylated side chain exhibits methyl signals at 17.3 and 19.2 ppm and the CH$_2$OH signal at 68.4 ppm.

Acetylation of **245** gave a pentaacetate in which the ^{1}H-NMR signal of 12-H has remained essentially constant at $\delta 3.58$, while the signal of 6-H has shifted from 3.72 ppm to 5.04 ppm. This provided evidence for location of the xylosyl residue at C-12.

Hydrolysis with 50% *aq.* acetic acid at 60°C of the major glucoside **246** gave 5α-cholestane-3α,6β,12β, 21-tetrol-3,6,12,21-tetraacetate. These compounds represent the first occurrence of steroidal glycosides in ophiuroids.

247

OCCURRENCE: *Ophioderma longicaudum* (*165*).
PHYSICAL DATA: selected ^{1}H-NMR signals: $\delta 0.75$ (s, 18-H$_3$), 0.85 (s, 19-H$_3$), 0.91 (6H, d, $J = 7.0$ Hz, 26- and 27-H$_3$), 3.97 (dd, $J = 9.5$, 5.3 Hz)–4.22 (1H, dd, $J = 9.5$, 3.8 Hz, 21-H$_2$), 4.64 (m, 3β-H).

248

OCCURRENCE: *Ophioderma longicaudum* (*165*).
ACTIVITY: Moderate cytotoxic activity (*36*).
PHYSICAL DATA: selected ^{1}H-NMR signals: $\delta 1.06$ (s, 19-H$_3$), 5.34 (br d, $J = 5.5$ Hz, 6-H).

249

OCCURRENCE: *Ophioderma longicaudum* (*165*).
PHYSICAL DATA: selected ^{1}H-NMR signals: $\delta 5.30$ (dd, $J = 16.0$, 10.0 Hz, 22-H), 5.43 (dt, $J = 16.0$, 7.5 Hz, 23-H).

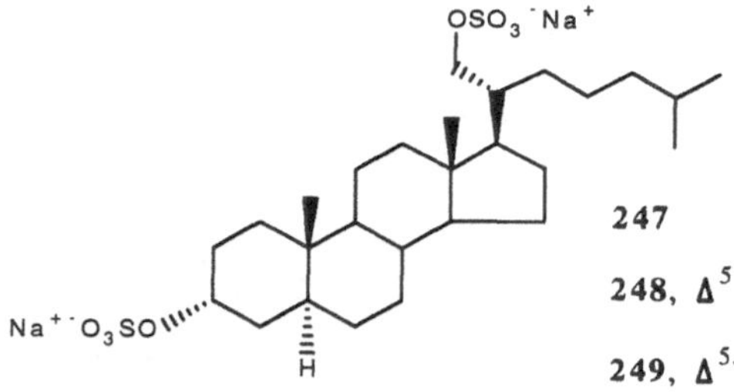

247

248, Δ^5

249, $\Delta^{5,22E}$

250, 24-Me, $\Delta^{5,24(28)}$

250

Occurrence: *Ophioderma longicaudum* (*165*).
Activity: Moderate cytotoxic activity (*36*).
Physical data: selected ^{1}H-NMR signals: δ 1.07 (6H, d, $J = 7.0$ Hz, 26- and 27-H$_3$), 4.70 (1H, br s) – 4.75 (1H, br s, 28-H$_2$).

The ^{1}H-NMR and mass spectra of the desulphated derivatives were also analysed. The ^{13}C-NMR spectra of the major cholestane **247** and its corresponding diol, obtained after solvolytic removal of sulphate, confirmed the 3α-OSO$_3^-$ 5α-cholestanol structure. The shifts of the nuclear carbons in the spectrum of 5α-cholestane-3α,21-diol were virtually identical with those published for 5α-cholestan-3α-ol except that the signal of C-17 was shifted to higher field because of the presence of the C-21 OH group.

251

Occurrence: *Ophioderma longicaudum* (*165*).
Activity: Moderate cytotoxic activity (*36*).
Physical data: m.p. 192–195°C; $[\alpha]_D = + 10.0°$ (MeOH); FAB-MS (− ve ion), m/z 649 [M$_{(SO_3K)(SO_3^-)}$], 633 [M$_{(SO_3Na)(SO_3^-)}$]; selected ^{1}H-NMR

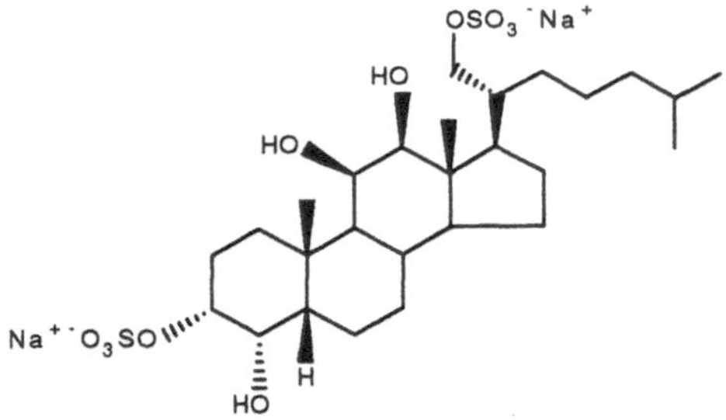

251

signals; δ 0.90 (s, 18-H$_3$), 1.19 (s, 19-H$_3$), 3.28 (d, $J = 3.0$ Hz, 12-H), 3.99 (t, $J = 3.0$ Hz, 11-H), 4.09 (1H, dd, $J = 9.0, 6.0$ Hz, 21-H), 4.19 (3H, m, 3-,4- and 21-H).

The ^{13}C-NMR signal at 27.0 ppm of the 19-methyl carbon suggested the cis-A/B ring fusion. Protons at C-11 and C-12 appeared in the ^{1}H-NMR spectrum as a distinct triplet at δ 3.99 ($J = 3.5$ Hz) and as a doublet at δ 3.28 ($J = 3.5$ Hz) coupled to each other. Protons at C-3 and C-4 overlapped at δ 4.19, but in the spectrum of the derived 5β-cholestane-3α,4α,11β,12β,21-pentaol 3,21-di-p-bromobenzoate, the protons at C-3 and C-4 appeared as a broad doublet at δ 4.99 (3-H) and as a sharp signal at δ 4.14 (4-H). Decoupling experiments demonstrated the presence of a structural element with vicinal hydroxyl groups, one adjacent to a carbon bearing one proton and the other adjacent to one bearing two protons, i.e. at the 3,4-positions. Treatment of the de-sulphated pentaol, m.p. 126–128°C, with Me$_2$CO and TsOH yields the bis-acetonide formed between the OH groups at C-3 and C-4 and C-11 and C-12.

252

OCCURRENCE: *Ophiocoma dentata* (*166*), *Ophioarthrum elegans* (*166*), *Ophiorachna incrassata* (*166*), *Ophiolepis superba* (*167*), *Ophiomastix annulosa* (*168*).

ACTIVITY: Moderate cytotoxic activity (*36*).

PHYSICAL DATA: $[\alpha]_D = +9.1°$ (MeOH); FAB-MS ($-$ ve ion), m/z 631 [M$_{(SO_3K)(SO_3^-)}$], 617 [M$_{(SO_3Na)(SO_3^-)}$]; selected ^{1}H-NMR signals: δ 0.95 (s, 18-H$_3$), 1.18 (s, 19-H$_3$), 3.99 (1H, dd, $J = 10.0, 6.3$ Hz, 21-H), 4.22 (3H, 3-,4- and 21-H).

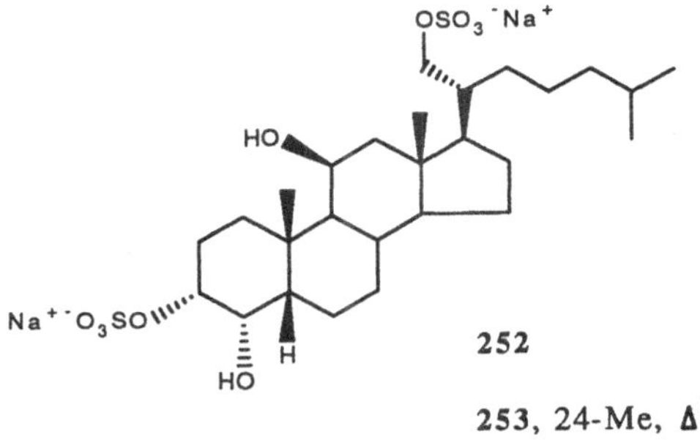

OSO$_3^-$Na$^+$

HO

Na$^+$ $^-$O$_3$SO

HO H

252

253, 24-Me, $\Delta^{24(28)}$

254, Δ^{22E}

253

Occurrence: *Ophiocoma dentata* (*166*).
Physical data: $[\alpha]_D = +15.8°$ (MeOH); FAB-MS (− ve ion), m/z 629 $[M_{(SO_3Na)(SO^-)}]^-$; selected ^{1}H-NMR signals: δ 1.06 (6H, d, $J = 7.0$ Hz, 26- and 27-H$_3$),^{3} 4.72 (1H, br s) − 4.74 (1H, br s, 28-H$_2$).

254

Occurrence: *Ophiocoma dentata* (*166*), *Ophioarthrum elegans* (*166*).
Physical data: $[\alpha]_D = -24.5°$ (MeOH); FAB-MS (− ve ion), m/z 615 $[M_{(SO_3Na)(SO_3^-)}]^-$; selected ^{1}H-NMR signals: δ 3.84 (1H, t, $J = 10.0$ Hz), − 4.25 (1H, m, 21-H$_2$), 5.24 (dd, $J = 15.0, 8.0$ Hz, 22-H), 5.45 (m, 23-H).

The disulphated tetraol **252** is the principal polar steroid constituent of the above ophiuroids, and was characterized by spectral studies. Protons at C-3, C-4, C-11 and C-21 overlap, but in the spectrum of the tetraol derived from **252** by solvolysis the resonances of hydroxy methine protons gave rise to five isolated signals: 3.62 (dt, $J = 12.0, 4.0$ Hz, 3β-H), 3.69 (1H, dd; $J = 11.0, 4.5$ Hz) − 3.74 (1H, dd, $J = 11.0, 3.0$ Hz, 21-H$_2$), 3.90 (t, $J = 4.0$ Hz, 4β-H), 4.22 (q, $J = 2.5$ Hz, 11α-H).

Oxidation of **252** with the Sarret reagent (chromium trioxide in dry pyridine) gave the 11-keto derivative **255**.

255

Occurrence: *Ophioarthrum elegans* (*166*).
Physical data: $[\alpha]_D = -25.0°$ (MeOH); CD, $\Theta_{300} + 765$ (positive max-imum); FAB-MS (− ve ion), m/z 629 $[M_{(SO_3K)(SO_3^-)}]$, 615 $[M_{(SO_3Na)(SO_3^-)}]$; selected ^{1}H-NMR signals: δ 0.68 (s, 18-H$_3$), 1.16 (s, 19-H$_3$), 2.47 (1H, d, $J = 12.5$ Hz), − 2.53 (1H, d, $J = 12.5$ Hz, 12-H$_2$), 3.18 (d, $J = 10.0$ Hz, 9-H).

Reduction of **255** with NaBH$_4$ in MeOH gave the 11β-hydroxy derivative **252**.

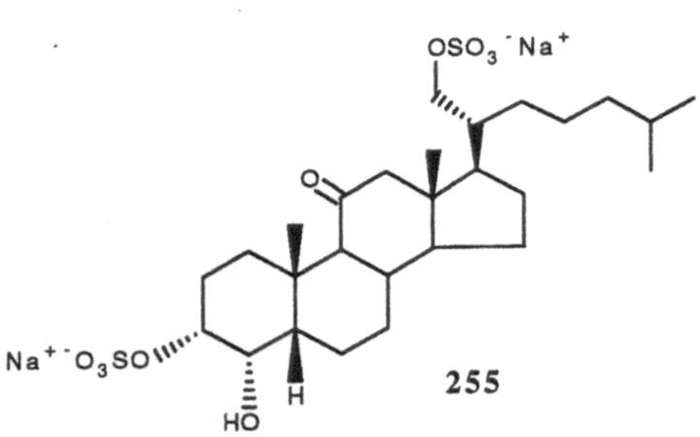

256

Occurrence: *Ophiorachna incrassata* (*166*).
Physical data: FAB-MS (− ve ion), m/z 719 [$M_{(SO_3K)(SO_3Na)(SO_3^-)}$], 703 (major peak) [$M_{(SO_3Na)_2(SO_3^-)}$], 697 [$M_{(SO_3K)(SO_3H)(SO_3^-)}$], 681 [$M_{(SO_3Na)(SO_3H)(SO_3^-)}$]; selected ^{1}H-NMR signals: δ 0.71 (s, 18-H$_3$), 0.96–0.98 (together 3H, d, J = 6.5 Hz, 27-H$_3$), 1.02 (s, 19-H$_3$), 3.80 (1H, dd, J = 10.0, 6.5 Hz, 26-H), 3.90–3.91 (1H, dd, J = 10.0, 5.0 Hz, 26-H), 4.72 (m, 2α-H or 3β-H), 4.74 (m, 3β-H or 2α-H).

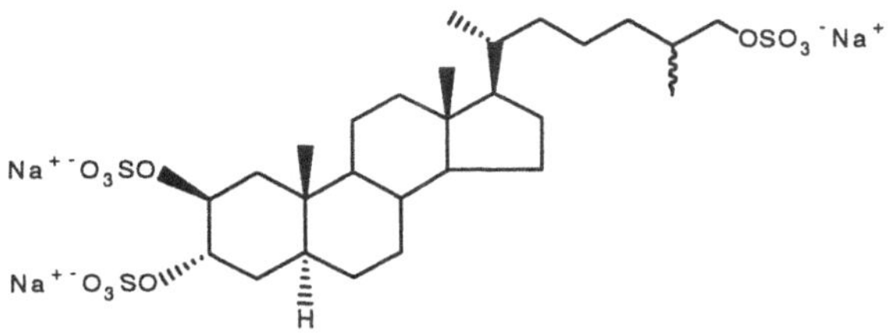

256, mixture of 25R- and 25S- isomers

257, mixture of 25R-and 25S- isomers, Δ^5

257

Occurrence: *Ophiorachna incrassata* (*166*).
Physical data: FAB-MS (− ve ion), m/z 701 [$M_{(SO_3Na)_2(SO_3^-)}$], 695 [$M_{(SO_3K)(SO_3H)(SO_3^-)}$], 679 [$M_{(SO_3Na)(SO_3H)(SO_3^-)}$]; selected ^{1}H-NMR signals: δ 0.75 (s, 18-H$_3$), 1.21 (s, 19-H$_3$), 4.75 (q, J = 2.5 Hz, 2α-H), 4.81 (q, J = 2.5 Hz, 3β-H).

The triol sulphates **256** and **257** represent the first occurrence of 26-hydroxylation in ophiuroids. The rare 2β,3α-disulphoxy feature which was established by sequential decoupling ^{1}H-NMR experiments on the Δ^5 analog **257**, thus revealing the J-connectivity path for C-1 to C-5, had been already found in a bioactive steroid, halistanol sulphate (24,25-dimethyl-5α-cholestane-2β,3α,6α-triyl sodium sulphate) from the sponge *Halichondria panicea* (*169*). The major support for the presence of both 25S- and 25R-epimers is found in the ^{13}C-NMR spectrum in which each side chain carbon signal is split in two peaks separated by 0.04–0.2 ppm.

258

Occurrence: *Ophiolepis superba* (*167*).
Physical data: $[\alpha]_D = +9.1°$ (MeOH); FAB-MS (− ve ion), m/z 615 $[M_{(SO_3K)(SO_3^-)}]$, 599 (major) $[M_{(SO_3Na)(SO_3^-)}]$, 577 $[M_{(SO_3H)(SO_3^-)}]$; selected ^{1}H-NMR signals: δ 0.75 (s, 18-H$_3$), 0.96 (s, 19-H$_3$), 1.74 (s, 27-H$_3$), 3.96 (1H, dd, $J = 10.5$, 6.5 Hz) – 4.22 (1H, dd, $J = 10.5$, 3.7 Hz, 21-H$_2$), 4.19 (2H, m, 3β- and 4β-H), 4.69 (2H, br s, 26-H$_2$).

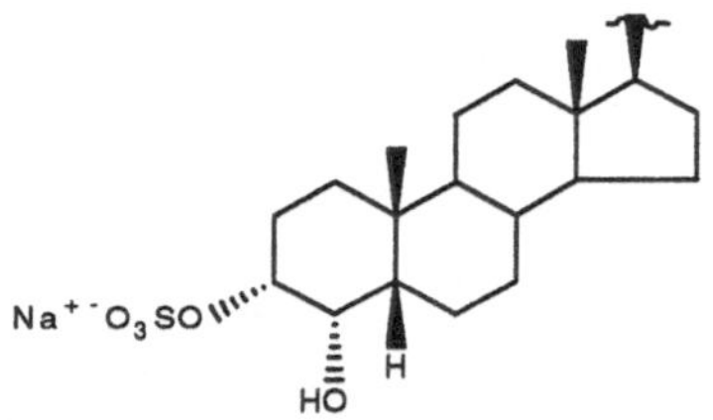

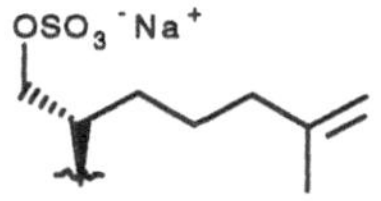

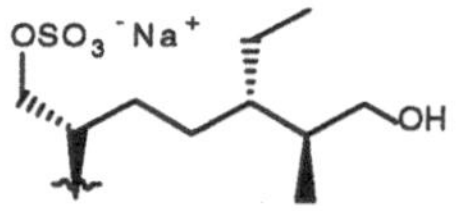

259

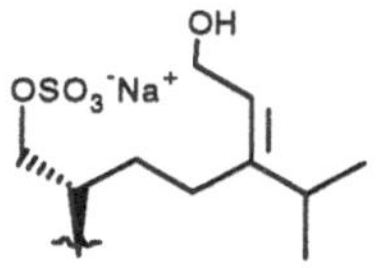

Occurrence: *Ophiolepis superba* (*167*).
Physical data: $[\alpha]_D = +21.9°$ (MeOH); FAB-MS (− ve ion), m/z 643 $[M_{(SO_3Na)(SO_3^-)}]$, 629 $[M_{(SO_3H)(SO_3^-)}]$; selected ^{1}H-NMR signals: δ 1.06 (6H, d, $J = 7.0$ Hz, 26- and 27-H$_3$), 4.18 (d, $J = 7.0$ Hz, 29-H$_2$), 5.34 (t, $J = 7.0$ Hz, 28-H).

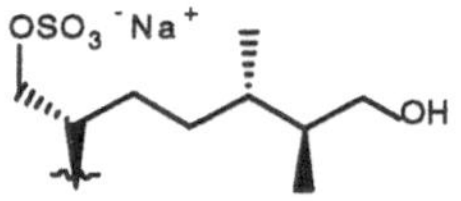

260

Occurrence: *Ophiolepis superba* (*167*).
Physical data: $[\alpha]_D = +31.5°$ (MeOH); FAB-MS (− ve ion), m/z 631 $[M_{(SO_3Na)(SO_3^-)}]$, 609 $[M_{(SO_3H)(SO_3^-)}]$; selected ^{1}H-NMR signals: δ 0.92 (6H, d, $J = 7.0$ Hz, 27- and 28-H$_3$), 3.34 (1H, dd, $J = 11.0$, 6.0 Hz) – 3.62 (1H, dd, $J = 11.0$, 6.0 Hz, 26-H$_2$).

261

OCCURRENCE: *Ophiolepis superba* (*167*).

PHYSICAL DATA: $[\alpha]_D = + 18.8°$ (MeOH); FAB-MS (− ve ion), m/z 661 $[M_{(SO_3K)(SO_3^-)}]$, 645 $[M_{(SO_3Na)(SO_3^-)}]$, 623 $[M_{(SO_3H)(SO_3^-)}]$; selected ^{1}H-NMR signals: δ 0.86 (t, J = 7.0 Hz, 29-H$_3$), 0.92 (d, J = 7.0 Hz, 27-H$_3$), 3.35 (1H, dd, J = 11.0, 7.0 Hz, 26-H$_2$), 3.58 (1H, dd, J = 11.0, 6.0 Hz, 26-H$_2$).

^{1}H and ^{13}C-NMR spectra indicated that compounds **258–261** possessed the same 3α,4α,21-triol-3,21-disulphated 5β-steroidal structure with different side chains. The ^{13}C-NMR signal at 24.0 ppm for the C-19 methyl carbon indicated the *cis* A/B ring fusion. Oxidation of **258** with chromium trioxide-pyridine reagent gave the corresponding 4-keto derivative which exhibited a weakly negative CD curve ($\Delta\varepsilon_{290}$ − 0.11) consistent with a *cis* A/B ring junction, and in whose ^{1}H-NMR spectrum the 19-methyl protons resonated at δ 1.16. In the alternative 3β-hydroxy-5α-cholestan-4-one structure the C-19 methyl protons should resonate at δ 0.79 according to ARNOLD et al. (*170*). The 24(28) double bond in **259** was assigned the E configuration based on the chemical shift of the C-25 proton at δ 2.32 ppm whereas the C-25 proton in the Z isomer should appear at around δ 3.00 ppm. In confirmation irradiation of H-25 proton in a NOEDS experiment produced enhancement of H-28.

After the synthesis of the stereoisomeric model steroids, **260** and **261** were assigned the 24S,25S-configuration (*140, 171*) (*cfr.* sections 9.5, 9.6).

262

OCCURRENCE: *Ophiolepis superba* (*167*).

PHYSICAL DATA: $[\alpha]_D = + 9.1°$ (MeOH); FAB-MS (− ve ion), m/z 631 $[M_{(SO_3K)(SO_3^-)}]$, 615 (major) $[M_{(SO_3Na)(SO_3^-)}]$, 593 $[M_{(SO_3H)(SO_3^-)}]$; selected ^{1}H-NMR signals: δ 0.74 (s, 18-H$_3$), 0.92 (s, 19-H$_3$), 3.99 (m, 4β-H), 4.69 (m, 3β-H).

The 3α,4α,5β-trihydroxycholestane structure was determined by

FAB mass spectroscopy and ^{1}H and ^{13}C-NMR techniques and verified by synthesis.

The synthesis of 5β-cholestane-3α,4α,5-triol started from cholest-4-en-3-one, and that of the alternative 5α-cholestane-3β,4β,5-triol started from 4α,5α-epoxy-cholestan-3β-ol (*167*). The ^{1}H-NMR spectrum of 5β-cholestane-3α,4α,5-triol showed the same shifts and coupling constants as the desulphated **262**, while substantial differences were observed with the spectrum of the isomeric 5α-cholestane-3β,4β,5-triol (*167*).

263

OCCURRENCE: *Ophiolepis superba* (*167*).
PHYSICAL DATA: $[\alpha]_D = + 9.0°$ (MeOH); FAB-MS (− ve ion), m/z 631 [$M_{(SO_3K)(SO_3^-)}$], 615 [$M_{(SO_3Na)(SO_3^-)}$], 593 [$M_{(SO_3H)(SO_3^-)}$]; selected ^{1}H-NMR signals: δ 0.75 (s, 18-H$_3$), 1.02 (s, 19-H$_3$), 4.03 (2H, m, 2α-H and 3β-H), 4.33 (t, $J = 3.4$ Hz, 4β-H).

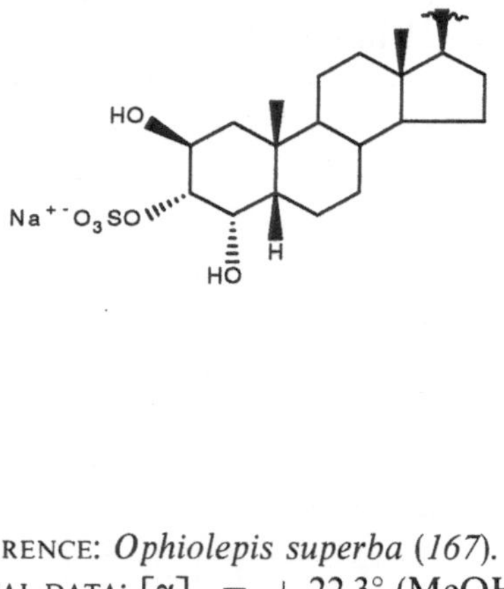

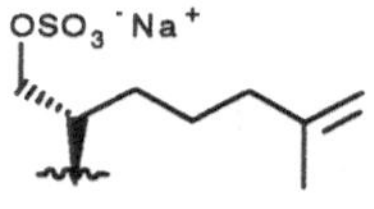

263

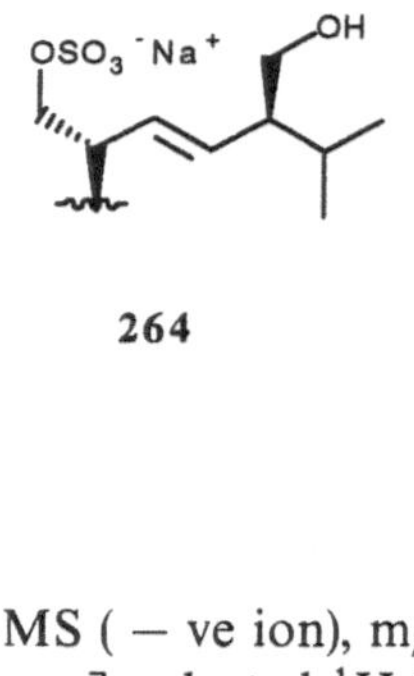

264

264

OCCURRENCE: *Ophiolepis superba* (*167*).
PHYSICAL DATA: $[\alpha]_D = + 22.3°$ (MeOH); FAB-MS (− ve ion), m/z 661 [$M_{(SO_3K)(SO_3^-)}$], 645 [$M_{(SO_3Na)(SO_3^-)}$], 623 [$M_{(SO_3H)(SO_3^-)}$]; selected ^{1}H-NMR signals: δ 0.89 (d, $J = 7.0$ Hz, 26- or 27-H$_3$), 0.95 (d, $J = 7.0$ Hz, 27- or 26-H$_3$), 3.49 (1H, dd, $J = 10.5, 7.5$ Hz) − 3.57 (1H, dd, $J = 10.5, 6.0$ Hz, 28-H$_2$), 5.29 (dd, $J = 14.0, 9.0$ Hz, 22-H), 5.34 (dd, $J = 14.0, 7.0$ Hz, 23-H).

Introduction of an additional hydroxyl group at C-2β of the 3α-sulphoxy-4α-hydroxy-5β-steroidal nucleus leads to the appearance in the ^{1}H-NMR spectrum of a signal at δ 4.03 which overlaps the 3-H signal, while the 4-H and 19-CH$_3$ signals are shifted slightly downfield to δ 4.33

(t, $J = 3.4$ Hz) and 1.02 (s), respectively. The three hydroxymethine protons appeared in the spectrum of the desulphated **263** as isolated signals at δ 3.81 (m), 3.23 (dd, $J = 9.5$, 3.5 Hz) and 3.94 (t, $J = 3.5$ Hz) ppm. Decoupling experiments proved that they were located in a sequential arrangement and allowed deduction of the 2β,3α,4α-trihydroxy-5β-steroidal structure. Comparison of the ^{1}H and ^{13}C-NMR data of **263** and **264** showed conclusively that both have the 2β,3α,4α,21-tetraol, 3,21-disulphate structure. The 24S configuration was assigned to **264** after stereoselective synthesis of (24R,22E)- and (24S,22E)-24-hydroxymethylcholesta-5,22-dien-3β-ol (*172*) (*cfr*. section 9.3).

265

Occurrence: *Ophiura sarsi* (*173*), *Ophiotrix fragilis* (*174*).
Physical data: m.p. 189–190°C; $[\alpha]_D = -15.6°$ (MeOH); selected ^{1}H-NMR signals: δ 0.78 (s, 18-H_3), 1.24 (s, 19-H_3), 4.19 (d, $J = 3.0$ Hz, 4α-H), 4.51 (q, $J = 3.0$ Hz, 3β-H), 5.65 (br s, 6-H).

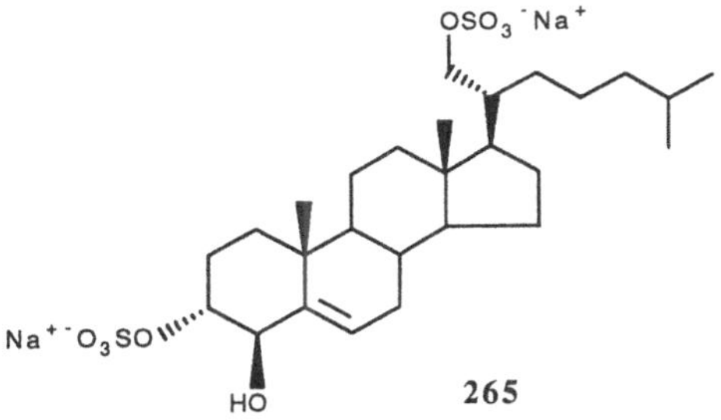

265

266

Occurrence: *Ophiomastix annulosa* (*168*).
Physical data: $[\alpha]_D = +53.4°$ (MeOH); FAB-MS (− ve ion), m/z 615 $[M_{(SO_3K)(SO_3^-)}]$, 599 $[M_{(SO_3Na)(SO_3^-)}]$; selected ^{1}H-NMR signals: δ 0.74 (s, 18-H_3), 1.02 (s, 19-H_3), 4.33 (2H, m, 3β- and 4β-H).

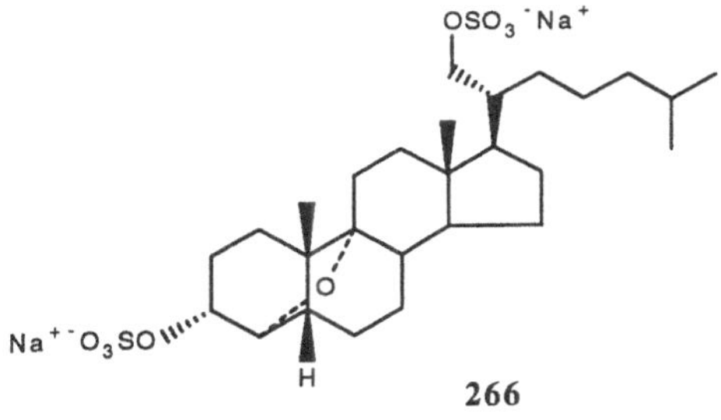

266

267

Occurrence: *Ophiomastix annulosa* (*168*).
Physical data: $[\alpha]_D = +46.0°$ (MeOH); FAB-MS (− ve ion), m/z 615 $[M_{(SO_3K)(SO_3^-)}]$, 599 $[M_{(SO_3Na)(SO_3^-)}]$, 577 $[M_{(SO_3H)(SO_3^-)}]$; selected ^{1}H-NMR signals: δ 0.69 (s, 18-H$_3$), 1.11 (s, 19-H$_3$), 4.12 (t, $J = 3.1$ Hz, 4β-H), 4.22 (m, 3β-H), 5.42 (d, $J = 5.8$ Hz, 11-H).

The unique 4α,9α-epoxide-5α-steroid structure of **266** was determined by detailed spectral analysis. High resolution mass spectrometry of the desulphated **266** indicated C$_{27}$H$_{46}$O$_3$ as the molecular formula which requires one additional unsaturation besides those of the tetracylic nucleus. Carbon signals shifted downfield to 84.5 (CH) and 87.5 (C) ppm, along with the absence of signals for sp^2 carbons indicated that one oxygen was involved in an ether bridge. Further analysis of the ^{13}C-NMR data provided structural information which led to location of the ether bridge between C-4 and C-9 in a 5β-steroid . Protons at C-3 and C-4 in **266** overlapped at δ 4.33 ppm, but in the spectrum of the desulphated **266** the two methine protons resonated at δ 3.62 (dd, $J = 9.6$, 6.4 Hz, 3β-H) and 3.98 (s, 4β-H). The absence of couplings between 3-H/4-H and 4-H/5-H indicated that their dihedral angles are close to 90°. Conformational analysis by MM2 calculations showed strain in the A and B rings.

8.1 Spectroscopy

Selected ^{1}H-NMR data for the polar sulphated steroids from ophiuroids are listed in the previous section. ^{13}C NMR chemical shifts for some representative 5β-H steroid are compiled in Table 7. The basis for the assignments was discussed in section 5.1

Table 7. ^{13}C nmr Shifts[a] of Selected 3α,21-Disulphoxy-4α-Hydroxy-5β-Steroids from Ophiuroids

C	3α,4α,21-OH's 258 (167)	3α,4α,5β,21-OH's 262 (167)	2β,3α,4α,21-OH's 263 (167)	3α,4α,11β,21-OH's 252 (166)	3α,4α,11β,12β,21-OH's 251 (165)
1	36.2	31.0	44.6	36.1	36.0
2	22.4	22.3	65.6	23.5	24.8
3	82.4	79.5	87.0	82.5	82.3
4	75.2	79.5	76.5	75.3	75.2
5	46.7	76.0	46.5	48.8	48.6
6	27.6	37.5	26.9	27.0	26.4
7	28.7	28.5	28.6	28.5	27.0
8	36.3	35.2	36.3	31.8	30.6
9	42.9	41.8	44.0	45.9	45.2
10	36.2	40.8	38.2	36.4	36.4
11	22.5	22.9	22.6	69.0	73.4
12	40.7	40.5	40.7	49.8	81.2
13	43.6	43.5	43.6	42.7	49.0
14	57.8	57.8	57.8	59.3	57.0
15	25.0	25.1	25.1	25.1	24.8
16	29.3	29.4	29.3	29.8	29.5
17	52.2	52.2	52.3	52.8	53.2
18	12.7	12.6	12.7	15.0	10.5
19	24.0	17.7	24.0	27.3	27.5
20	41.3	41.3	41.2	41.5	39.1
21	69.7	69.9	69.9	69.8	72.3
22	30.5	30.5	30.6	31.0	30.1
23	24.8	25.0	25.0	24.7	23.5
24	39.3	39.3	39.3	40.7	40.7
25	147.1	147.2	147.2	29.0	29.1
26	110.3	110.2	110.2	23.1	23.0
27	22.4	22.4	22.4	23.0	23.1

[a] Data (in ppm) were mostly obtained at 62.9 MHz from solution in [$^2H_4$]-methanol and are referred to the central line of the solvent signal (49.0 ppm)

9. Assignment of absolute configurations to stereogenic carbons in the side chain of polyhydroxysteroids

The occurrence of a large variety of steroids with multiple oxygen functionalities and different alkylation patterns in the side chains has often required the assignment of absolute configurations to stereogenic centers. In most cases this involved synthesis of appropriate models and analysis of their ^{1}H and ^{13}C-NMR spectra and of those of their derivatives with a chiral reagent. Details concerning the assignment of absolute configurations are reviewed below.

9.1 24-Hydroxysteroids

Hydroxylation at C-24 is the most common functionality found in the aglycone side chain of glycosides of polyhydroxysteroids and is also found to a minor extent in polyhydroxysterols themselves. The need for determining the absolute configuration arose in the case of nodososide (**146**), the first representative of this group of compounds to be isolated from a starfish. Although (24R)- and (24S)-hydroxycholesterol have been synthesized (*175, 176*), the differences in their ^{1}H and ^{13}C-NMR spectra are so small that a direct comparison of both stereoisomers is necessary for unequivocal determination of stereochemistry. We assigned the 24S configuration to nodososide (**146**) by applying the gas-chromatographic modification of Horeau's method to the (24S) 3β,5,6β,15α-tetrametoxy-5α-cholest-8(9)-en-24-ol obtained by methylation of nodososide and subsequent acid methanolysis (*147*). More recently use of the Mosher method for determining of absolute configuration of secondary carbinols (*177, 178*) was found to be more convenient. This method is based on the observation that ^{1}H-NMR chemical shifts of selected signals of dia-stereotopic groups contiguous to the chiral carbinol become non-equi-valent upon derivatization with chiral α-methoxy-α-(trifluoro-methyl)phenylacetic acid (MTPA, the Mosher reagent). According to the NMR configurational correlation scheme developed by DALE and MO-SHER, in the ^{1}H-NMR spectra of diastereomeric MTPA esters of the chiral carbinol, signals due to the diastereotopic groups will typically be shifted upfield in one diastereomer and downfield in the other in consequence of their interaction with the phenyl ring of the MTPA moiety. We tested the applicability of this useful configurational correla-tion scheme, to the determination of absolute configuration in 24-hydroxysteroids by measuring the ^{1}H-NMR spectra of (R)-(+)-MTPA esters [the term R-(+)- or S-(−)-MTPA ester refers to an ester obtained using the acid chloride prepared from R-(+)- and S-(−)-α methoxy-α-(trifluoromethyl)phenylacetic acid, respectively] of authentic (24S)- and (24R)-6β-methoxy-3α,5-cyclo-5α-cholestan-24-ol. As expected the signals due to the 26- and 27-methyl protons appeared as two upfield doublets at δ 0.83 ($J = 6.5$ Hz) and 0.85 ($J = 6.5$ Hz) in the spectrum of the 24S-isomer and as a downfield six proton doublet at δ 0.91 in the spectrum of the 24R-isomer. From this it follows that in the presence of a single stereoisomer, as is usual in the case of natural products, assignment of configuration at C-24 can be made by comparing the resonances of the 26- and 27-methyl protons in the R-(+)- and (S)-(−)-MTPA derivatives. In 24S-stereoisomers these signals will appear at higher field in the spectrum of the (R)-(+)-MTPA ester and at lower field in that of the (S)-(−)-MTPA ester, while this behaviour will be reversed for 24R-stereo-isomers in which the 26- and 27-methyl protons will resonate at higher field in the spectrum of the (S)-(−)-MTPA derivative. This method has been used to confirm the 24S configuration in several 24-oxygenated steroids from echinoderms and its application is here exemplified by the assignment of configuration at C-24 in amurensoside A (**61**) (Fig. 11). The (24S) 3β,6α,15α-trimethoxy-5α-cholestan-24-ol, obtained by methylation of **61** and successive acid methanolysis, was converted into the diastereomeric (R)-(+)- and (S)-(−)-MTPA esters which showed the signals of the isopropyl methyls significantly upfield (δ 0.84 and 0.86) in the NMR spectrum of the (R)-(+)-MTPA ester and downfield (δ 0.89 and 0.91) in that of the (S)-(−)-MTPA, thus establishing the 24S-configuration in amurensoside A (**61**) (*43*). Recently X-ray analysis has been applied to desulphated asterosaponin P-1 (**64**) confirming the 24S absolute configuration of the steroid aglycone (*193*).

A small number of 24-hydroxylated steroids contain a 22,23(E)-double bond. The configuration at C-24 in these compounds might conceivable be derived in a somewhat simpler fashion because of small but significant differences observed for the 22-, 23- and 24-proton signals in the ^{1}H-NMR spectra of the two synthetic stereoisomers (Fig. 12). However caution has to be used in deducing configurations solely on chemical shift grounds, since substituents on ring D consistently exert an influence on the chemical shifts of side chain signals. Greater confidence in the stereochemical assignment can be had by relying on ^{1}H-NMR spectral data of the R-(+)- and (S)-(−)-MTPA derivatives; indeed chemical shift differences among side chain signals in the two derivatives still comply with the above mentioned correlation as shown in Fig. 13. Alternatively it is possible to make use of the exciton chirality method using allylic benzoates devised by K. NAKANISHI (*179*). The CD curves of

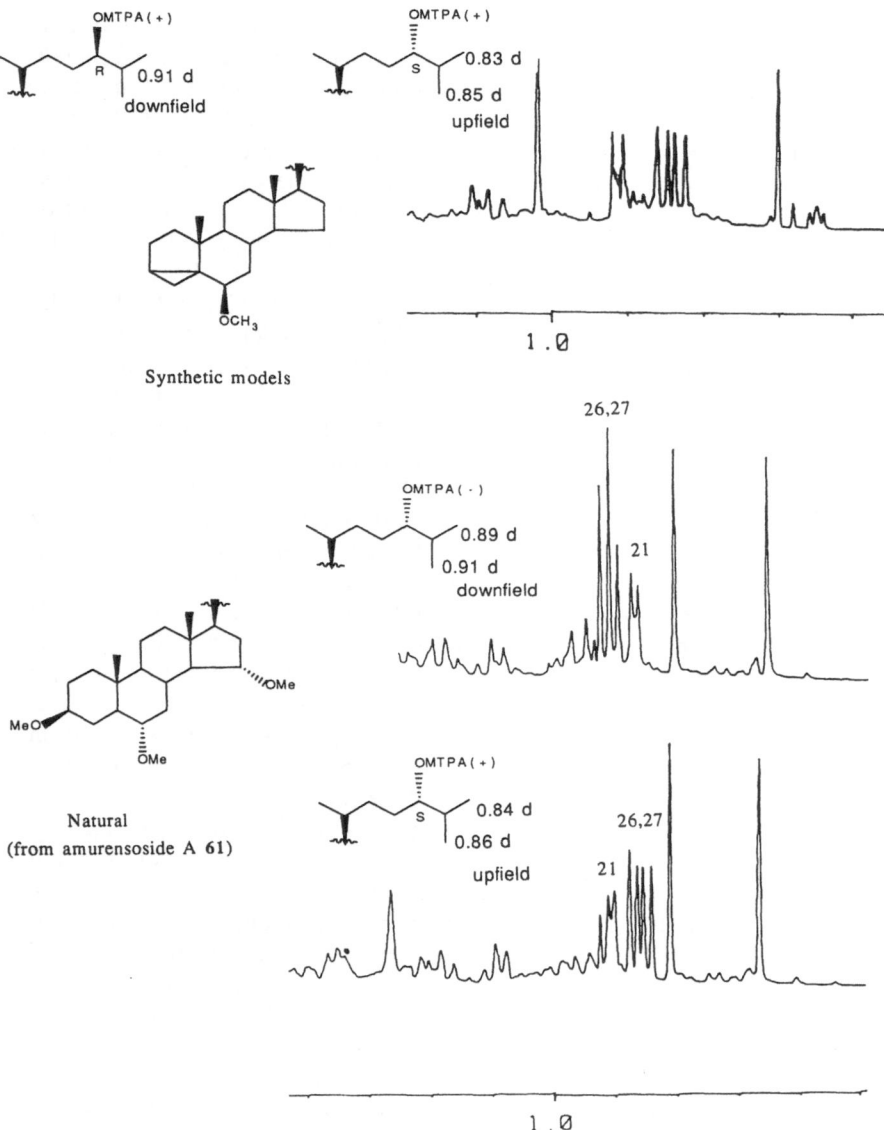

Fig. 11. Configuration at C-24 in 24-hydroxysteroids by ^{1}H NMR data (250 MHz, CD$_3$OD) of (R)-(+)- and (S)-(−)-24-MTPA derivatives

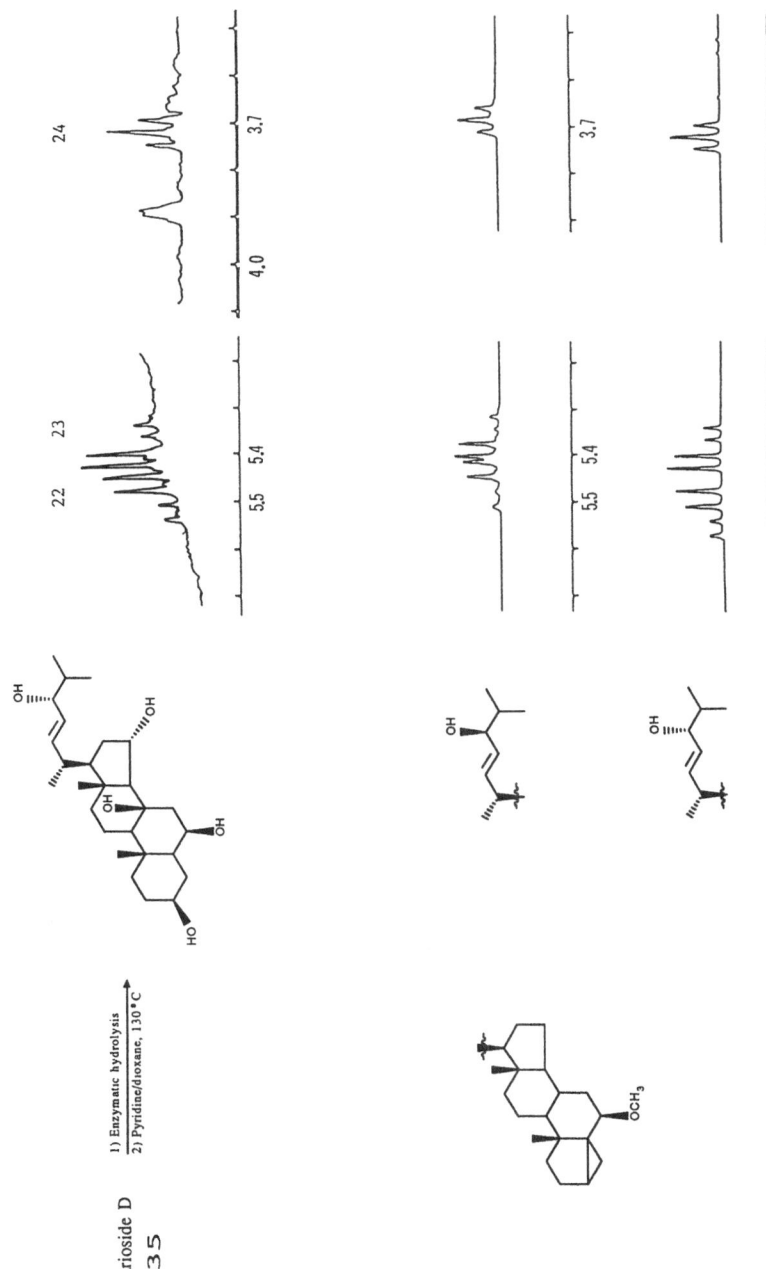

Fig. 12. Selected ^{1}H NMR data (250 MHz, CD$_3$OD) of (24R)- and (24S)-Δ^{22E}-24- hydroxysteroids

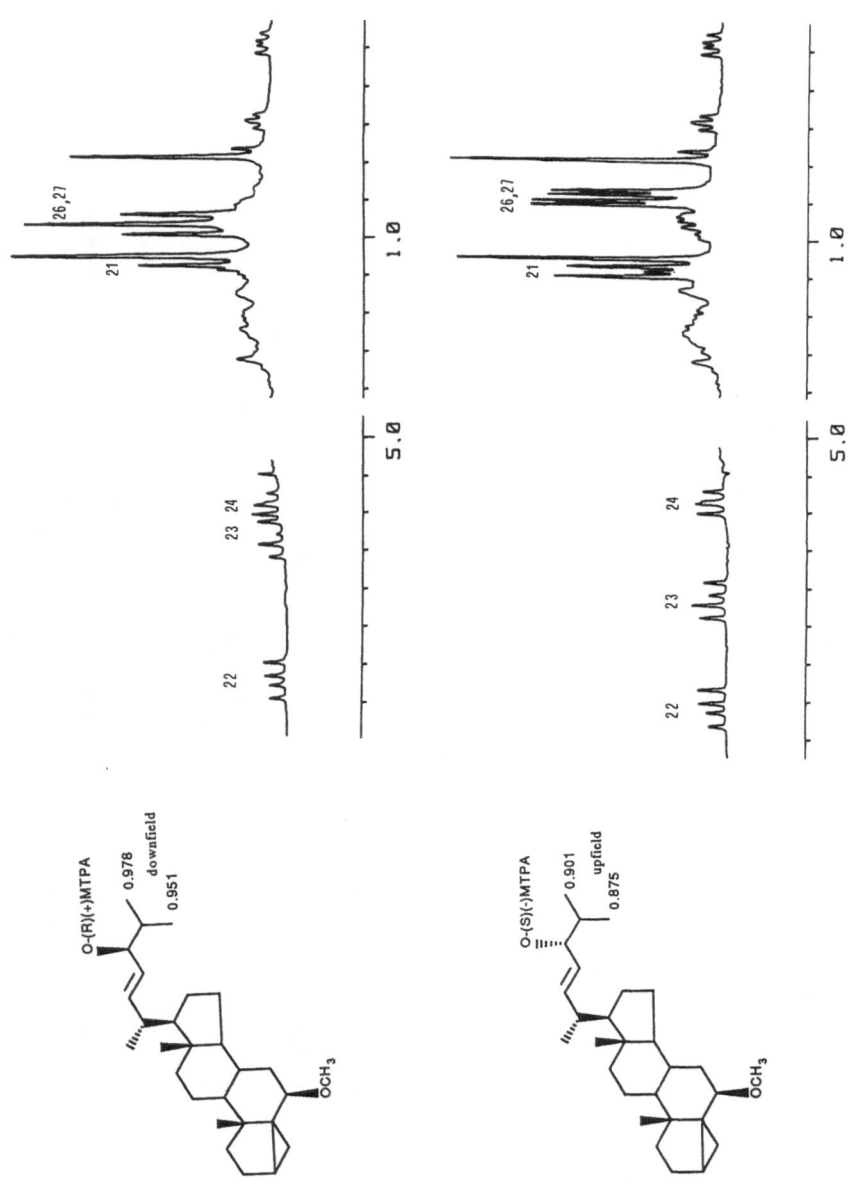

Fig. 13. Configuration at C-24 in Δ^{22E}-24-hydroxysteroids by ^{1}H NMR data (250 MHz, CD$_3$OD) of (R)-(+)- and (S)-(–)-24-MTPA derivatives

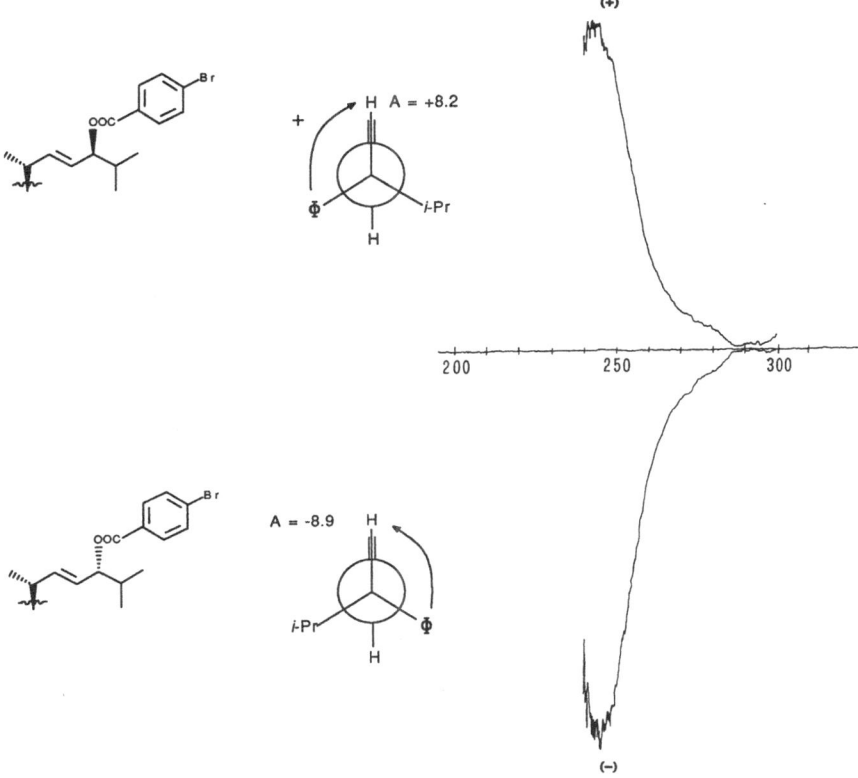

Fig. 14. Exciton split CD curves of the 24-allylic *p*-bromobenzoates of (24R)- and (24S)-Δ^{22E}-24-hydroxysteroids

the 24-O-*p*-bromobenzoates of synthetic Δ^{22}-24-hydroxysteroids are in full agreement with data reported for acyclic allylic alcohols (Fig. 14).

9.2 26-Hydroxysteroids

Hydroxylation at C-26 is a common feature in the large majority of polyhydroxysterols from starfishes. As in the case of 24-hydroxysteroids, the spectra of epimeric 26-hydroxysteroids show only very small differences (*180*) and assignment of configuration requires derivatization with a chiral reagent which induces significant differences in the spectra of 25R and 25S diastereomeric derivatives. The absolute configuration at C-25 in this group of compounds was first determined in the case of 5α-cholestane-3β,6α,8,15α,16β,25-hexaol (**173**), a polyhydroxysterol isolated from the starfish *Protoreaster nodosus* by using a method developed by Yasuhara and Yamaguchi. This method allows determination of the absolute configuration of primary carbinols with a chiral center at C-2 on the basis of lanthanide induced shifts of the OMe signal of the MTPA moiety in the ^{1}H-NMR spectra of the (R)-(+)- and (S)-(−)-MTPA ester derivatives (*178, 181*). We observed that a clearly different pattern was shown by the 26-H$_2$ signals in the ^{1}H-NMR spectra of the (R)-(+)- and (S)-(−)-MTPA esters. These appeared as two overlapping double doublets, at δ 4.15 (J = 11.0 and 7.0 Hz) and 4.17 (J = 11.0 and 5.5 Hz), in the

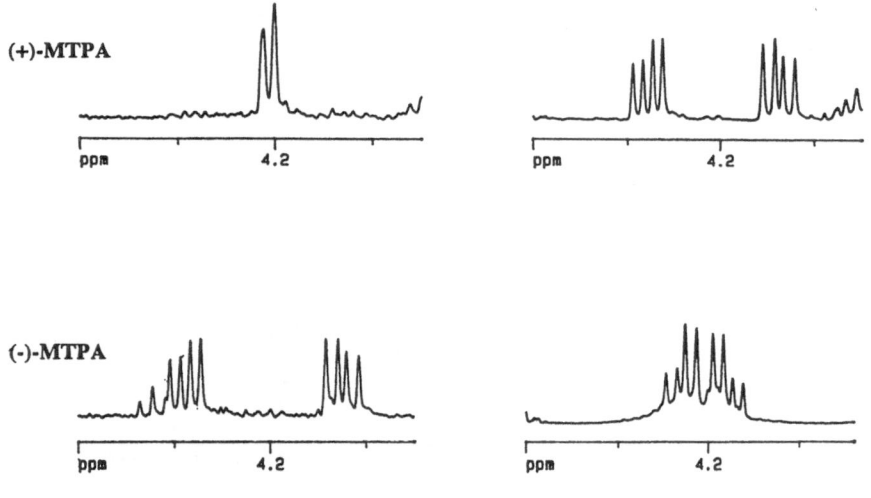

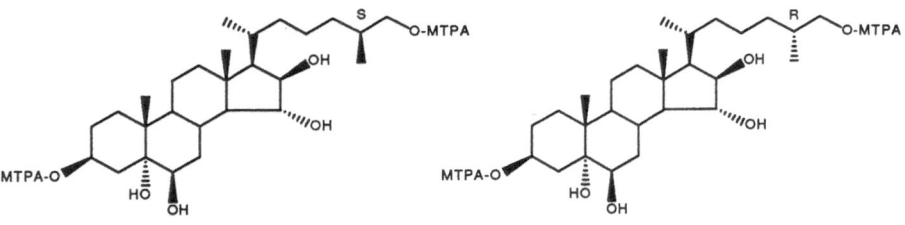

From natural **194**   From natural **197**

Fig. 15. Configuration at C-25 in 26-hydroxysteroids by ^{1}H NMR data (250 MHz, CD$_3$OD) of (R)-(+)- and (S)-(−)-26-MTPA derivatives

References, pp. 297–308

spectrum of the (R)-(+)-MTPA ester and as two well separated double doublets, at δ 4.06 (J = 10.8 and 6.8 Hz) and 4.25 (J = 10.8 and 5.5 Hz), in the spectrum of the (S)-(−)-MTPA ester. Because the inverse behaviour had been observed by TACHIBANA and NAKANISHI during the configurational analysis of a 25R-hydroxysteroid (11), where the 26-H$_2$ signals appeared as well separated signals in the spectrum of the (R)-(+)-MTPA ester and as a doublet in the spectrum of the (S)-(−)-MTPA ester, a direct correlation could be deduced between the pattern of the diastereotopic 26-H$_2$ signals in the spectrum of the (R)-(+)- and (S)-(−)-MTPA esters and the configuration at C-25. Thus, in the case of a 25S-isomer the 26-methylene proton signals are much closer together in the spectrum of the (R)-(+)-MTPA ester than in that of the (S)-(−)-MTPA derivative, while the reverse occurs for MTPA esters of a 25R isomer, the resonances being closer together in the (S)-(−)-MTPA ester and more separated in the (R)-(+)-MTPA ester. Because it has also been observed that derivatization on ring D can effect separation of the 26-methylene signal, it is always advisable to assign the configuration by comparing the spectra data of the (R)-(+)- and (S)-(−)-MTPA derivatives (85). This correlation has since been used as a basis for the assignment of configuration at C-25 in most of the 26-hydroxysteroids isolated from starfishes (Fig. 15).

9.3 24-Hydroxymethylsteroids

Stereochemical analysis of such compounds required the stereoselective synthesis of (24S)- and (24R)-24-(hydroxymethyl)cholesta-5,22-(E)-dien-3β-ols and of their side-chain saturated derivatives (268–271), which permitted acquisition of a set of spectral data suitable for stereochemical assignments in natural steroids possessing a 24-hydroxymethyl side chain (Table 8) (172). Synthesis of model compounds was performed *via* a Claisen rearrangement reaction on a *cis*-allylic C-22 alcohol, a method developed by SUCROW and co-workers (182) which has grown into a general and effective method for the stereospecific functionalization of steroidal side chains (183–187). In series with a saturated side chain the two C-24 epimers can be differentiated directly by their ^{1}H-NMR spectra (Fig. 16): the C-28 methylene protons resonate as a broad doublet (δ 3.52 br d, J = 5.0 Hz) in the ^{1}H-NMR spectrum of the 24R synthetic model (268) and as two well separated signals (δ 3.47 dd, J = 6.5, 10.0 Hz and 3.56 dd, J = 6.5, 10.0 Hz) in the spectrum of its 24S epimer (269). Additional useful information can also be obtained from the ^{1}H- and ^{13}C-NMR chemical shifts of the isopropyl methyl's; these appear more separated in the spectrum of the 24R isomer (19.2, 20.4 ppm) than in that

268   **269**

270, R=H   **271, R=H**

270a, R=(+)MTPA   **271a, R=(+)MTPA**

270b, R=(-)MTPA   **271b, R=(-)MTPA**

272 (from natural **115**)

of the 24S isomer (19.3, 19.9 ppm). The 24S configuration of culcitoside C$_6$ (**119**) was assigned on these grounds by comparing of the ^{1}H-NMR spectrum of the 24-hydroxymethylene steroid **119a**, derived from culcitoside C$_6$ (**119**) by acid treatment with 2M HCl/MeOH, with those of the synthetic models (Fig. 16).

The ^{1}H-NMR spectra of the two synthetic and epimeric Δ^{22}-24-hydroxymethyl models (**270, 271**) were virtually identical and the ^{13}C-NMR spectra were equally useless for differentiating between the C-24 epimers. However the ^{1}H-NMR spectra of the (R)-(+)- and (S)-(−)-MTPA esters showed a number of diagnostic differences in the side chain

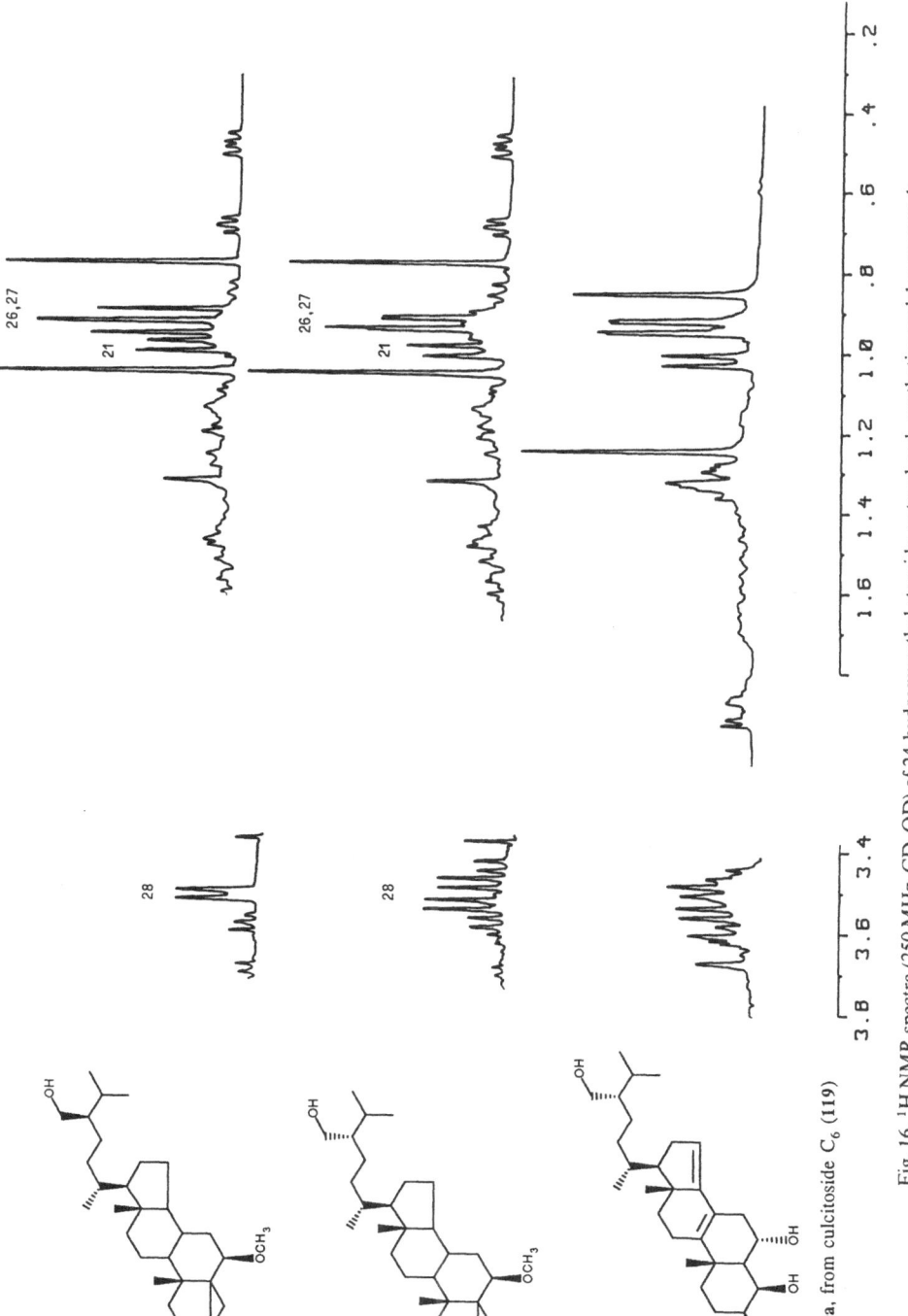

119a, from culcitoside C₆ (119)

Fig. 16. ¹H NMR spectra (250 MHz, CD₃OD) of 24-hydroxymethyl steroids: natural and synthetic model compounds

Table 8. *Selected* ^{1}*H-NMR Data for Side Chain Signals in (24S)-, (24R)-24-Hydroxymethylsteroids and Their (R)-(+)- and (S)-(−)-MTPA Derivatives*

Compound	21-Me	26,27-Me	28-H$_2$
268 (24R)	0.98	0.91, 0.94	3.52 br d (5.0)
269 (24S)	0.99	0.93, 0.92	3.47 dd (6.5, 10.0) 3.56 dd (6.5, 10.0)
270 (24S)	1.07	0.88, 0.94	3.49 dd (6.5, 10.0) 3.57 dd (6.5, 10.0)
270a (24S), (R)-(+)-MTPA	0.96	0.87, 0.94	4.25 dd (7.5, 10.5) 4.39 dd (5.5, 10.5)
270b (24S), (S)-(−)-MTPA	0.96	0.88, 0.92	4.32 br d (6.2)
271 (24R)	1.08	0.87, 0.93	3.51 dd (6.5, 10.0) 3.58 dd (6.5, 10.0)
271a (24R), (R)-(+)-MTPA	1.03	0.87, 0.91	4.34 d (6.5)
271b (24R), (S)-(−)-MTPA	1.01	0.87, 0.93	4.23 br dd (7.0, 10.5) 4.43 dd (5.5, 10.5)
272 (24R), (R)-(+)-MTPA	1.06	0.87, 0.91	4.37 d (6.5)

250 MHz (CD$_3$OD), chemical shift values are given in δ ppm and are referred to CHD$_2$OD (3.34). Coupling constants (in parentheses) are given in Hz

signals which can be used for assignment of configuration (Table 8). The most noticeable feature deals with the diastereotopic C-28 methylene proton signals, which in the spectra of (R)-(+)-MTPA esters appear more separated in the (24S)- than in the (24R)-isomer. In the spectra of the (S)-(−)-MTPA esters the differences, as expected, are reversed. The chemical shift of the C-21 methyl protons is also noteworthy: the signal is shifted significantly upfield in both MTPA derivatives of the (24S)-isomer, whereas in the MTPA esters of the (24R)-isomer it is affected to a minor extent.

Such data made it possible to assign the 24R configuration to coscinasteroside C (**115**) (*42*), pisasteroside A (**116**) (*72*) and laeviuscolosides C (**150**) and D (**151**) (*94*). The 24-methyl-5α-cholest-22(E)-ene-3β,6α,8,15β,16β,28-hexol, derived from coscinasteroside C (**115**) after removing the glucose unit by enzymatic hydrolysis, was converted to a 3β,6α,28-(R)-(+)-MTPA derivative (**272**). In the ^{1}H-NMR spectrum (CD$_3$OD) of **272** (Table 8) the resonances of the C-28 protons (a doublet at δ 4.37, J = 6.5 Hz), and of the 21-Me protons (a doublet at δ 1.06, J = 7 Hz), this last essentially at the same chemical shifts as in the underivatized steroid, were in good agreement with the corresponding resonances observed in the (R)-(+)-MTPA ester of the 24R-synthetic model (**271a**), thus establishing the stereochemistry 24R in coscinasteroside C (**115**).

9.4 24-(β-Hydroxyethyl)- and 24-(carboxymethyl)steroids

The configuration at C-24 in 24-(β-hydroxyethyl)steroids can be determined by comparison of ^{1}H and ^{13}C-NMR spectral data. It has

References, pp. 297–308

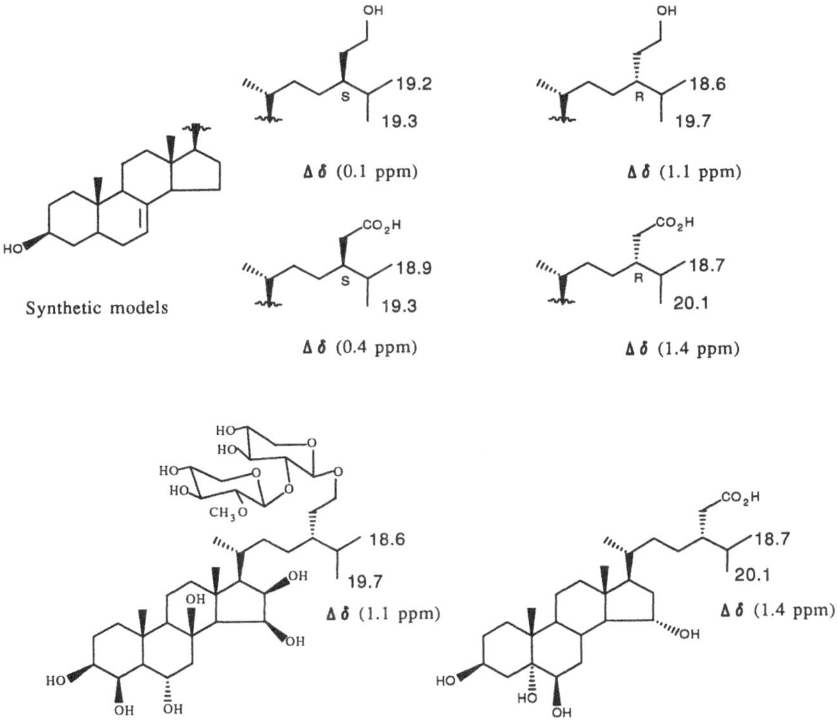

126 (halityloside A) **215** (from Myxoderma platiacanthum)

Fig. 17. Configuration at C-24 in 24-(β-hydroxyethyl)- and 24-(carboxymethyl)steroids. ^{13}C NMR data of synthetic models and natural compounds

been reported that a small but diagnostic difference can be observed in the C-26 and C-27 proton signals of the epimeric 29-hydroxyclionasterol (24R) and 29-hydroxysitosterol (24S) (*185*): in the 220 MHz ^{1}H-NMR spectrum of the (24R)-isomer the isopropyl signal is observed as an apparent triplet at δ 0.84 because of coincidental overlap of the low-field arm of one doublet (δ 0.83) with the high field arm of the other (δ 0.86), while in the spectrum of the (24S)-isomer the isopropyl methyls give rise to two overlapping doublets at δ 0.84 and 0.85. These findings were confirmed in a study aimed at investigating the diagnostic utility of ^{13}C-NMR spectroscopy within a series of synthetic 24R- and 24S-(β-hydroxyethyl)- and -(carboxymethyl)steroids, for assignment of configuration at C-24 in unknown natural steroids (*188*). Assignment is possible on the basis of the chemical shift differences (Δδ) observed between the C-26 and C-27 carbon signals in the 24R- and 24S-isomers. Indeed, in C-24 epimeric pairs with saturated side chains, the difference in chemical shift ($\Delta\delta$) between the C-26 and C-27 carbon resonances ranges between 1.1 and 1.4 ppm in the case of the 24R-isomers and between 0.1 and 0.4 ppm for the 24S-isomers (fig. 17). In conclusion ^{1}H-NMR spectroscopy allows differentiation between pairs of epimers, but comparison of spectra of both compounds is advisable. ^{13}C-NMR spectroscopy, owing to the larger difference of $\Delta\delta$ values, permits unambiguous assignment of the C-24 stereochemistry of 29-oxygenated steroids with a saturated side chain even in the presence of only a single epimer. In the case of Δ^{22} side-chains the chemical shifts of the isopropyl methyl carbons are very similar and ^{13}C-NMR spectra fail to differentiate between 24R- and 24S-epimers. However assignment is always possible in principle by examining the spectrum of the saturated derivative.

The 24R configuration was assigned to a number of 24-(β-hydroxyethyl)steroidal glycosides on the basis of the above arguments. For example, in halityloside A (126) signals due to the isopropyl methyls, at 18.6 and 19.7 ppm ($\Delta\delta$ 1.1 ppm) in the ^{13}C-NMR spectrum, and at δ 0.88, 0.91, in the ^{1}H-NMR spectrum, allowed the 24R stereochemistry to be assigned (112). Similarly the 24R configuration was assigned to 3β,5,6β,15α-tetrahydroxy-5α-stigmastan-29-oic acid (215) isolated from *Myxoderma platiacanthum* (85).

9.5 24-Methyl-26-hydroxy- and 24-methyl-26-oic steroidal side-chains

In order to acquire a set of spectral data allowing the assignment of configuration at C-24 and C-25 of 24-methyl-26-hydroxysteroids, model compounds with all possible configurations at C-24 and C-25 were synthesized by a scheme involving a Claisen rearrangement of *cis*-allylic C-22 alcohols (273, 274 in Fig. 18) (140). These were converted by reaction with triethyl orthopropionate into a mixture of two Δ^{22E}-26-oic esters epimeric at C-25 which, upon reduction by lithium aluminium hydride followed by HPLC separation afforded the four possible Δ^{22E}-24-methyl-26-hydroxy stereoisomers. These were then converted by catalytic hydrogenation into the saturated derivatives (Fig. 18). The Claisen rearrangement of the individual allylic alcohols gave, in both cases, a major amount of the *erythro* product, in agreement with the expected predominance of the more stable (E) isomers of the intermediate ketene acetals formed during the reaction when ethanol is lost from the orthoester intermediates (189, 190). The configuration at C-24 of all synthetic models followed from the stereoselectivity of the Claisen

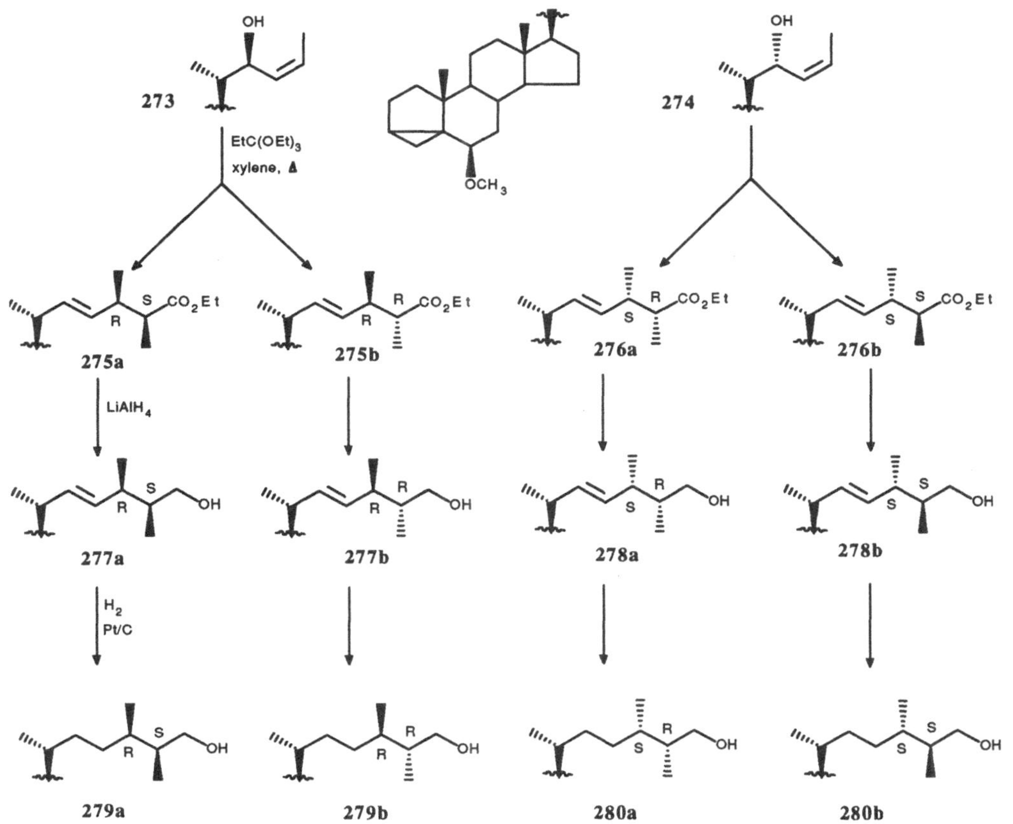

Fig. 18. Synthesis of model 24-methyl-26-hydroxysteroids

rearrangement reaction, while the stereochemistry at C-25 was confirmed by application of the lanthanide-induced chemical shift (LIS) non-equivalence method described by YASUHARA and YAMAGUCHI (*178*, *181*). The chemical shifts of the C-26, C-27 and C-28 protons and those of carbons 24, 27 and 28 are virtually identical in the Δ^{22}-*threo* pair (**277a–278a**) and significantly different from those, likewise identical, of the alternative *erythro* pair (**277b–278b**). The same is true in the series containing a saturated side chain, with a more differentiated pattern exhibited by the carbon signals (Table 9).

Thus, NMR spectroscopy makes assignment of relative stereochemistry in the side chain straightforward, while differentiation between the

Table 9. Selected NMR Data for Side Chain Signals in Natural and Synthetic 24-Methyl-26-Hydroxysteroids

Compound	^{13}C			^{1}H	27-H$_3$	28-H$_3$	(R)-(+)-MTPA esters
	C-24	C-27	C-28	26-H$_2$			26-H$_2$
Δ^{22}-series							
277a (24R,25S) *threo*	39.7	13.8	17.1	3.28 dd, 3.60 dd	0.90 d	0.95 d	4.19 dd, 4.31 dd
278a (24S,25R) *threo*	39.5	13.8	16.8	3.29 dd, 3.59 dd	0.90 d	0.97 d	4.13 dd, 4.38 dd
278b (24S,25S) *erythro*	39.2	13.6	19.0	3.34 dd, 3.53 dd	0.87 d	1.02 d	4.21 br d
277b (24R,25R) *erythro*	39.3	13.6	19.2	3.34 dd, 3.52 dd	0.88 d	1.02 d	4.16 dd, 4.21 dd
Saturated side-chain series							
279a (24R,25S) *threo*	35.1	14.8	12.0	3.38 dd, 3.55 dd	0.83 d	0.81 d	4.23 br d
280a (24S,25R) *threo*	35.1	15.1	11.6	3.38 dd, 3.49 dd	0.81 d	0.81 d	4.14 dd, 4.34 dd
280b (24S,25S) *erythro*	36.8	17.5	14.4	3.36 dd, 3.58 dd	0.93 d	0.92 d	4.22 dd, 4.32 dd
279b (24R,25R) *erythro*	36.1	17.4	14.1	3.37 dd, 3.57 dd	0.91 d	0.91 d	4.16 dd, 4.38 dd
Natural steroids							
echinasteroside A (**152**) [Δ^{22}-side chain, (24R,25S) *threo*]	40.4	14.5	17.5	3.46 dd, 3.58 dd	0.91 d	0.97 d	4.17 dd, 4.33 dd
							[(S)-(−)-MTPA; 4.06 dd, 4.46 dd]
(**260**) [saturated side chain, (24S,25S) *erythro*]	36.7	17.4	14.3	3.34 dd, 3.62 dd	0.92 d	0.92 d	4.24 dd, 4.32 dd
							[(S)-(−)-MTPA; 4.16 dd, 4.37 dd]

250 MHz (CD$_3$OD), chemical-shift values are given in ppm and are referred to central CHD$_2$OD (δ_H 3.34) and CD$_3$OD (δ_C 49.0) signals

individual stereoisomers of every *threo* or *erythro* pair can be achieved by analysis of ^{1}H-NMR spectra of their (R)-(+)- and (S)-(–)-MTPA derivatives. The following strategy has been formulated for the stereochemical assignment of an unknown natural 24-methyl-26-hydroxy steroid: (a) identification of relative stereochemistry by comparison of NMR spectral data with those of reference models (through a typical *threo* or *erythro* pattern); (b) assignment of absolute configuration at C-25, and hence to C-24, by the shape of the C-26 methylene protons signal in the ^{1}H-NMR spectrum of the (R)-(+)- and (S)-(–)-MTPA derivatives. In this connection it has been established that the C-26 proton signals always appear as two double doublets, but that, in every pair they are closer to each other in the spectrum of the (R)-(+)-MTPA derivative of the (25S)-isomer than in that of the (25R)-isomer and such behaviour is reversed in the (S)-(–)-MTPA derivatives, the signals being now closer together in the spectrum of the (25R)-isomer. Thus, a lesser difference in the positions of the C-26 methylene proton signals in the spectrum of the (R)-(+)-MTPA derivative will be indicative of 25S stereochemistry while on the contrary a lesser difference in the positions of the (S)-(–)-MTPA derivative will point to 25R stereochemistry. These arguments were utilised for the stereochemical analysis of echinasteroside A (**152**) and of the steroid **260** from the ophiuroid *Ophiolepis superba* (*140*).

The *threo*-24R,25S stereochemistry was assigned to steroid **213** with a Δ^{22},24-methyl-26-oic side chain encountered in *Myxoderma platyacanthum*, by spectral comparison with the spectra of the synthetic models **275a, b** and **276a, b** (Table 10). As expected the *erythro* isomers can be differentiated easily from the *threo* isomers by ^{1}H-NMR spectroscopy, especially by comparing the shifts of the olefinic protons. More subtle differences are observed in the shifts of the side chain methyl protons between each pair. In order to assign the absolute configuration more

213

213a

Table 10. *Selected* ^{1}*H-NMR Data of Synthetic and Natural 24-methyl-26-oic Steroidal Side-Chains*

Compound	Olefinic H's	Methyl doublets
275a (24R,25S) *threo*	5.27 m	1.00, 1.03, 1.10
276a (24S,25R) *threo*	5.28 m	1.01, 1.03, 1.11
275b (24R,25R) *erythro*	5.16 dd, 5.31 dd	0.99, 1.05, 1.09
276b (24S,25S) *erythro*	5.16 dd, 5.33 dd	1.00, 1.06, 1.08
Natural **213** (24R,25S) *threo*	5.26 m	0.99, 1.01, 1.10
C-25 epimer **213a** (24R,25R) *erythro*	5.15 dd, 5.30 dd	0.98, 1.04, 1.09

250 MHz (CD$_3$OD), chemical-shift values are given in δ ppm and are referred to central CHD$_2$OD (δ$_H$ 3.34) signal

confidently, the natural amide **213** was equilibrated with 10% KOH thus affording the epimer at C-25 (**213a**). This was followed by a comparison of the spectra data of **213** with those of the *threo* models and the spectrum of the epimer at C-25 **213a** with the spectra of the *erythro* models. The pattern of the methyl doublet in the ^{1}H-NMR spectrum of **213** compared more closely with that of the (24R, 25S)-isomer, and accordingly the pattern of **213a** compared better with that of the (24R, 25R)-isomer.

9.6 24-Ethyl-26-hydroxysteroids

Model 24-ethyl-26-hydroxysteroids with all possible configurations at C-24 and C-25 have been synthesized (*171, 187*). In the Δ^{22}-series the ^{1}H and ^{13}C-NMR spectra of the pair **281a–282a** are significantly different from those of the pair **281b–282b**, major differences relating to the chemical shifts of the C-26 methylene and C-27 methyl protons and of the C-24, C-27 and C-28 carbons. In the series with a saturated side chain the ^{1}H-NMR spectra of all isomers are very similar, while the ^{13}C-NMR spectra of the pair **283a–284a** are significantly different from those of the pair **283b–284b**, with major differences relating to the C-23 and C-28 carbon signals (Tab. 11). Thus NMR analysis permits discrimination of the relative stereochemistry (24R, 25S)/(24S, 25R) from the alternative (24R, 25R)/(24S, 25S). The absolute configuration can then be derived by ^{1}H-NMR analysis of the (R)-(+)- and (S)-(−)-MTPA derivatives. Indeed, as mentioned earlier, in the (R)-(+)-MTPA esters of each pair with the same relative stereochemistry the C-26 methylene protons of the 25S-isomer appear as signals resonating much more closely than those in the corresponding 25R-isomer and the opposite is true for the (S)-(−)-

281a   281b   282a   282b

283a   283b   284a   284b

OCH₃

MTPA esters. Here again more proximate signals in the ¹H-NMR spectrum of the (R)-(+)-MTPA ester point to the 25S configuration, while the inverse behaviour is an indication of the 25R configuration. By using these arguments we have assigned the stereochemistry at C-24 and C-25 of the steroid **261** from the ophiuroid *Ophiolepis superba* (*171*).

9.7 24-Methyl-25,26-dihydroxysteroids.

A Δ^{22}-24-methyl-25,26-dihydroxysteroidal side-chain was found in a highly polyhydroxylated steroid (**220**) isolated from the extracts of the New Caledonian starfish *Archaster typicus* (159). Simpler side chain models were synthesized by epoxidation of (*E*)-2-methyl-2-pentenol followed by reaction with lithium dimethylcuprate to give the (2R,3R)/(2S,3S)-2, 3-dimethylpentane-1,2-diol enantiomeric pair (**285a, b**). These were converted to the (2S,3R)/(2R,3S)-enantiomeric pair (**286a, b**) by tosylation, alkaline treatment and opening of the resulting 1,2-epoxide with diluted aqueous sulphuric acid. Enantiomeric pairs of model compounds exhibited typical and easily recognizable ¹H-NMR spectra which, when compared with that of the 22,23-dihydroderivative of the natural steroid (**288**), allowed recognition of its relative stereo-

Table 11. *Selected NMR Data for Side-Chain Signals in Synthetic and Natural 24-Ethyl-26-Hydroxysteroids*

Compound	^{13}C C-23	C-24	C-27	C-28	^1H 26-H$_2$	27-H$_3$	(R)-(+)-MTPA esters 26-H$_2$
Δ^{22}-side chain series							
281a (24R,25S)		48.7	15.2	25.5	3.25 dd, 3.64 dd	0.93 d	4.10 dd, 4.32 dd
282a (24S,25R)		48.5	15.2	25.5	3.25 dd, 3.63 dd	0.94 d	4.04 dd, 4.38 dd
282b (24S,25S)		47.1	12.9	26.9	3.32 dd, 3.48 dd	0.83 d	4.15 br d
281b (24R,25R)		47.1	12.9	26.9	3.34 dd, 3.48 dd	0.85 d	4.08 dd, 4.19 dd
Saturated side-chain series							
283a (24R,25S)	28.3	43.1	13.6	23.6	3.37 dd, 3.56 dd	0.89 d	4.17 dd, 4.26 dd
284a (24S,25R)	28.4	42.8	13.0	23.7	3.38 dd, 3.52 dd	0.84 d	4.11 dd, 4.32 dd
284b (24S,25S)	27.3	43.2	13.5	25.2	3.37 dd, 3.55 dd	0.89 d	4.22 br d
283b (24R,25R)	27.0	42.5	13.0	24.9	3.39 dd, 3.53 dd	0.87 d	4.12 dd, 4.31 dd
Natural (**261**)	27.6	43.4	13.4	25.1	3.35 dd, 3.58 dd	0.92 d	4.26 br d
[saturated side chain, (24S,25S)]							[(S)-(−)-MTPA; 4.16 dd, 4.39 dd]

250 MHz (CD$_3$OD), chemical-shift values are given in ppm and are referred to central CH$_2$OD (δ_H 3.34) and CD$_3$OD (δ_C 49.0) signals

chemistry on the basis of the good agreement between its C-26,C-27 and C-28 proton signals and the corresponding signals of the (2R,3R)/(2S,3S)-2,3-dimethylpentane-1,2-diol enantiomeric pair (**285a, b**). In order to establish the absolute configuration, the optically active (2R,3R)-2,3-dimethylpentane-1,2-diol (**285a**) was synthesized by using the titanium tartrate-catalysed asymmetric epoxidation of allylic alcohols devised by KATSUKI and SHARPLESS (*191, 192*). (R)-(+)-MTPA Esters were then prepared from the (2R,3R)/(2S,3S)-2,3-dimethylpentane-1,2-diol enantiomeric pair (**285a, b**), the single enantiomer (2R,3R)-2,3-dimethylpentane-1,2-diol (**285a**), and the 22,23-dihydro derivative (**288**) and their ^{1}H-NMR spectra compared. In the ^{1}H-NMR spectrum (250 MHz, CDCl$_3$) of the (R)-(+)-MTPA ester of the (2R,3R)-2,3-dimethylpentane-1,2-diol (**287a**), the signal due to the C-1 methylene protons appeared as an AB quartet centred at δ 4.20 (J = 11.0 Hz) with

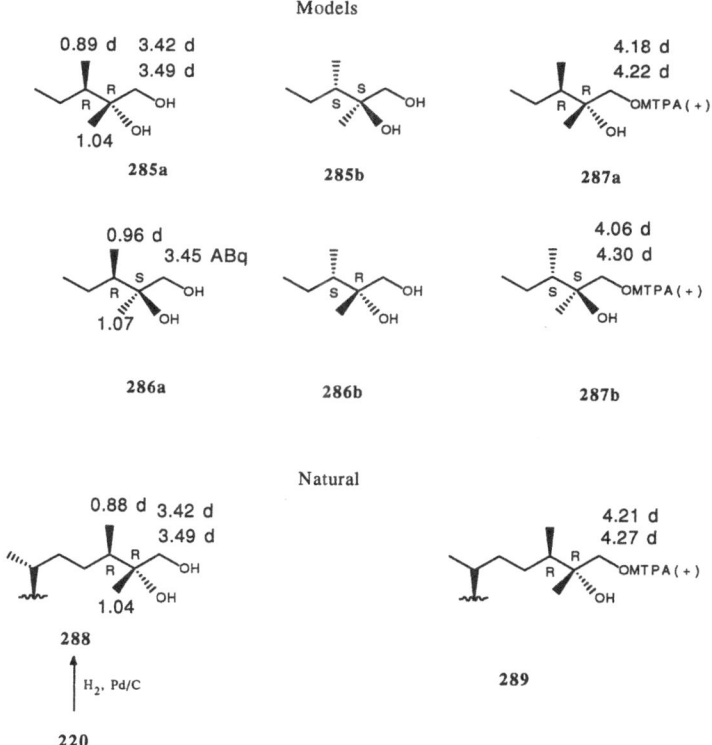

Fig. 19. Configuration at C-24 and C-25 in 24-methyl-25,26-dihydroxysteroids. ^{1}H NMR data of synthetic models and natural compounds

the two central lines separated by *ca.* 4.0 Hz, closely resembling the signal of the C-26 methylene protons of the (R)-(+)-MTPA ester **289**, also an AB quartet with central lines separated by *ca.* 7.0 Hz. An identical signal was present in the spectrum of the (R)-(+)-MTPA ester mixture of the epimeric (2R,3R)/(2S,3S)-2,3-dimethylpentane-1,2-diols (**287a, b**) together with two additional well separated doublets (J = 11.0 Hz) at δ 4.06 and 4.30 (J = 11.0 Hz, central lines separated by *ca.*45 Hz), which were consequently assigned to the C-1 methylene protons of the (2S,3S)-stereoisomer (**287b**) (Fig. 19). On this basis the configurations of C-24 and C-25 in **288** and accordingly in the naturally occurring **220** were determined as 24R, 25R.

Aknowledgements

Our studies described in this chapter were supported by grants from Ministero Pubblica Istruzione (M.P.I.), Roma, Ministero dell'Università e della Ricerca Scientifica e Tecnologica (MURST), Roma and Consiglio Nazionale delle Ricerche (C.N.R.), Roma, through the "Progetto Finalizzato Chimica Fine e Secondaria", "Progetto Finalizzato Chimica Fine II" and various contributions from the "Comitato Scienze Chimiche".

A number of studies described here are part of the projects SNOM "Substances Naturelles d' Origine Marine" and SMIB "Substances Marines d' Interêt Biologique", ORSTOM-CNRS, Nouméa, New Caledonia. We wish to express our thanks to Dr. P. Potier (Institute de Chimie de Substances Naturelles, CNRS, Gif sur Yvette, France) for having promoted the collaboration between our group in Napoli and the groups of CNRS and ORSTOM in Nouméa. We wish also to express our warm thanks to the collaborators engaged in the projects on the chemistry of echinoderms in Napoli: Dr.s M. V. D' Auria, M. Iorizzi, O. Squillace Greco, I. Bruno, E. Finamore, F. De Riccardis and L. Gomez Paloma, and in Nouméa: T. Sevenet, J. Pusset, D. Laurent and C. Debitus, for their extensive intellectual, creative and technical contributions. We are grateful to Prof.'s W. Fenical (Scripps Institution of Oceanography, La Jolla, CA, USA), T. Yasumoto (Faculty of Agriculture, Tohoku University, Sendai, Japan) and T. Higa (University of the Ryukyus, Okinawa, Japan) for their help (hospitality, scientific discussion, collection of material) provided during our work. We also thank Prof. M. Jangoux (Universite' Libre de Bruxelles, Belgium) for the identification of many of the starfish species collected off Nouméa, New Caledonia.

References, pp. 297–308

Name Index of Asterosponins

Name	Structure	M.W.	Molecular Formula
Acanthaglycoside A	24	1248	$C_{56}H_{89}O_{27}S_1Na_1$
Acanthaglycoside B	25	1264	$C_{56}H_{89}O_{28}S_1Na_1$
Acanthaglycoside C	7	1266	$C_{56}H_{91}O_{28}S_1Na_1$
Acanthaglycoside D	26	1264	$C_{56}H_{89}O_{28}S_1Na_1$
Acanthaglycoside F	46	1426	$C_{63}H_{103}O_{32}S_1Na_1$
Asteroside A	35	1266	$C_{56}H_{91}O_{28}S_1Na_1$
Asteroside B	36	1252	$C_{55}H_{89}O_{28}S_1Na_1$
Asteroside C	45	1280	$C_{57}H_{93}O_{28}S_1Na_1$
Asteroside D	42	1264	$C_{57}H_{93}O_{27}S_1Na_1$
Co-aris II	40	1260	$C_{57}H_{89}O_{27}S_1Na_1$
Forbeside E	52	684	$C_{27}H_{42}O_{13}S_2Na_2$
Forbeside F	21	1134	$C_{51}H_{83}O_{24}S_1Na_1$
Forbeside G	22	988	$C_{45}H_{73}O_{20}S_1Na_1$
Forbeside H	23	958	$C_{44}H_{71}O_{19}S_1Na_1$
Glycoside B2 (forbeside B)	5	1266	$C_{56}H_{91}O_{28}S_1Na_1$
Henricioside A	50	1266	$C_{56}H_{91}O_{28}S_1Na_1$
Laevigatoside	10	1250	$C_{56}H_{91}O_{27}S_1Na_1$
Luidiaglycoside C	28	1262	$C_{57}H_{91}O_{27}S_1Na_1$
Luidiaglycoside D	30	1264	$C_{57}H_{93}O_{27}S_1Na_1$
Maculatoside (luidiaglycoside B)	8	1250	$C_{56}H_{91}O_{27}S_1Na_1$
Marthasteroside A1	14	1412	$C_{62}H_{101}O_{32}S_1Na_1$
Marthasteroside A2 (luidiaglycoside A)	15	1396	$C_{62}H_{101}O_{31}S_1Na_1$
Marthasteroside B	27	1262	$C_{57}H_{91}O_{27}S_1Na_1$
Marthasteroside C	29	1264	$C_{57}H_{93}O_{27}S_1Na_1$
Myxodermoside A	20	1120	$C_{50}H_{81}O_{24}S_1Na_1$
Ophidianoside B	31	1252	$C_{55}H_{89}O_{28}S_1Na_1$
Ophidianoside C	32	1222	$C_{54}H_{87}O_{27}S_1Na_1$
Ophidianoside F	9	1236	$C_{55}H_{89}O_{27}S_1Na_1$
Ovarian asterosaponin I (Co-Aris I; forbeside C)	6	1280	$C_{57}H_{93}O_{28}S_1Na_1$
Ovarian asterosaponin-4 (Co-Aris III)	38	1250	$C_{56}H_{91}O_{27}S_1Na_1$
Patirioside A	49	1412	$C_{62}H_{101}O_{32}S_1Na_1$
Patirioside B	19	1442	$C_{63}H_{103}O_{33}S_1Na_1$
Pectinioside A	12	1280	$C_{57}H_{93}O_{28}S_1Na_1$
Pectinioside B	48	1442	$C_{63}H_{103}O_{33}S_1Na_1$
Pectinioside C	51	1440	$C_{64}H_{105}O_{32}S_1Na_1$
Pectinioside E	16	1412	$C_{62}H_{101}O_{32}S_1Na_1$
Pectinioside F	17	1428	$C_{62}H_{101}O_{33}S_1Na_1$
Pectinioside G	18	1412	$C_{62}H_{101}O_{32}S_1Na_1$
Protoreasteroside	37	1250	$C_{56}H_{91}O_{27}S_1Na_1$
Regularoside A	47	1294	$C_{58}H_{95}O_{28}S_1Na_1$
Regularoside B	11	1250	$C_{56}H_{91}O_{27}S_1Na_1$
Solasteroside A	39	1234	$C_{56}H_{91}O_{26}S_1Na_1$
Tenuispinoside A	33	1266	$C_{56}H_{91}O_{28}S_1Na_1$
Tenuispinoside B	34	1250	$C_{56}H_{91}O_{27}S_1Na_1$
Tenuispinoside C	41	1282	$C_{56}H_{91}O_{29}S_1Na_1$
Thornasteroside A	4	1266	$C_{56}H_{91}O_{28}S_1Na_1$
Versicoside A (forbeside A)	13	1428	$C_{62}H_{101}O_{33}S_1Na_1$
Versicoside B	44	1442	$C_{63}H_{103}O_{33}S_1Na_1$
Versicoside C (thornasteroside B)	43	1280	$C_{57}H_{93}O_{28}S_1Na_1$

Molecular Weight Index of Asterosaponins

Name	Structure	M.W.	Molecular Formula
Acanthaglycoside A	24	1248	$C_{56}H_{89}O_{27}S_1Na_1$
Acanthaglycoside B	25	1264	$C_{56}H_{89}O_{28}S_1Na_1$
Acanthaglycoside C	7	1266	$C_{56}H_{91}O_{28}S_1Na_1$
Acanthaglycoside D	26	1264	$C_{56}H_{89}O_{28}S_1Na_1$
Acanthaglycoside F	46	1426	$C_{63}H_{103}O_{32}S_1Na_1$
Asteroside A	35	1266	$C_{56}H_{91}O_{28}S_1Na_1$
Asteroside B	36	1252	$C_{55}H_{89}O_{28}S_1Na_1$
Asteroside C	45	1280	$C_{57}H_{93}O_{28}S_1Na_1$
Asteroside D	42	1264	$C_{57}H_{93}O_{27}S_1Na_1$
Co-Aris II	40	1260	$C_{57}H_{89}O_{27}S_1Na_1$
Forbeside E	52	684	$C_{27}H_{42}O_{13}S_2Na_2$
Forbeside F	21	1134	$C_{51}H_{83}O_{24}S_1Na_1$
Forbeside G	22	988	$C_{45}H_{73}O_{20}S_1Na_1$
Forbeside H	23	958	$C_{44}H_{71}O_{19}S_1Na_1$
Glycoside B2 (forbeside B)	5	1266	$C_{56}H_{91}O_{28}S_1Na_1$
Henricioside A	50	1266	$C_{56}H_{91}O_{28}S_1Na_1$
Laevigatoside	10	1250	$C_{56}H_{91}O_{27}S_1Na_1$
Luidiaglycoside C	28	1262	$C_{57}H_{91}O_{27}S_1Na_1$
Luidiaglycoside D	30	1264	$C_{57}H_{93}O_{27}S_1Na_1$
Maculatoside (luidiaglycoside B)	8	1250	$C_{56}H_{91}O_{27}S_1Na_1$
Marthasteroside A1	14	1412	$C_{62}H_{101}O_{32}S_1Na_1$
Marthasteroside A2 (luidiaglycoside A)	15	1396	$C_{62}H_{101}O_{31}S_1Na_1$
Marthasteroside B	27	1262	$C_{57}H_{91}O_{27}S_1Na_1$
Marthasteroside C	29	1264	$C_{57}H_{93}O_{27}S_1Na_1$
Myxodermoside A	20	1120	$C_{50}H_{81}O_{24}S_1Na_1$
Ophidianoside B	31	1252	$C_{55}H_{89}O_{28}S_1Na_1$
Ophidianoside C	32	1222	$C_{54}H_{87}O_{27}S_1Na_1$
Ophidianoside F	9	1236	$C_{55}H_{89}O_{27}S_1Na_1$
Ovarian asterosaponin I (Co-Aris I; forbeside C)	6	1280	$C_{57}H_{93}O_{28}S_1Na_1$
Ovarian asterosaponin-4 (Co-Aris III)	38	1250	$C_{56}H_{91}O_{27}S_1Na_1$
Patirioside A	49	1412	$C_{62}H_{101}O_{32}S_1Na_1$
Patirioside B	19	1442	$C_{63}H_{103}O_{33}S_1Na_1$
Pectinioside A	12	1280	$C_{57}H_{93}O_{28}S_1Na_1$
Pectinioside B	48	1442	$C_{63}H_{103}O_{33}S_1Na_1$
Pectinioside C	51	1440	$C_{64}H_{105}O_{32}S_1Na_1$
Pectinioside E	16	1412	$C_{62}H_{101}O_{32}S_1Na_1$
Pectinioside F	17	1428	$C_{62}H_{101}O_{33}S_1Na_1$
Pectinioside G	18	1412	$C_{62}H_{101}O_{32}S_1Na_1$
Protoreasteroside	37	1250	$C_{56}H_{91}O_{27}S_1Na_1$
Regularoside A	47	1294	$C_{58}H_{95}O_{28}S_1Na_1$
Regularoside B	11	1250	$C_{56}H_{91}O_{27}S_1Na_1$
Solasteroside A	39	1234	$C_{56}H_{91}O_{26}S_1Na_1$
Tenuispinoside A	33	1266	$C_{56}H_{91}O_{28}S_1Na_1$
Tenuispinoside B	34	1250	$C_{56}H_{91}O_{27}S_1Na_1$
Tenuispinoside C	41	1282	$C_{56}H_{91}O_{29}S_1Na_1$
Thornasteroside A	4	1266	$C_{56}H_{91}O_{28}S_1Na_1$
Versicoside A (forbeside A)	13	1428	$C_{62}H_{101}O_{33}S_1Na_1$
Versicoside B	44	144	$C_{63}H_{103}O_{33}S_1Na_1$
Versicoside C (thornasteroside B)	43	1280	$C_{57}H_{93}O_{28}S_1Na_1$

Name Index of Polyhydroxysteriod Glycosides

Name	Structure	M.W.	Molecular Formula
–	77	614	$C_{33}H_{58}O_{10}S_0Na_0$
–	65	686	$C_{32}H_{55}O_{12}S_1Na_1$
–	78	630	$C_{33}H_{58}O_{11}S_0Na_0$
–	79	628	$C_{38}H_{60}O_{10}S_0Na_0$
–	162	626	$C_{34}H_{58}O_{10}S_0Na_0$
Amurensoside A	61	670	$C_{32}H_{55}O_{11}S_1Na_1$
Amurensoside B	62	686	$C_{32}H_{55}O_{12}S_1Na_1$
Amurenoside C	63	668	$C_{32}H_{53}O_{11}S_1Na_1$
Amurensoside D	98	584	$C_{32}H_{56}O_9S_0Na_0$
Aphelasteroside A	137	686	$C_{32}H_{55}O_{12}S_1Na_1$
Aphelasteroside B	155	698	$C_{32}H_{51}O_{13}S_1Na_1$
Asterosaponin P-1	64	700	$C_{33}H_{57}O_{12}S_1Na_1$
Asterosaponin P-2	81	760	$C_{35}H_{61}O_{14}S_1Na_1$
Attenuatoside A-I	91	730	$C_{38}H_{66}O_{13}S_0Na_0$
Attenuatoside A-II	82	584	$C_{32}H_{56}O_9S_0Na_0$
Attenuatoside B-1	100	746	$C_{38}H_{66}O_{14}S_0Na_0$
Attenuatoside B-2	99	600	$C_{32}H_{56}O_{10}S_0Na_0$
Attenuatoside C	74	732	$C_{37}H_{64}O_{14}S_0Na_0$
Attenuatoside S-I	157	730	$C_{34}H_{59}O_{13}S_1Na_1$
Attenuatoside S-II (22-dehydro-attenuatoside S-I)	158	728	$C_{34}H_{57}O_{13}S_1Na_1$
Attenuatoside S-III	159	730	$C_{34}H_{59}O_{13}S_1Na_1$
Borealoside A	69	832	$C_{38}H_{65}O_{16}S_1Na_1$
Borealoside B	70	818	$C_{37}H_{63}O_{16}S_1Na_1$
Borealoside C	71	598	$C_{33}H_{58}O_9S_0Na_0$
Borealoside D	75	614	$C_{33}H_{58}O_{10}S_0Na_0$
Coscinasteroside A	110	730	$C_{33}H_{55}O_{14}S_1Na_1$
Coscinasteroside B	134	686	$C_{32}H_{55}O_{12}S_1Na_1$
Coscinasteroside C	115	744	$C_{34}H_{57}O_{14}S_1Na_1$
Coscinasteroside D	113	730	$C_{33}H_{55}O_{14}S_1Na_1$
Coscinasteroside E	109	616	$C_{32}H_{56}O_{11}S_0Na_0$
Coscinasteroside F	108	600	$C_{32}H_{56}O_{10}S_0Na_0$
Crossasteroside A	72	760	$C_{39}H_{68}O_{14}S_0Na_0$
Crossasteroside B	67	744	$C_{39}H_{68}O_{13}S_0Na_0$
Crossasteroside C	68	730	$C_{38}H_{66}O_{13}S_0Na_0$
Crossasteroside D	73	746	$C_{38}H_{66}O_{14}S_0Na_0$
Crossasteroside P1	164	804	$C_{41}H_{72}O_{15}S_0Na_0$
Crossasteroside P2	165	820	$C_{41}H_{72}O_{16}S_0Na_0$
Culcitoside C2	118	790	$C_{40}H_{70}O_{15}S_0Na_0$
Culcitoside C3	117	774	$C_{40}H_{70}O_{14}S_0Na_0$
Culcitoside C4	92	730	$C_{38}H_{66}O_{13}S_0Na_0$
Culcitoside C5	101	746	$C_{38}H_{66}O_{14}S_0Na_0$
Culcitoside C6	119	774	$C_{40}H_{70}O_{14}S_0Na_0$
Culcitoside C7	120	876	$C_{40}H_{69}O_{17}S_1Na_1$
Culcitoside C8	112	758	$C_{39}H_{66}O_{14}S_0Na_0$
22-Dehydrohalityloside D	103	758	$C_{39}H_{66}O_{14}S_0Na_0$
22-Dehydrohalityloside E (poranoside A)	94	742	$C_{39}H_{66}O_{13}S_0Na_0$
5-Deoxyisonodososide	138	730	$C_{38}H_{66}O_{13}S_0Na_0$
Distolasteroside D1	95	716	$C_{37}H_{64}O_{13}S_0Na_0$
Distolasteroside D2	96	714	$C_{37}H_{62}O_{13}S_0Na_0$
Distolasteroside D3	97	746	$C_{38}H_{66}O_{14}S_0Na_0$
Echinasteroside A	152	726	$C_{34}H_{55}O_{13}S_1Na_1$

Name	Structure	M.W.	Molecular Formula
Echinasteroside B	153	742	$C_{35}H_{59}O_{13}S_1Na_1$
Echinasteroside B1	148	788	$C_{40}H_{68}O_{15}S_0Na_0$
Echinasteroside B2	147	746	$C_{38}H_{66}O_{14}S_0Na_0$
6-Epi-nodososide	76	746	$C_{38}H_{66}O_{14}S_0Na_0$
Forbeside K	161	758	$C_{39}H_{66}O_{14}S_0Na_0$
Forbeside L	160	640	$C_{35}H_{60}O_{10}S_0Na_0$
Glacialoside A	89	686	$C_{32}H_{55}O_{12}S_1Na_1$
Glacialoside B	104	702	$C_{32}H_{55}O_{13}S_1Na_1$
Gomophioside A	167	760	$C_{39}H_{68}O_{14}S_0Na_0$
Gomophioside B	122	788	$C_{41}H_{72}O_{14}S_0Na_0$
Granulatoside A	142	746	$C_{38}H_{66}O_{14}S_0Na_0$
Granulatoside B	139	714	$C_{38}H_{66}O_{12}S_0Na_0$
Halityloside A	126	790	$C_{40}H_{70}O_{15}S_0Na_0$
Halityloside B	125	774	$C_{40}H_{70}O_{14}S_0Na_0$
Halityloside D (culcitoside C1)	102	760	$C_{39}H_{68}O_{14}S_0Na_0$
Halityloside E	93	744	$C_{39}H_{68}O_{13}S_0Na_0$
Halityloside F	166	744	$C_{39}H_{68}O_{13}S_0Na_0$
Halityloside H	127	804	$C_{41}H_{72}O_{15}S_0Na_0$
Halityloside H, 6-O-sulphate	128	906	$C_{41}H_{71}O_{18}S_1Na_1$
Halityloside I	111	860	$C_{39}H_{65}O_{17}S_1Na_1$
Imbricatoside A	106	906	$C_{41}H_{71}O_{18}S_1Na_1$
Imbricatoside B	105	890	$C_{41}H_{71}O_{17}S_1Na_1$
Indicoside A	121	674	$C_{35}H_{62}O_{12}S_0Na_0$
Indicoside B	85	700	$C_{33}H_{57}O_{12}S_1Na_1$
Indicoside C	107	716	$C_{33}H_{57}O_{13}S_1Na_1$
Isonodososide	145	746	$C_{38}H_{66}O_{14}S_0Na_0$
Laeviuscoloside A	58	826	$C_{39}H_{63}O_{15}S_1Na_1$
Laeviuscoloside B	149	714	$C_{34}H_{59}O_{12}S_1Na_1$
Laeviuscoloside C	150	728	$C_{34}H_{57}O_{13}S_1Na_1$
Laeviuscoloside D	151	726	$C_{34}H_{55}O_{13}S_1Na_1$
Laeviuscoloside E [4(5)-dihydro-Echinasteroside B]	154	744	$C_{35}H_{61}O_{13}S_1Na_1$
Laeviuscoloside F	144	744	$C_{39}H_{68}O_{13}S_0Na_0$
Laeviuscoloside G (forbeside J)	143	760	$C_{39}H_{68}O_{14}S_0Na_0$
Laeviuscoloside H (22-dehydro-laeviuscoloside I)	141	626	$C_{34}H_{58}O_{10}S_0Na_0$
Laeviuscoloside I (forbeside I)	140	628	$C_{34}H_{60}O_{10}S_0Na_0$
Miniatoside A	80	744	$C_{35}H_{61}O_{13}S_1Na_1$
Miniatoside B	66	818	$C_{37}H_{63}O_{16}S_1Na_1$
Moniloside A	59	562	$C_{33}H_{54}O_7S_0Na_0$
Moniloside B	60	578	$C_{33}H_{54}O_8S_0Na_0$
Moniloside C	169	598	$C_{33}H_{58}O_9S_0Na_0$
Moniloside D	168	614	$C_{33}H_{58}O_{10}S_0Na_0$
Moniloside E	129	790	$C_{40}H_{70}O_{15}S_0Na_0$
Moniloside F (22-dehydro-moniloside E)	130	788	$C_{40}H_{68}O_{15}S_0Na_0$
Moniloside G	131	936	$C_{46}H_{80}O_{19}S_0Na_0$
Moniloside H (22-dehydro-moniloside G)	132	934	$C_{46}H_{78}O_{19}S_0Na_0$
Nodososide	146	746	$C_{38}H_{66}O_{14}S_0Na_0$
Pisasteroside A	116	744	$C_{34}H_{57}O_{14}S_1Na_1$
Pisasteroside B	136	686	$C_{32}H_{55}O_{12}S_1Na_1$
Pisasteroside C	124	744	$C_{34}H_{57}O_{14}S_1Na_1$
Pisasteroside D	156	684	$C_{32}H_{53}O_{12}S_1Na_1$
Pisasteroside E	133	684	$C_{32}H_{53}O_{12}S_1Na_1$
Pisasteroside F	123	760	$C_{35}H_{61}O_{14}S_1Na_1$
Placentoside A	114	744	$C_{38}H_{64}O_{14}S_0Na_0$
Pycnopodioside A	86	584	$C_{32}H_{56}O_9S_0Na_0$

References, pp. 297–308

Name	Structure	M.W.	Molecular Formula
Pycnopodioside B	87	686	$C_{32}H_{55}O_{12}S_1Na_1$
Pycnopodioside C	88	716	$C_{33}H_{57}O_{13}S_1Na_1$
Scoparioside A	83	686	$C_{32}H_{55}O_{12}S_1Na_1$
Scoparioside B	84	686	$C_{32}H_{55}O_{12}S_1Na_1$
Scoparioside C	90	700	$C_{33}H_{57}O_{12}S_1Na_1$
Scoparioside D (22-dehydro-coscinasteroside B)	135	684	$C_{32}H_{53}O_{12}S_1Na_1$
Thromidioside	163	642	$C_{34}H_{58}O_{11}S_0Na_0$

Molecular Weight Index of Polyhydroxysteroid Glycosides

Name	Structure	M.W.	Molecular Formula
Amurensoside A	61	670	$C_{32}H_{55}O_{11}S_1Na_1$
Amurensoside B	62	686	$C_{32}H_{55}O_{12}S_1Na_1$
Amurensoside C	63	668	$C_{32}H_{53}O_{11}S_1Na_1$
Amurensoside D	98	584	$C_{32}H_{56}O_9S_0Na_0$
Aphelasteroside A	137	686	$C_{32}H_{55}O_{12}S_1Na_1$
Aphelasteroside B	155	698	$C_{32}H_{51}O_{13}S_1Na_1$
Asterosaponin P-1	64	700	$C_{33}H_{57}O_{12}S_1Na_1$
Asterosaponin P-2	81	760	$C_{35}H_{61}O_{14}S_1Na_1$
Attenuatoside A-1	91	730	$C_{38}H_{66}O_{13}S_0Na_0$
Attenuatoside A-II	82	584	$C_{32}H_{56}O_9S_0Na_0$
Attenuatoside B-1	100	746	$C_{38}H_{66}O_{14}S_0Na_0$
Attenuatoside B-2	99	600	$C_{32}H_{56}O_{10}S_0Na_0$
Attenuatoside C	74	732	$C_{37}H_{64}O_{14}S_0Na_0$
Attenuatoside S-I	157	730	$C_{34}H_{59}O_{13}S_1Na_1$
Attenuatoside S-II (22-dehydro-attenuatoside S-I)	158	728	$C_{34}H_{57}O_{13}S_1Na_1$
Attenuatoside S-III	159	730	$C_{34}H_{59}O_{13}S_1Na_1$
Borealoside A	69	832	$C_{38}H_{65}O_{16}S_1Na_1$
Borealoside B	70	818	$C_{37}H_{63}O_{16}S_1Na_1$
Borealoside C	71	598	$C_{33}H_{58}O_9S_0Na_0$
Borealoside D	75	614	$C_{33}H_{58}O_{10}S_0Na_0$
Coscinasteroside A	110	730	$C_{33}H_{55}O_{14}S_1Na_1$
Coscinasteroside B	134	686	$C_{32}H_{55}O_{12}S_1Na_1$
Coscinasteroside C	115	744	$C_{34}H_{57}O_{14}S_1Na_1$
Coscinasteroside D	113	730	$C_{33}H_{55}O_{14}S_1Na_1$
Coscinasteroside E	109	616	$C_{32}H_{56}O_{11}S_0Na_0$
–	162	626	$C_{34}H_{58}O_{10}S_0Na_0$
Coscinasteroside F	108	600	$C_{32}H_{56}O_{10}S_0Na_0$
–	77	614	$C_{33}H_{58}O_{10}S_0Na_0$
Crossasteroside A	72	760	$C_{39}H_{68}O_{14}S_0Na_0$
Crossasteroside B	67	744	$C_{39}H_{68}O_{13}S_0Na_0$
Crossasteroside C	68	730	$C_{38}H_{66}O_{13}S_0Na_0$
Crossasteroside D	73	746	$C_{38}H_{66}O_{14}S_0Na_0$
Crossasteroside P1	164	804	$C_{41}H_{72}O_{15}S_0Na_0$
Crossasteroside P2	165	820	$C_{41}H_{72}O_{16}S_0Na_0$
Culcitoside C2	118	790	$C_{40}H_{70}O_{15}S_0Na_0$
Culcitoside C3	117	774	$C_{40}H_{70}O_{14}S_0Na_0$
Culcitoside C4	92	730	$C_{38}H_{66}O_{13}S_0Na_0$
Culcitoside C5	101	746	$C_{38}H_{66}O_{14}S_0Na_0$
Culcitoside C6	119	774	$C_{40}H_{70}O_{14}S_0Na_0$
Culcitoside C7	120	876	$C_{40}H_{69}O_{17}S_1Na_1$
Culcitoside C8	112	758	$C_{39}H_{66}O_{14}S_0Na_0$
22-Dehydrohalityloside D	103	758	$C_{39}H_{66}O_{14}S_0Na_0$
22-Dehydrohalityloside E (poranoside A)	94	742	$C_{39}H_{66}O_{13}S_0Na_0$
5-Deoxyisonodososide	138	730	$C_{38}H_{66}O_{13}S_0Na_0$

Name	Structure	M.W.	Molecular Formula
Distolasteroside D1	95	716	$C_{37}H_{64}O_{13}S_0Na_0$
Distolasteroside D2	96	714	$C_{37}H_{62}O_{13}S_0Na_0$
Distolasteroside D3	97	746	$C_{38}H_{66}O_{14}S_0Na_0$
Echinasteroside A	152	726	$C_{34}H_{55}O_{13}S_1Na_1$
Echinasteroside B	153	742	$C_{35}H_{59}O_{13}S_1Na_1$
Echinasteroside B1	148	788	$C_{40}H_{68}O_{15}S_0Na_0$
Echinasteroside B2	147	746	$C_{38}H_{66}O_{14}S_0Na_0$
6-Epi-Nodososide	76	746	$C_{38}H_{66}O_{14}S_0Na_0$
Forbeside K	161	758	$C_{39}H_{66}O_{14}S_0Na_0$
Forbeside L	160	640	$C_{35}H_{60}O_{10}S_0Na_0$
Glacialoside A	89	686	$C_{32}H_{55}O_{12}S_1Na_1$
Glacialoside B	104	702	$C_{32}H_{55}O_{13}S_1Na_1$
Gomophioside A	167	760	$C_{39}H_{68}O_{14}S_0Na_0$
Gomophioside B	122	788	$C_{41}H_{72}O_{14}S_0Na_0$
Granulatoside A	142	746	$C_{38}H_{66}O_{14}S_0Na_0$
Granulatoside B	139	714	$C_{38}H_{66}O_{12}S_0Na_0$
Halityloside A	126	790	$C_{40}H_{70}O_{15}S_0Na_0$
Halityloside B	125	774	$C_{40}H_{70}O_{14}S_0Na_0$
Halityloside D (culcitoside C1)	102	760	$C_{39}H_{68}O_{14}S_0Na_0$
Halityloside E	93	744	$C_{39}H_{68}O_{13}S_0Na_0$
Halityloside F	166	744	$C_{39}H_{68}O_{13}S_0Na_0$
Halityloside H	127	804	$C_{41}H_{72}O_{15}S_0Na_0$
Halityloside H, 6-O-sulphate	128	906	$C_{41}H_{71}O_{18}S_1Na_1$
Halityloside I	111	860	$C_{39}H_{65}O_{17}S_1Na_1$
Imbricatoside A	106	906	$C_{41}H_{71}O_{18}S_1Na_1$
Imbricatoside B	105	890	$C_{41}H_{71}O_{17}S_1Na_1$
Indicoside A	121	674	$C_{35}H_{62}O_{12}S_0Na_0$
Indicoside B	85	700	$C_{33}H_{57}O_{12}S_1Na_1$
Indicoside C	107	716	$C_{33}H_{57}O_{13}S_1Na_1$
Isonodososide	145	746	$C_{38}H_{66}O_{14}S_0Na_0$
Laeviuscoloside A	58	826	$C_{39}H_{63}O_{15}S_1Na_1$
Laeviuscoloside B	149	714	$C_{34}H_{59}O_{12}S_1Na_1$
Laeviuscoloside C	150	728	$C_{34}H_{57}O_{13}S_1Na_1$
Laeviuscoloside D	151	726	$C_{34}H_{55}O_{13}S_1Na_1$
Laeviuscoloside E [4(5)-dihydro-echinasteroside B]	154	744	$C_{35}H_{61}O_{13}S_1Na_1$
Laeviuscoloside F	144	744	$C_{39}H_{68}O_{13}S_0Na_0$
Laeviuscoloside G (forbeside J)	143	760	$C_{39}H_{68}O_{14}S_0Na_0$
Laeviuscoloside H (22-dehydro-laeviuscoloside I)	141	626	$C_{34}H_{58}O_{10}S_0Na_0$
–	79	628	$C_{34}H_{60}O_{10}S_0Na_0$
Laeviuscoloside I (forbeside I)	140	628	$C_{34}H_{60}O_{10}S_0Na_0$
–	78	630	$C_{33}H_{58}O_{11}S_0Na_0$
Miniatoside A	80	744	$C_{35}H_{61}O_{13}S_1Na_1$
Miniatoside B	66	818	$C_{37}H_{63}O_{16}S_1Na_1$
Moniloside A	59	562	$C_{33}H_{54}O_7S_0Na_0$
Moniloside B	60	578	$C_{33}H_{54}O_8S_0Na_0$
Moniloside C	169	598	$C_{33}H_{58}O_9S_0Na_0$
Moniloside D	168	614	$C_{33}H_{58}O_{10}S_0Na_0$
Moniloside E	129	790	$C_{40}H_{70}O_{15}S_0Na_0$
Moniloside F (22-dehydro-moniloside E)	130	788	$C_{40}H_{68}O_{15}S_0Na_0$
Moniloside G	131	936	$C_{46}H_{80}O_{19}S_0Na_0$
Moniloside H (22-dehydro-moniloside G)	132	934	$C_{46}H_{78}O_{19}S_0Na_0$
Nodososide	146	746	$C_{38}H_{66}O_{14}S_0Na_0$
Pisasteroside A	116	744	$C_{34}H_{57}O_{14}S_1Na_1$
Pisasteroside B	136	686	$C_{32}H_{55}O_{12}S_1Na_1$

References, pp. 297–308

Name Index of Polyhydroxysteroids

Name	Structure	M.W.	Molecular Formula
5β-Cholestane-3α,4α,11β,21-tetraol,3,21-sodium disulphate	252	640	$C_{27}H_{46}O_{10}S_2Na_2$
(22R)-5α-Cholestane-3β,6α,22-triol,22-acetate-3,6-sodium disulphate	236	666	$C_{29}H_{48}O_{10}S_2Na_2$
(22R)-5α-Cholestane-3β,6α,22-triol,3,6-sodium disulphate	235	624	$C_{27}H_{46}O_9S_2Na_2$
5α-Cholestane-2β,3α,26-triol, 2,3,26-sodium tri-sulphate	256	726	$C_{27}H_{45}O_{12}S_3Na_3$
(22E)-5α-Cholest-22-ene-3β,21-diol	240	402	$C_{27}H_{46}O_2S_0Na_0$
Cholest-5-ene-3β,21-diol	243	402	$C_{27}H_{46}O_2S_0Na_0$
Cholest-5-ene-3β,21-diol,3,21-sodium disulphate	248	606	$C_{27}H_{44}O_8S_2Na_2$
(22E,24R)-5α-Cholest-22-ene-3β,4β,6α,8,15β,16β,24-heptaol, 3-sodium sulphate	231	584	$C_{27}H_{45}O_{10}S_1Na_1$
(22E)-5α-Cholest-24(25)-ene-3β,4β,6α8,14,15α,26-heptaol-15-sodium sulphate	217	584	$C_{27}H_{45}O_{10}S_1Na_1$
(22E,25S)-5α-Cholest-22-ene-3β,6α,8,15β,16β,26-hexaol	185	466	$C_{27}H_{46}O_6S_0Na_0$
(24E)-5α-Cholest-24-(25)-ene-3β,6α,8,14,15α,26-hexaol-15-sodium sulphate	216	568	$C_{27}H_{45}O_9S_1Na_1$
(24E)-5α-Cholest-24-ene-3β,4β,5,6α,7β,8,14,15α,26-nonaol,6-sodium sulphate	224	616	$C_{27}H_{45}O_{12}S_1Na_1$
(22E,25S)-5α-Cholest-22-ene-3β,4β,6β,7α,8,15β,16β,26-octaol	208	498	$C_{27}H_{46}O_8S_0Na_0$
(22E,25S)-5α-Cholest-22-ene-3β,6α,8,15β,26-pentaol	184	450	$C_{27}H_{46}O_5S_0Na_0$
(22E,24R)-5α-Cholest-22-ene-3β,6α,8,15β,24-pentaol	230	450	$C_{27}H_{46}O_5S_0Na_0$
(22E,24R)-5α-Cholest-22-ene-3β,6β,8,15α,24-pentaol,15-sodium sulphate	232	552	$C_{27}H_{45}O_8S_1Na_1$
5β-Cholest-25(26)-ene-2β,3α,4α,21-tetraol,3,21-sodium disulphate	263	638	$C_{27}H_{44}O_{10}S_2Na_2$
(22E)-5α-Cholest-22-ene-3α,4α,11β,21-tetraol,3,21-sodium disulphate	254	638	$C_{27}H_{44}O_{10}S_2Na_2$
5β-Cholest-25(26)-ene-3β,4α,5,21-tetraol,3,21-sodium disulphate	262	638	$C_{27}H_{44}O_{10}S_2Na_2$
(22R)-5α-Cholest-9(11)-ene-3β,6α,22-triol,22-acetate-3,6-sodium disulphate	234	664	$C_{29}H_{46}O_{10}S_2Na_2$
(22R)-5α-Cholest-9(11)-ene-3β,6α,22-triol,3,6-sodium disulphate	233	622	$C_{27}H_{44}O_9S_2Na_2$
(23S)-5α-Cholest-9(11)-ene-3β,6α,23-triol,3,6-sodium disulphate	237	622	$C_{27}H_{44}O_9S_2Na_2$
Cholest-5-ene-3α,4β,21-triol,3,21-sodium disulphate	265	622	$C_{27}H_{44}O_9S_2Na_2$
5β-Cholest-25(26)-ene-3α,4α,21-triol,3,21-sodium disulphate	258	622	$C_{27}H_{44}O_9S_2Na_2$
5β-Cholest-9(11)-ene-3α,4α,21-triol,3,21-sodium disulphate	267	622	$C_{27}H_{44}O_9S_2Na_2$
5α-Cholest-5-ene-2β,3α,26-triol,2,3,26-sodium trisulphate	257	724	$C_{27}H_{43}O_{12}S_3Na_3$
Ergosta-5,24(28)-diene-3β-21-diol	244	414	$C_{28}H_{46}O_2S_0Na_0$
Ergosta-5,24(28)-diene-3α,21-diol,3,21-sodium disulphate	250	618	$C_{28}H_{44}O_8S_2Na_2$
(24R,25S)-5α-Ergostane-3β,6α,8,15α,16β,26-hexaol	177	482	$C_{28}H_{50}O_6S_0Na_0$
(25S)-5α-crgostane-3β,4β,6β,7α,8,15β,16β,26-octaol	209	514	$C_{28}H_{50}O_8S_0Na_0$
(24S,25S)-5β-Ergostane-3α,4α,21,26-tetraol,3,21-sodium disulphate	260	654	$C_{28}H_{48}O_{10}S_2Na_2$

References, pp. 297–308

Name	Structure	M.W.	Molecular Formula
5α-Ergost-24(28)-ene-3β,21-diol	241	416	$C_{28}H_{48}O_2S_0Na_0$
(22E)-5α-Ergost-22-ene-3β,21-diol	242	416	$C_{28}H_{48}O_2S_0Na_0$
(25S)-5α-Ergost-24(28)-ene-3β,4β,6α,8,15β,16β,26-heptaol	187	496	$C_{28}H_{48}O_7S_0Na_0$
(22E,24R,25S)-5α-Ergost-22-ene-3β,4β,6α,8,15α,16β,26-heptaol	178	496	$C_{28}H_{48}O_7S_0Na_0$
(25S)-5α-Ergost-24(28)-ene-3β,6α,8,14,15β,16β,26-heptaol	190	496	$C_{28}H_{48}O_7S_0Na_0$
(25S)-5α-Ergost-24(28)-ene-3β,6β,8,15α,16β,26-hexaol	202	480	$C_{28}H_{48}O_6S_0Na_0$
(25S)-5α-Ergost-24(28)-ene-3β,6α,8,15β,16β,26-hexaol	186	480	$C_{28}H_{48}O_6S_0Na_0$
(22E,24R,25R)-5α-Ergost-22-ene-3β,4β,5,6α,8,14,15α,25,26-Nonaol	220	528	$C_{28}H_{48}O_9S_0Na_0$
(22E,24S)-5α-Ergost-22-ene-3β,4β,5,6α,8,14,15α,25,28-nonaol	221	528	$C_{28}H_{48}O_9S_0Na_0$
(22E,24R,25R)⁵5α-Ergost-22-ene-3β,4β,6α,7α,8,15β,16β,25,26-nonaol,6-sodium sulphate	191	630	$C_{28}H_{47}O_{12}S_1Na_1$
(25S)-5α-Ergost-24(28)-ene-3β,4β,6α,7α,8,15β,16β,26-octaol	188	512	$C_{28}H_{48}O_8S_0Na_0$
(25S)-5α-Ergost-24(28)-ene-3β,6β,15α,16β,26-pentaol	193	464	$C_{28}H_{48}O_5S_0Na_0$
(24S)-5β-Ergost-22-ene-2β,3α,4α,21,28-pentaol,3,21-sodium disulphate	264	668	$C_{28}H_{46}O_{11}S_2Na_2$
5β-Ergost-24(28)-ene-3α,4α,11β,21-tetraol,3,21-sodium disulphate	253	652	$C_{28}H_{46}O_{10}S_2Na_2$
(25S)-N-(2'-Ethane sodium sulphonate)-3β,5,6β,15α-tetrahydroxy-5α-cholestan-27-amide	212	595	$C_{29}H_{50}N_1O_8S_1Na_1$
(24R,25S)-N-(2'-Ethane sodium sulphonate)-3β,5,6β,15α-tetrahydroxy-5α-ergost-22-en-27-amide	213	607	$C_{30}H_{50}N_1O_8S_1Na_1$
3β,4β,5,6α,8,14,15α,24-Heptahydroxy-27-nor-5α-cholestan-24-one	219	484	$C_{26}H_{44}O_8S_0Na_0$
(22E)-24-Methyl-26,27-bisnor-5α-cholest-22-ene-3β,4β,6β,8,15α,16β,25-heptaol	205	468	$C_{26}H_{44}O_7S_0Na_0$
24-Methyl-27-nor-5α-cholestane-3β,6β,8,15α,16β,25-heptaol	203	468	$C_{27}H_{48}O_6S_0Na_0$
24-Methyl-27-nor-5α-cholestane-3β,4β,6β,8,15α,16β,26-heptaol	204	484	$C_{27}H_{48}O_7S_0Na_0$
(24S)-24-Methyl-27-nor-3β,5,6β,15α-tetrahydroxy-5α-cholest-22-en-26-oic acid	214	464	$C_{27}H_{44}O_6S_0Na_0$
(24R)-27-Nor-5α-cholestane-3β,4β,5,6α,7β,8,14,15α,24α-nonaol	222	502	$C_{26}H_{46}O_9S_0Na_0$
(24R)-27-Nor-5α-cholestane-3β,4β,5,6α,7β,8,14,15α,24α-nonaol,6-sodium sulphate	223	604	$C_{26}H_{45}O_{12}S_1Na_1$
(24R)-27-Nor-5α-cholestane-3β,4β,5,6α,8,14,15α,24-octaol	218	486	$C_{26}H_{46}O_8S_0Na_0$
(24R)-5α-Stigmastane-3β,6α,8,15α,16β,29-hexaol,29-sodium sulphate	179	598	$C_{29}H_{51}O_9S_1Na_1$
(24S,25S)-5β-Stigmastane-3α,4α,21,26-tetraol,3,21-sodium disulphate	261	668	$C_{29}H_{50}O_{10}S_2Na_2$
(22R)-5α-Stigmastane-3β,6α,22-triol,22-acetate,6-sodium phosphate,3-sodium sulphate	238	694	$C_{31}H_{53}O_{10}P_1S_1Na_2$
(24E)-5β-Stigmast-24(28)-ene-3α,4α,21,29-tetraol,3,21-sodium disulphate	259	666	$C_{29}H_{48}O_{10}S_2Na_2$
(24R)-3β,5,6β,15α-Stigmastan-29-oic acid	215	494	$C_{29}H_{50}O_6S_0Na_0$
3α,4α,21-Trihydroxy-5β-cholestan-11-one,3,21-sodium disulphate	255	638	$C_{38}H_{44}O_{10}S_2Na_2$

Molecular Weight Index of Polyhydroxysteroids

Name	Structure	M.W.	Molecular Formula
(22E)-Cholesta-5,22-diene-3α,21-diol,3,21-sodium disulphate	249	604	$C_{27}H_{42}O_8S_2Na_2$
5α-Cholestane-3β-21-diol	239	404	$C_{27}H_{48}O_2S_0Na_0$
5α-Cholestane-3α,21-diol,3,21-sodium disulphate	247	608	$C_{27}H_{46}O_8S_2Na_2$
(25S)-5α-Cholestane-3β,4β,6α,8,15β,16β,26-heptaol	181	484	$C_{27}H_{48}O_7S_0Na_0$
(25S)-5α-Cholestane-3β,4β,6α,8,15α,16β,26-heptaol	176	484	$C_{27}H_{48}O_7S_0Na_0$
(25S)-5α-Cholestane-3β,5,6β,7α,15α,16β,26-heptaol	196	484	$C_{27}H_{48}O_7S_0Na_0$
(25S)-5α-Cholestane-3β,6β,7α,8,15β,16β,26-heptaol	206	484	$C_{27}H_{48}O_7S_0Na_0$
(25S)-5α-Cholestane-3β,6β,7α,8,15α,16β,26-heptaol	200	484	$C_{27}H_{48}O_7S_0Na_0$
(25S)-5α-Cholestane-3β,6α,7α,8,15β,16β,26-heptaol	182	484	$C_{27}H_{48}O_7S_0Na_0$
(25S)-5α-Cholestane-3β,6α,7α,8,15α,16β,26-heptaol	174	484	$C_{27}H_{48}O_7S_0Na_0$
(25S)-5α-Cholestane-3β,6α,8,14,15β,16β,26-heptaol	189	484	$C_{27}H_{48}O_7S_0Na_0$
(24S)-5α-Cholestane-3β,4β,6α,8,15α,24-hexaol	229	468	$C_{27}H_{48}O_6S_0Na_0$
(25S)-5α-Cholestane-3β,5,6β,15α,16β,26-hexaol	194	468	$C_{27}H_{48}O_6S_0Na_0$
(25R)-5α-Cholestane-3β,5,6β,15α,16β,26-hexaol	197	468	$C_{27}H_{48}O_6S_0Na_0$
(25S)-5α-Cholestane-3β,6β,7α,15α,16β,26-hexaol	195	468	$C_{27}H_{48}O_6S_0Na_0$
(25S)-5α-Cholestane-3β,6β,8,15α,16β,26-hexaol	199	468	$C_{27}H_{48}O_6S_0Na_0$
(25S)-5α-Cholestane-3β,6α,8,15β,16β,26-hexaol	180	468	$C_{27}H_{48}O_6S_0Na_0$
(25S)-5α-Cholestane-3β,6α,8,15α,16β,26-hexaol	173	468	$C_{27}H_{48}O_6S_0Na_0$
(25S)-5α-Cholestane-3β,4β,6β,7α,8,15β,16β,26-octaol	207	500	$C_{27}H_{48}O_8S_0Na_0$
(25S)-5α-Cholestane-3β,4β,6α,7α,8,15β,16β,26-octaol	183	500	$C_{27}H_{48}O_8S_0Na_0$
(25S)-5α-Cholestane-3β,4β,6α,7α,8,15α,16β,26-octaol	175	500	$C_{27}H_{48}O_8S_0Na_0$
(25S)-5α-Cholestane-3β,4β,6β,7α,8,15α,16β,26-octaol	201	500	$C_{27}H_{48}O_8S_0Na_0$
(25S)-5α-Cholestane-3β,5,6β,15α,26-pentaol	210	452	$C_{27}H_{48}O_5S_0Na_0$
(25S)-5α-Cholestane-3β,6β,15α,16β,26-pentaol	192	452	$C_{27}H_{48}O_5S_0Na_0$
(24S)-5α-Cholestane-3β,6α,8,15β,24-pentaol	227	452	$C_{27}H_{48}O_5S_0Na_0$
(24S)-5α-Cholestane-3β,6α,8,15α,24-pentaol	225	452	$C_{27}H_{48}O_5S_0Na_0$
5β-Cholestane-3α,4α,11β,12β,21-pentaol,3,21-sodium disulphate	251	656	$C_{27}H_{46}O_{11}S_2Na_2$
(24S)-5α-Choleatane-3β,6α,8.15β,24-pentaol,24-sodium sulphate	228	554	$C_{27}H_{47}O_8S_1Na_1$
(24S)-5α-Choleatane-3β,6α,8,15α,24-pentaol,24-sodium sulphate	226	554	$C_{27}H_{47}O_8S_1Na_1$
(24S)-5α-Choleatane-3β,5,6β,15α,26-pentaol,15-sodium sulphate	211	554	$C_{27}H_{47}O_8S_1Na_1$
5-β-Cholestan-4α,9α-epoxy-3α,21-diol,3,21-sodium disulphate	266	622	$C_{27}H_{44}O_9S_2Na_2$
5β-Cholestane-3α,4α,11β,21-tetraol,3,21-sodium disulphate	252	640	$C_{27}H_{46}O_{10}S_2Na_2$
(22R)-5α-Cholestane-3β,6α,22-triol,22-acetate-3,6-sodium disulphate	236	666	$C_{29}H_{48}O_{10}S_2Na_2$
(22R)-5α-Cholestane-3β,6α,22-triol,3,6-sodium disulphate	235	624	$C_{27}H_{46}O_9S_2Na_2$
(25R)-5β-Cholestane-3α,6β,15α,16β,26-pentaol	198	452	$C_{27}H_{48}O_5S_0Na_0$
5α-Cholestane-2β,3α,26-triol,2,3,26-sodium trisulphate	256	726	$C_{27}H_{45}O_{12}S_3Na_3$
(22E)-5α-Cholest-22-ene-3β-21-diol	240	402	$C_{27}H_{46}O_2S_0Na_0$
Cholest-5-ene-3β-21-diol	243	402	$C_{27}H_{46}O_2S_0Na_0$
Cholest-5-ene-3α,21-diol,3,21-sodium disulphate	248	606	$C_{27}H_{44}O_8S_2Na_2$
(22E,24R)-5α-Cholest-22-ene-3β,4β,6α,8,15β,-16β,24-heptaol,3-sodium sulphate	231	584	$C_{27}H_{45}O_{10}S_1Na_1$
(24E)-5α-Cholest-24(25)-ene-3β,4β,6α,8,14,15α,26-heptaol-15-sodium sulphate	217	584	$C_{27}H_{45}O_{10}S_1Na_1$

References, pp. 297–308

Name	Structure	M.W.	Molecular Formula
(22E,25S)-5α-Cholest-22-ene-3β,6α,8,15β,16β,26-hexaol	185	466	$C_{27}H_{46}O_6S_0Na_0$
(24E)-5α-Cholest-24(25)-ene-3β,6α,8,14,15α,26-hexaol-15-sodium sulphate	216	568	$C_{27}H_{45}O_9S_1Na_1$
(24E)-5α-Cholest-24-ene-3β,4β,5,6α,7β,8,14,15α,26-nonaol,6-sodium sulphate	224	616	$C_{27}H_{45}O_{12}S_1Na_1$
(22E,25S)-5α-Cholest-22-ene-3β,4β,6β,7α,8,15β,-16β,26-octaol	208	498	$C_{27}H_{46}O_8S_0Na_0$
(22E,25S)-5α-Cholest-22-ene-3β,6α,8,15β,26-pentaol	184	450	$C_{27}H_{46}O_5S_0Na_0$
(22E,24R)-5α-Cholest-22-ene-3β,6α,8,15β,24-pentaol	230	450	$C_{27}H_{46}O_5S_0Na_0$
(22E,24R)-5α-Cholest-22-ene-3β,6β,8,15α,24-pentaol,15-sodium sulphate	232	552	$C_{27}H_{45}O_8S_1Na_1$
5β-Cholest-25(26)-ene-2β,3α,4α,21-tetraol,3,21-sodium disulphate	263	638	$C_{27}H_{44}O_{10}S_2Na_2$
(22E)-5β-Cholest-22-ene-3α,4α,11β,21-tetraol,3,21-sodium disulphate	254	638	$C_{27}H_{44}O_{10}S_2Na_2$
5β-Cholest-25(26)-ene-3α,4α,5,21-tetraol,3,21-sodium disulphate	262	638	$C_{27}H_{44}O_{10}S_2Na_2$
(22R)-5α-Cholest-9(11)-ene-3β,6α,22-triol,22-acetate-3,6-sodium disulphate	234	664	$C_{29}H_{46}O_{10}S_2Na_2$
(22R)-5α-Cholest-9(11)-ene-3β,6α,22-triol,3,6-sodium disulphate	233	622	$C_{27}H_{44}O_9S_2Na_2$
(23S)-5α-Cholest-9(11)-ene-3β,6α,23-triol,3,6-sodium disulphate	237	622	$C_{27}H_{44}O_9S_2Na_2$
Cholest-5-ene-3α-4β,21-triol,3,21-sodium disulphate	265	622	$C_{27}H_{44}O_9S_2Na_2$
5β-Cholest-25(26)-ene-3α,4α,21-triol,3,21-sodium disulphate	258	622	$C_{27}H_{44}O_9S_2Na_2$
5β-Cholest-9(11)-ene-3α,4α,21-triol,3,21-sodium disulphate	267	622	$C_{27}H_{44}O_9S_2Na_2$
5α-Cholest-5-ene-2β,3α,26-triol,2,3,26-sodium trisulphate	257	724	$C_{27}H_{43}O_{12}S_3Na_3$
Ergosta-5,24(28)-diene-3β-21-diol	244	414	$C_{28}H_{46}O_2S_0Na_0$
Ergosta-5,24(28)-diene-3α,21-diol,3,21-sodium disulphate	250	618	$C_{28}H_{44}O_8S_2Na_2$
(24R,25S)-5α-Ergostane-3β,6α,8,15α,16β,26-hexaol	177	482	$C_{28}H_{50}O_6S_0Na_0$
(25S)-5α-Ergostane-3β,4β,6β,7α,8,15β,16β,26-octaol	209	514	$C_{28}H_{50}O_8S_0Na_0$
(24S,25S)-5β-Ergostane-3α,4α,21,26-tetraol,3,21-sodium disulphate	260	654	$C_{28}H_{48}O_{10}S_2Na_2$
5α-Ergost-24(28)-ene-3β-21-diol	241	416	$C_{28}H_{48}O_2S_0Na_0$
(22E)-5α-Ergost-22-ene-3β-21-diol	242	416	$C_{28}H_{48}O_2S_0Na_0$
(25S)-5α-Ergost-24(28)-ene-3β,4β,6α,8,15β,16β,26-heptaol	187	496	$C_{28}H_{48}O_7S_0Na_0$
(22E,24R,25S)-5α-Ergost-22-ene-3β,4β,6α,8,15α,-16β,26-heptaol	178	496	$C_{28}H_{48}O_7S_0Na_0$
(25S)-5α-Ergost-24(28)-ene-3β,6α,8,14,15β,16β,26-heptaol	190	496	$C_{28}H_{48}O_7S_0Na_0$
(25S)-5α-Ergost-24(28)-ene-3β,6β,8,15α,16β,26-hexaol	202	480	$C_{28}H_{48}O_6S_0Na_0$
(25S)-5α-Ergost-24(28)-ene-3β,6α,8,15β,16β,26-hexaol	186	480	$C_{28}H_{48}O_6S_0Na_0$
(24E,24R,25R)-5α-Ergost-22-ene-3β,4β,5,6α,-8,14,15α,25,26-nonaol	220	514	$C_{28}H_{50}O_8S_0Na_0$
(24E,24S)-5α-Ergost-22-ene-3β,4β,5,6α,-8,14,15α,25,28-nonaol	221	528	$C_{28}H_{48}O_9S_0Na_0$
(22E,24R,25R)-5α-Ergost-22-ene-3β,4β,6α,7α,8,15β,-16β,25,26-nonaol,6-sodium sulphate	191	630	$C_{28}H_{47}O_{12}S_1Na_1$

Name	Structure	M.W.	Molecular Formula
(25S)-5α-Ergost-24(28)-ene-3β,4β,6α,7α,8,15β,-16β,26-octaol	188	512	$C_{28}H_{48}O_8S_0Na_0$
(25S)-5α-Ergost-24(28)-ene-3β,6β,15α,16β,26-pentaol	193	464	$C_{28}H_{48}O_5S_0Na_0$
(24S)-5β-Ergost-22-ene-2β,3α,4α,21,28-pentaol,3,21-sodium disulphate	264	668	$C_{28}H_{46}O_{11}S_2Na_2$
5β-Ergost-24(28)-ene-3α,4α,11β,21-tetraol,3,21-sodium disulphate	253	652	$C_{28}H_{46}O_{10}S_2Na_2$
(25S)-N-(2'-Ethane sodium sulphonate)-3β,5,6β,-15-tetrahydroxy-5α-Cholestan-27-amide	212	595	$C_{29}H_{50}N_1O_8S_1Na_1$
(24R,25S)-N-(2'-Ethane sodium sulphonate)-3β,5,6β,15-tetrahydroxy-5α-ergost-22-en-27-amide	213	607	$C_{30}H_{50}N_1O_8S_1Na_1$
3β,4β,5,6α,8,14,15α,24-Heptahydroxy-27-nor-5α-cholestan-24-one	219	484	$C_{26}H_{44}O_8S_0Na_0$
(22E)-24-Methyl-26,27-bisnor-5α-Cholest-22-ene-3β,4β,6β,8,15α,16β,25-heptaol	205	468	$C_{26}H_{44}O_7S_0Na_0$
24-Methyl-27-nor-5α-Cholestane-3β,6β,8,15α,16β,26-hexaol	203	468	$C_{27}H_{48}O_6S_0Na_0$
24-Methyl-27-nor-5α-cholestane-3β,4β,6β,8,15α,-16β,26-heptaol	204	484	$C_{27}H_{48}O_7S_0Na_0$
(24S)-24-Methyl-27-nor-3β,5,6β,15α-tetrahydroxy-5α-cholest-22-en-26-oic acid	214	464	$C_{27}H_{44}O_6S_0Na_0$
(24R)-27-Nor-5α-Cholestane-3β,4β,5,6α,7β,-8,14,15α,24-nonaol	222	502	$C_{26}H_{46}O_9S_0Na_0$
(24R)-27-Nor-5α-cholestane-3β,4β,5,6α,7β,-8,14,15α,24α-nonaol,6-sodium sulphate	223	604	$C_{26}H_{45}O_{12}S_1Na_1$
(24R)-27-Nor-5α-cholestane-3β,4β,5,6α,8,14,15α,-24-octaol	218	486	$C_{26}H_{46}O_8S_0Na_0$
(24R)-5α-Stigmastane-3β,6α,8,15α,16β,29-hexaol,29-sodium sulphate	179	598	$C_{29}H_{51}O_9S_1Na_1$
(24S,25S)-5β-Stigmastane-3α,4α,21,26-tetraol,3,21-sodium disulphate	261	668	$C_{29}H_{50}O_{10}S_2Na_2$
(22R)-5β-Stigmastane-3β,6α,22-triol,22-acetate,6-sodium phosphate,3-sodium sulphate	238	694	$C_{31}H_{53}O_{10}P_1S_1Na_2$
(24E)-5β-Stigmast-24(28)-ene-3α,4α,21,29-tetraol,3,21-sodium disulphate	259	666	$C_{29}H_{48}O_{10}S_2Na_2$
(24R)-3β,5,6β,15α-tetrahydroxy-5α-stigmastan-29-oic acid	215	494	$C_{29}H_{50}O_6S_0Na_0$
3α,4α,21-Trihydroxy-5β-cholestan-11-one,3,21-sodium disulphate	255	638	$C_{27}H_{44}O_{10}S_2Na_2$

References

1. BURNELL, D. J. and J. W. APSIMON: Echinoderm Saponins. In: Marine Natural Products, Chemical and Biological Perspectives (SCHEUER, P. J., ed.) Vol. V, p. 287. New York: Academic Press 1983.
2. KOBAYASHI, M., Y. KIYOTA, O. SATOMI, Y. KYOGOKU, and I. KITAGAWA: Five New Steroidal Glycosides, Pregnedioside-A, -B, and their three Monoacetals, from an Okinawan Soft Coral of *Alcyonium* sp. Tetrahedron Lett., **25**, 3731 (1984).
3. BANDURRAGA, M. M., and W. FENICAL: Isolation of Muricins. Evidence of a Chemical Adaptation against Fouling in the Marine Octocoral *Murina fruticosa* (Gorgonacea). Tetrahedron, **41**, 1057 (1985).
4. FUSETANI, N., K. YASUKAWA, S. MATSUNAYA, and K. HASHIMOTO: Dimorphosides A and B, Novel Steroid Glycosides from the Gorgonian *Anthoplexaura dimorpha*. Tetrahedron Lett., **28**, 1187 (1987).
5. WASYLYK, J. M., G. E. MARTIN, A. J. WEINHEIMER, and M. ALAM: Isolation and Structure Identification of a new Pregnane Glycoside from the Gorgonian *Pseudoplexaura wayenaori*. J. Nat. Prod., **52**, 391 (1989).
6. KITAGAWA, I., M. KOBAYASHI, Y. OKAMOTO, M. YOSHIKAWA, and Y. HAMAMOTO: Structures of Sarasinosides A_1, B_1, and C_1; New Norlanostane Triterpenoid Oligoglycosides from the Paluan Marine Sponge *Asteropus sarasinosum*. Chem. Pharm. Bull., **35**, 5036 (1987).
7. SCHMITZ, F. J., M. B. KSEBATI, S. P. GUNASEKERA, and S. AGARWAL: Sarasinoside A_1: A Saponin Containing Amino Sugars Isolated from a Sponge. J. Org. Chem., **53**, 594 (1988).
8. CARMELY, S., M. ROLL, Y. LOYA, and Y. KASHMAN: The Structure of Eryloside, a New Antitumor and Antifungal 4-Methylated Steroidal Glycoside from the Sponge *Erylus lendenfeldi*. J. Nat. Prod., **52**, 167 (1989).
9. HIROTA, H., S. TAKAYAMA, S. MIYASHIRO, Y. OZAKI, and S. IKEGAMI: Structure of a Novel Steroidal Saponin, Pachastrelloside A, Obtained from a Marine Sponge of the Genus *Pachastrella*. Tetrahedron Lett., **31**, 3321 (1990).
10. D' AURIA, M. V., L. GOMEZ PALOMA, L. MINALE, and R. RICCIO, and C. DEBITUS: Structure Characterization, by Two-Dimensional NMR Spectroscopy, of Two Marine Triterpene Oligoglycosides from a Pacific Sponge of the Genus *Erylus*. Tetrahedron, **48**, 491 (1992).
11. TACHIBANA, K., M. SAKAITANI, and K. NAKANISHI: Pavonins, Shark-Repelling and Ichthyotoxic Steroid N-Acetylglucosaminides from the Defense secretion of the Sole *Pardachirus Pavininus* (Soleidae). Tetrahedron, **41**, 1027 (1985).
12. GOAD, L. J.: The Sterols of Marine Invertebrates: Composition, Biosynthesis and Metabolites. In: Marine Natural Products. Chemical and Biological Perspectives (SCHEUER, P. J., ed.), Vol. II, p. 75. New York: Academic Press. 1978.
13. STONIK, V. A., and G. B. ELYAKOV: Secondary Metabolites from Echinoderms as Chemotaxonomic Markers. In: Bioorganic Marine Chemistry (SCHEUER, P. J., ed.), Vol. 2, p. 43. Berlin Heidelberg New York Tokyo: Springer. 1988.
14. RICCIO, R., M. V. D'AURIA, and L. MINALE: Two New Steroidal Glycoside Sulfates, Longicaudoside-A and -B, from the Mediterranean Ophiuroid *Ophioderma longicaudum*. J. Org. Chem., **51**, 533 (1986).
15. HASHIMOTO, Y. : Marine Toxins and Other Bioactive Marine Metabolites, pp. 268–288. Tokyo: Japan Scientific Societies Press, 1979.
16. KREBS, H. CHR.: Recent Developments in the Field of Marine Natural Products with Emphasis on Biologically Active Compounds. In: Progress in the Chemistry of

Organic Natural Products (Herz, W., H. Grisebach, G. W. Kirby, and Ch. Tamm eds.) Vol. 50, p. 151. Wien, New York: Springer. 1986.

17. Minale, L., R. Riccio, C. Pizza, and F. Zollo: Steroidal Olygoglycosides of Marine Origin. In: Natural Products and Biological Activities. A Naito Foundation Symposium (Imura, H., T. Goto, T. Murachi, and T. Nakajima, eds.), p. 59. Tokyo: University of Tokyo Press. 1986.

18. Quinn, R. J. : Chemistry of Aqueous Marine Extracts: Isolation Techniques. In: Biorganic Marine Chemistry (Scheuer, P. J. ed.), Vol. 2, p.1. Berlin Heidelberg New York Tokyo: Springer. 1988.

19. Habermehl, G. G., and H. Chr. Krebs: Toxins of Echinoderms. In: Studies in Natural Products Chemistry, (Atta-Ur-Rahman, ed.), Vol. 7, p. 265. Amsterdam: Elsevier Science Publishers B. V. 1990.

20. Hashimoto, Y., and T. Yasumoto: Confirmation of Saponin as a Toxic Principle of Starfish. Bull. Japn. Soc. Scient. Fish., 26, 1132 (1960).

21. Yasumoto, T., T. Wanatabe, and Y. Hashimoto: Physiological Activities of Starfish Saponin. Bull. Japn. Soc. Sci. Fish., 30, 357 (1964).

22. Owellen, R. J., R. G. Owellen, M. A. Gorof, and D. Klein: Cytolytic Saponin Fraction from Asterias vulgaris. Toxicon, 11, 319 (1973).

23. Mackie, A. M., R. Lasker, and P. T. Grant: Avoidance Reactions of Mollusc Buccinum undatum to Saponin-like Surface-active Substances in Extracts of the Starfish Asterias rubens and Marthasterias glacialis. Comp. Biochem. Physiol., 36, 415 (1968).

24. Ikegami, S., Y. Kamiya, and S. Tamura: Isolation and Characterization of Spawning Inhibitors in Ovary of the Starfish Asterias amurensis. Agr. Biol. Chem., 36, 2005 (1972).

25. Fujimoto, Y., T. Yamada, N. Ikekawa, I. Nishiyama, T. Matsui, and M. Hoshi: Structure of Acrosome Reaction-Inducing Steroidal Saponins from the Egg Jelly of the Starfish Asterias amurensis. Chem. Pharm. Bull., 35, 1829 (1987).

26. Nigrelli, R. F., M. F. Stempien, Jr., G. D. Ruggeri, V. R. Liguori, and J. T. Cecil: Substances of Potential Biomedical Importance from Marine Organisms. Fed. Proceed., 26, 1197 (1967).

27. Ruggeri, G. D., and R. F. Nigrelli: Physiologically Active Substances from Echinoderms. In: Bioactive Compounds from the Sea (Humm, H. J., and C. E. Lane, eds.), Marine Science Series, Vol. 1, p. 183. New York: Marcel Dekker 1974.

28. Pettit, G. R., J. F. Day, J. L. Hartwell, H. B. Wood: Antineoplastic Components of Marine Animals. Nature, 227, 962 (1970).

29. Shimizu, Y.: Antiviral Substances in Starfish. Experientia, 27, 1188 (1971).

30. Friess, S. L., R. C. Durant, and J. D. Chanley: Further Studies on Biological Actions of Steroidal Saponins Produced by Poisonous Echinoderms. Toxicon, 6, 81 (1968).

31. Friess, S. L. : Mode of Action of Marine Saponins on Neuromuscolar Tissue. Fed. Proceed., 31, 1146 (1972).

32. Goldsmith, L. A., and G. P. Carlson: Pharmacological Evolution of an Asterosaponin from Asterias forbesi. In: Food-Drugs from the Sea, Proc. 1974 (Webber, H. H., and G. D. Ruggeri eds.), p. 354. Washington: Marine Technol. Soc. 1976.

33. Ruggeri G. D., and R. F. Nigrelli: Effects of Extracts of the Sea Star Acanthaster planci on the Developping Sea Urchin. Am. Soc. Biol., 6, 592 (1966).

34. Fusetani, N., Y. Kato, K. Hashimoto, T. Komori, Y. Itakura, and T. Kawasaki: Biological Activities of Asterosaponins with Special reference to Structure-Activity Relationships. J. Nat. Prod., 47, 997 (1984).

35. BRUNO, I: Isolamento, Determinazione Strutturale e Attività Farmacologica di Nuovi Steroidi Poliossidrilati e Oligoglicosidi da Echinodermi. Tesi di Dottorato di Ricerca (Ph. D. Thesis), Facoltà di Farmacia, Università degli Studi di Napoli "Federico II", 1990, Biblioteca Nazionale, Roma, Firenze.

36. ANDERSSON, L., L. BOHLIN, M. IORIZZI, R. RICCIO, L. MINALE, and W. MORENO-LOPEZ: Biological Activity of Saponins and Saponin-like Compounds from Starfishs and Brittle-stars. Toxicon, 27, 179 (1989).

37. DUBOIS, M. A., R. HIGUCHI, T. KOMORI, and T. SASAKI: Structures of Two New Oligoglycoside Sulfates, Pectiniosides E and F, and Biological Activities of the Six New Pectiniosides. Liebigs Ann. Chem., 845 (1988).

38. VERBIST, J. F.: Pharmacological Effects of Compounds from Echinoderms. In: Echinoderm Studies (Jangoux, M., ed.), Vol. I, Bolkema Rotterdam, 1991,

39. MINALE, L., C. PIZZA, R. RICCIO, and F. ZOLLO: Steroidal Glycosides from Starfishs. Pure and Appl. Chem., 54, 1935 (1982).

40. DE RICCARDIS, F., M. IORIZZI, L. MINALE, R. RICCIO, and C. DEBITUS: The First Occurrence of Polyhydoxylated Steroids with Phosphate Conjugation from the Starfish *Tremaster novaecaledoniae*. Tetrahedron Lett., 33, 1097 (1992).

41. MARSTON, A., and K. HOSTETTMANN: Modern Separation Methods. Nat. Prod. Rep., 8, 391 (1991).

42. RICCIO, R., M. IORIZZI, and L. MINALE: Starfish Saponins XXX. Isolation of Sixteen Steroidal Glycosides and Three Polyhydroxysteroids from the Mediterranean Starfish *Coscinasterias tenuispina*. Bull. Soc. Chim. Belg., 95, 869 (1986).

43. RICCIO, R., M. IORIZZI, L. MINALE, Y. OSHIMA, and T. YASUMOTO: Starfish Saponins. Part 34. Novel Steroidal Glycosides Sulphates from the Starfish *Asterias amurensis*. J. Chem. Soc. Perkin Trans I, 1337 (1988).

44. IORIZZI, M., L. MINALE, R. RICCIO, T. HIGA, and J. TANAKA: Starfish Saponins. Part 46. Steroidal Glycosides and Polyhydroxysteroids from the Starfish *Culcita novaeguineae*. J. Nat. Prod., 54, 1254 (1991).

45. OKANO, K., and S. IKEGAMI: Separation of Ovarian Asterosaponins in the Starfish *Asterias amurensis*. Agric. Biol. Chem., 45, 801 (1981).

46. NOGUCHI, Y., R. HIGUCHI, N. MARUBAYASHI, and T. KOMORI: Steroidal Oligoglycosides from the Starfish *Asterina pectinifera* Muller and Troschel, 1. Structures of Two New Saponins and Two New Oligoglycoside Sulfates: Pectinioside A and Pectinioside B. Liebigs Ann. Chem., 341 (1987).

47. DUBOIS, M. A., Y. NOGUCHI, R. HIGUCHI, and T. KOMORI: Structures of Two New Oligoglycoside sulfates: Pectinioside C and Pectinioside D. Liebig Ann. Chem., 495 (1988).

48. FINDLAY, J. A., ZHENG-QUAN HE, and B. BLACKWELL: Minor Saponins from the Starfish *Asterias forbesi*. Can. J. Chem., 68, 1215 (1990).

49. D'AURIA, M. V., E. FINAMORE, L. MINALE, C. PIZZA, R. RICCIO, F. ZOLLO, M. PUSSET, and P. TIRARD: Steroids from the Starfish *Euretaster insignis*: a Novel Group of Sulphated 3α,21-Dihydroxysteroids. J. Chem. Soc. Perkin Trans I, 2277 (1984).

50. SHEIKH, Y. M., B. M. TURSCH, and C. DJERASSI: 5α-Pregn-9(11)-ene-3β,6α-diol-20-one and 5α-Cholesta-9(11).20(22)-diene-3β,6α-diol-23-one. Two Novel Steroids from the Starfish *Acanthaster planci*. J. Am. Chem. Soc., 94, 3278 (1972).

51. SHIMIZU, Y.: Characterization of an Acid Hydrolysis Product of Starfish Toxins as a 5α-Pregnane Derivative. J. Am. Chem. Soc., 94, 4051 (1972).

52. APSIMON, J. W., J. A. BUCCINI, and S. BADRIPERSAUD: Marine Organic Chemistry I. Isolation of 3β,6α-Dihydroxy-5α-pregn-9(11)-en-20-one from the Saponins of the Starfish *Asterias forbesi*. Can. J. Chem., 51, 850 (1973).

53. KITAGAWA, I ., K. KOBAYASHI, and T. SUGAWARA: Saponin and Sapogenol. XXV. Steroidal Saponins from the Starfish *Acanthaster planci* L. (Crown of the Thorns). (1). Structures of Two Genuine Sapogenols, Thornasterol A and Thornasterol B, and Their Sulfates. Chem. Pharm. Bull., **26**, 1852 (1978).

54. DE SIMONE, F., A. DINI, L. MINALE, C. PIZZA, and R. RICCIO: Starfish Saponins III. A Novel Steroidal Sapogenin, 17β-Methyl-3β,6α-dihydroxy-18-nor-5α-cholesta-9(11),13-dien-23-one from the Starfish *Astropecten aurantiacus*. Tetrahedron Lett., 959 (1979).

55. DE SIMONE, F., A. DINI, E. FINAMORE, L. MINALE, C. PIZZA, and R. RICCIO: Starfish Saponins IV. Sapogenins from the Starfish *Astropecten aurantiacus and Marthasterias glacialis*. Comp. Biochem. Physiol., **64B**, 25 (1979).

56. YASUMOTO, T., and Y. HASHIMOTO: Properties and Sugar Components of Asterosaponin A, Isolated from Starfish. Agric. Biol. Chem., **29**, 804 (1965).

57. IKEGAMI, S., Y. HIROSE, Y. KAMIYA, and S. TAMURA: Structure of Carbohydrate Moiety of Asterosaponin A. Agr. Biol. Chem., **36**, 2453 (1972).

58. IKEGAMI, S., Y. KAMIYA, and S. TAMURA: New Steroidal Sulfate Obtained from a Starfish Saponin, Asterosaponin A. Tetrahedron Lett., 731 (1973).

59. OKANO, K., T. NAKAMURA, Y. KAMIYA, and S. IKEGAMI: Structure of Ovarian Asterosaponin-1 in the Starfish *Asterias amurensis*. Agric. Biol. Chem., **45**, 805 (1981).

60. KITAGAWA, I., and M. KOBAYASHI: Saponin and Sapogenol. XXVI. Steroidal Saponins from the Starfish *Acanthaster planci* L. (Crown of the Thorns). (2). Structure of the Major Saponin Thornasteroside A. Chem. Pharm. Bull., **23**, 1864 (1978).

61. IKEGAMI, S., K. OKANO, and H. MURAGAKI: Structure of Glycoside B₂. A Steroidal Saponin in the Ovary of the Starfish *Asterias amurensis*. Tetrahedron Lett., 1769 (1979).

62. RICCIO, R., O. SQUILLACE GRECO, L. MINALE, J. PUSSET, and J. L. MENOU: Starfish Saponins, Part 18. Steroidal Glycoside Sulfates from the Starfish *Linkia laevigata*. J. Nat. Prod., **48**, 97 (1985).

63. IORIZZI, M., L. MINALE, and R. RICCIO: Starfish Saponins. Part 39. Steroidal Oligoglycoside Sulphates and Polyhydroxysteroids from the Starfish *Asterina pectinifera*. Gazz. Chim. It., **120**, 147 (1990).

64. D'AURIA, M. V., M. IORIZZI, L. MINALE, and R. RICCIO: Starfish Saponins. Part 40. Structures of Two New "Asterosaponins" from the Starfish *Patiria miniata*: Patirioside A and Patirioside B. J. Chem. Soc. Perkin Trans I, 1019 (1990).

65. ITAKURA, Y., T. KOMORI, and T. KAWASAKI: Steroid Oligoglycosides from the Starfish *Asterias amurensis* [*cf.*] *versicolor* Sladen, 1. Structural Elucidation of a New Oligoglycoside Sulfate, Liebig. Ann. Chem., 2079 (1983).

66. MINALE, L., R. RICCIO, O. SQUILLACE GRECO, J. PUSSET, and J. L. MENOU: Starfish Saponins-XVI. Composition of the Steroidal Glycoside Sulphates from the Starfish *Luidia maculata*. Comp. Biochem. Physiol., **80B**, 113 (1985).

67. RICCIO, R., C. PIZZA, O. SQUILLACE GRECO, and L. MINALE: Starfish Saponins. Part 17. Steroidal Glycoside Sulfates from the Starfish *Ophidiaster ophidianus* (Lamark) and *Hacelia attenuata* (Gray). J. Chem. Soc. Perkin Trans I, 655 (1985).

68. RICCIO, R., F. ZOLLO, E. FINAMORE, L. MINALE, D. LAURENT, G. BARGIBANT, and J. PUSSET: Starfish Saponins, 19. A Novel Steroidal Glycoside Sulfate from the Starfishs *Protoreaster nodosus* and *Pentaceraster alveolatus*. J. Nat. Prod., **48**, 266 (1985).

69. RICCIO, R., M. IORIZZI, O. SQUILLACE GRECO, L. MINALE, M. DEBRAY, and J. L. MENOU: Starfish Saponins, 22., Asterosaponins from the Starfish *Halityle regularis*. A Novel 22,23-Epoxysteroidal Glycoside Sulfate. J. Nat. Prod., **48**, 756 (1985).

70. RICCIO, R., O. SQUILLACE GRECO, L. MINALE, D. DUHET, D. LAURENT, J. PUSSET, G. CHAUVIERE, and M. PUSSET: Starfish Saponins, Part 28. Steroidal Glycosides from the Pacific Starfish of the Genus *Nardoa*. J. Nat. Prod., **49**, 1141 (1986).

71. RICCIO, R., O. SQUILLACE GRECO, L. MINALE, S. LA BARRE, and D. LAURENT: Starfish Saponins, Part 36. Steroidal Oligoglycosides from the Pacific Starfish *Tromidia catalai*. J. Nat. Prod., **51**, 1003 (1988).

72. ZOLLO, F., E. FINAMORE, R. RICCIO, and L. MINALE: Starfish Saponins, Part 37. Steroidal Glycoside Sulfates from the Starfish of the Genus *Pisaster*. J. Nat. Prod., **52**, 693, (1989).

73. ZOLLO, F., E. FINAMORE, C. MARTUCCIO, and L. MINALE: Starfish Saponins, Part 44. Steroidal Glycosides from the Starfish *Pisaster giganteus*. J. Nat. Prod., **53**, 1000 (1990).

74. BRUNO, I., L. MINALE, and R. RICCIO: Starfish Saponins, Part 38. Steroidal Glycosides from the Starfish *Pycnopodia helianthoides*. J. Nat. Prod., **52**, 1022 (1989).

75. HONDA, M., and T. KOMORI: Structures of Thornasterols A and B (Biologically Active Glycosides from Asteroidea, XI). Tetrahedron Lett., **27**, 3369 (1986).

76. FINDLAY, J. A., M. JASEJA, D. J. BURNELL, and J. R. BRISSON: Major Saponins from the Starfish *Asterias forbesi*. Complete Structures by Nuclear Magnetic Resonance Methods. Can. J. Chem., **65**, 1386 (1987).

77. ITAKURA, Y., and T. KOMORI: Structure Elucidation of Two New Oligoglycoside Sulfates, Versicoside B and Versicoside C. Liebigs Ann. Chem., 359 (1986).

78. FINDLAY, J. A., M. JASEJA, and J. R. BRISSON: Forbeside C, a Saponin from *Asterias forbesi*. Complete Structures by Nuclear Magnetic Resonance Methods. Can. J. Chem., **65**, 2605 (1987).

79. ITAKURA, Y., T. KOMORI, and T. KAWASAKI: Steroid Oligoglycosides from Starfish *Astropecten latespinosus* Meissner. Liebigs Ann. Chem., 56 (1983).

80. ITAKURA, Y., and T. KOMORI : Structures of Four New Oligoglycosides Sulfates. Liebig Ann. Chem., 499 (1986).

81. KOMORI, T., H. Chr. KREBS, Y. ITAKURA, R. HIGUCHI, K. SAKAMOTO, S. TAGUCHI, and T. KAWASAKI: Steroid Oligoglycoside aus dem Seestern *Luidia maculata* Muller et Troschel, 1. Die Structuren Eines Neuen Aglyconsulfates und von Zwei Neuen Oligoglycosidsulfaten. Liebigs Ann. Chem., 2092 (1983).

82. DINI, A., F. A. MELLON, L. MINALE, C. PIZZA, R. RICCIO, R. SELF, and F. ZOLLO: Starfish Saponins - XI. Isolation and Partial Characterization of the Saponins from the Starfish *Marthasterias glacialis*. Comp. Biochem. Physiol., **76B**, 839 (1983).

83. BRUNO, I., L. MINALE, C. PIZZA, F. ZOLLO, R. RICCIO, and F. A. MELLON: Starfish Saponins. Part 14. Structures of the Steroidal Glycosides Sulphates from the Starfish *Marthasterias glacialis*. J. Chem. Soc. Perkin Trans I, 1875 (1984).

84. IORIZZI, M., L. MINALE, R. RICCIO, and H. KAMIYA: Starfish Saponins, 45. Novel Sulfated Steroidal Glycosides from the Starfish *Astropecten scoparius*. J. Nat. Prod., **53**, 1225 (1990).

85. FINAMORE, E., L. MINALE, R. RICCIO, G. RINALDO, and F. ZOLLO: Novel Marine Polyhydroxylated Steroids from the Starfish *Myxoderma platyacanthum*. J. Org. Chem., **56**, 1146 (1991).

86. KOMORI, T., H. NANRI, Y. ITAKURA, K. SAKAMOTO, S. TAGUCHI, R. HIGUCHI, T. KAWASAKI, and T. HIGUCHI: Structures of Two Newly Characterized Genuine Sapogenins and an Oligoglycoside Sulfate. Liebigs Ann. Chem., 37 (1983).

87. MACKIE, A. M., and A. B. TURNER: Partial Characterization of a Biologically Active Steroidal Glycoside Isolated from the Starfish *Marthasterias glacialis*. Biochem. J., **117**, 543 (1970).

88. Smith, D. H. S., A. B. Turner, and A. M. Mackie: Marine Steroid Part I. Structure of the Principal Aglycones from the Saponins of the Starfish *Marthasterias glacialis.* J. Chem. Soc. Perkin Trans I, 1745 (1973).

89. Nicholson, S. H., and A. B. Turner: Marine Steroids. Part III. On the Structure of Marthasterone Glucoside, from the Starfish *Martasterias glacialis.* J. Chem. Soc. Perkin Trans I, 1357 (1976).

90. Krebs, H. Chr., T. Komori, and T. Kawasaki: Steroid-Oligoglycoside aus dem Seestern *Luidia maculata* Muller et Troschel, 2. Die Strukturen von Zwei Neuen Oligoglycosidsulfaten. Liebig Ann. Chem., 296 (1984).

91. Hikino, H., T. Okuyama, S. Arihara, Y. Hikino, T. Takemoto, H. Mori, and K. Shibata: Shidasterone, an Insect Metamorphosing Substance from *Bleehnum niponicum*: Structure. Chem. Pharm. Bull., **23**, 1458 (1975).

92. Okano, K., N. Ohkawa, and S. Ikegami: Structure of Ovarian Asterosaponin-4, an Inhibitor of Spontaneous Oocyte Maturation from the Starfish *Asterias amurensis.* Agric. Biol. Chem., **49**, 2823 (1985).

93. Iorizzi, M., L. Minale, R. Riccio, and T. Yasumoto: Starfish Saponins, 48. Isolation of Fifteen Steroid Constituents (Six Glycosides and Nine polyhydroxysteroids) from the Starfish *Solaster borealis.* J. Nat. Prod., **55**, 866 (1992).

94. D' Auria, M. V., A. Fontana, L. Minale, and R. Riccio: Starfish Saponins. Part XLII. Isolation of Twelve Steroidal Glycosides from the Pacific Ocean Starfish *Henricia laeviuscula.* Gazz. Chim. It., **120**, 155 (1990).

95. Honda, M., T. Igarashi, and T. Komori: Structure of Pectinioside C: Determination of the Stereochemistry of the C-17 Side Chain of the Steroidal Aglycone. Liebigs. Ann. Chem., 547 (1990).

96. Findlay, J. A., Zheng-quan He, and M. Jaseja: Forbeside E: a Novel Sulfated Glycoside from *Asterias forbesi.* Can. J. Chem., **67**, 2078 (1989).

97. Apsimon, J. W., S. Badripersaud, J. A. Buccini, J. Eenkhoorn, and M. W. Gilgan: Marine Organic Chemistry. IV. Isolation of (20R)-5α-pregn-9(11)-3β,6α,20-triol from the Saponins of the Starfish, *Asterias forbesi* and *Asterias vulgaris,* and its Synthesis. Can. J. Chem., **58**, 2703 (1980).

98. Gorin, P. A., and M. Mazurek: Further Studies on the Assignment of Signals in ^{13}C Magnetic Resonance Spectra of Aldoses and Derived Methyl Glycosides. Can. J. Chem., **53**, 1212 (1975).

99. Breitmaier, E., and W. Voelter: In: Carbon-13 NMR Spectroscopy, ed. VCH Verlagsgesellshaft GmbH, D-6940, Weinheim (Federal Republic of Germany), pp. 380–393 (1987).

100. Petrakova, E., and P. Kovac: ^{13}C-NMR Spectra of Isomeric D-Xylobioses. Chem. Zvesti, **35**, 551 (1981).

101. Tori, K., S. Shujino, Y. Yoshimura, M. Arita, and Y. Tomita: Glycosidation Shifts in Carbon-13 NMR Spectroscopy: Carbon-13 Signal Shifts from Aglycone and Glucose to Glucoside. Tetrahedron Lett., 179 (1977).

102. Komori, T., T. Kawasaki, and H. R. Schulten: Field Desorption and Fast Atom Bombardment Mass Spectrometry of Biologically Active Natural Oligoglycosides. Mass Spectrom. Reviews, **4**, 255 (1985).

103. De Simone, F., A. Dini, E. Finamore L. Minale, C. Pizza, R. Riccio, and F. Zollo: Starfish Saponins. Part 5. Structure of Sepositoside A, a Novel Steroidal Cyclic Glycoside from the Starfish *Echinaster sepositus.* J. Chem. Soc. Perkin Trans I, 1855 (1981).

104. Riccio, R., F. De Simone, A. Dini, L. Minale, C. Pizza, F. Senatore, and F. Zollo: Starfish Saponins VI - Unique 22,23-Epoxysteroidal Cyclic Glycosides Minor Constituents from *Echinaster sepositus.* Tetrahedron Lett., **22**, 1557 (1981).

105. Riccio, R., A. Dini, L. Minale, C. Pizza, F. Zollo, and T. Sevenet: Starfish Saponins VII. Structure of Luzonicoside, a further Steroidal Cyclic Glycoside from the Pacific Starfish *Echinaster luzonicus*. Experientia **38**, 68 (1982).

106. De Riccardis, F., M. Iorizzi, L. Minale, and R. Riccio, unpublished.

107. Minale, L., R. Riccio, F. De Simone, A. Dini, C. Pizza, and E. Ramundo: Starfish Saponins I. 3β-Hydroxy-5α-cholesta-8,14-dien-23-one, the Major Genin from the Starfish *Echinaster sepositus*. Tetrahedron Lett., 2609 (1978).

108. Minale, L., R. Riccio, F. De Simone, A. Dini, and C. Pizza: Starfish Saponins II. 22,23-Epoxy Steroids, Minor Genins from the Starfish *Echinaster sepositus*. Tetrahedron Lett., 645 (1979).

109. Casapullo, A., E. Finamore, L. Minale, F. Zollo, J. B. Carre, C. Debitus, D. Laurent, A. Folgore, F. Galdieri, Starfish Saponins, Port 49. New Cytotoxic Steroidal glycosides from the starfish *Fromia monilis*. J. Nat. Prod., **55**, 0000 (1992).

110. Riccio, R., L. Minale, C. Pizza, F. Zollo, and J. Pusset: Starfish Saponins. Part 8: Structure of Nodososide, a Novel Type of Steroidal Glycoside from the Starfish *Protoreaster nodosus*. Tetrahedron Lett., **23**, 2899 (1982).

111. Minale, L., C. Pizza, R. Riccio, F. Zollo: Starfish Saponins. Part 9. A Novel 24-O-Glycosidated Steroid from the Starfish *Hacelia attenuata*. Experientia **39**, 567 (1983).

112. Iorizzi, M., L. Minale, R. Riccio, M. Debray, and J. L. Menou: Starfish Saponins, Part 23. Steroidal Glycosides from the Starfish *Halytile regularis*. J. Nat. Prod., **49**. 67 (1986).

113. Eggert, H., C. L. Van Antwerp, N. S. Bhacca, and C. Djerassi: Carbon-13 Nuclear Magnetic Resonance Spectra of Hydroxysteroids. J. Org. Chem., **41**, 71 (1976).

114. Van Antwerp, C. L., H. Eggert, G. D. Meakins, J. O. Miners, and C. Djerassi: Additivity Relationships in Carbon-13 Nuclear Magnetic Resonance Spectra of Dihydroxysteroids. J. Org. Chem., **42**, 789 (1977).

115. Kicha, A. A., A. J. Kalinosky, E. V. Levina, V. A. Stonik, G. B. Elyakov: Asterosaponin P_1 from the Starfish *Patiria pectinifera*. Tetrahedron Lett., **24**, 3893 (1983).

116. Segura De Correa, R., R. Riccio, L. Minale, and C. Duque: Starfish Saponins, Part 21. Steroidal Glycosides from the Starfish *Oreaster reticulatus*. J. Nat. Prod., **48**, 751 (1985).

117. D' Auria, M. V., M. Iorizzi, L. Minale, R. Riccio, and E. Uriarte: Starfish Saponins, Part 41. Structure of Two New Steroidal Glycoside Sulfates (Miniatosides A and B) and Two New Polyhydroxylated from the Starfish *Patiria miniata*. J. Nat. Prod., **53**, 94 (1990).

118. Andersson, L., L. Bohlin, R. Riccio, and L. Minale: Studies of Swedish Marine Organisms. Part 8: Three Novel Minor Polyhydroxysteroids Steroid Glycosides from the Starfish *Crossaster papposus*. J. Chem. Research (S), 246 (1987), (M), 2085 (1987).

119. Higuchi, R., Y. Noguchi, T. Komori, and T. Sasaki: ^{1}H -NMR Spectroscopy and Biological Activities of Polyhydroxylated Steroids from the Starfish. *Asterina pectinifera* Muller et Trochel. Liebigs Ann. Chem., 1185 (1988).

120. Pizza, C., P. Pezzullo, L. Minale, E. Breitmaier, J. Pusset, and P. Tirard: Starfish Saponins. Part 20. Two Novel Steroidal Glycosides from the Starfish *Acanthaster planci* L. J. Chem. Research, (S) 76 (1985); (M) 969 (1985).

121. Andersson, L., S. Bano, L. Bohlin, R. Riccio, and L. Minale: Studies of Swedish Marine Organisms VII. A Novel Bioactive Steroidal Glycoside from the Starfish *Crossaster papposus*. J. Chem. Research (S), 366 (1985); (M), 3873 (1985).

304   L. MINALE, R. RICCIO, and F. ZOLLO

122. MINALE, L., C. PIZZA, R. RICCIO, and F. ZOLLO: Starfish Saponins. Part 10. Further 24-O-Glycosidated Steroids from the Starfish *Hacelia attenuata.* Experientia, **39**, 569 (1983).

123. HARADA, N., and K. NAKANISHI: In: Circular Dichroic Spectroscopy- Exciton Coupling in Organic Stereochemistry. University Sceince Books, Mill Valley, U.S.A., pp. 164–189 (1983).

124. ZOLLO, F., E. FINAMORE, L. MINALE, D. LAURENT, and G. BARGIBANT: Starfish Saponins, 29. A Novel Steroidal Glycoside from the Starfish *Pentaceraster alveolatus.* J. Nat. Prod., **49**, 919 (1986).

125. RICCIO, R., L. MINALE, S. PAGONIS, C. PIZZA, F. ZOLLO, and J. PUSSET: A Novel Group of Highly Hydroxylated Steroids from the Starfish *Protoreaster nodosus.* Tetrahedron, **38**, 3615 (1982).

126. KICHA, A. A., A. I. KALINOVSKII, E. V. LEVINA, Ya. B. RASHKES, V. A. STONIK, and G. B. ELYAKOV: A New Steroid Glycoside from the Starfish *Patiria pectinifera.* Chem. Nat. Compd. (English) 332 (1985).

127. MINALE, L., C. PIZZA A. PLOMITALLO, R. RICCIO, F.ZOLLO, F. A. MELLON: Starfish Saponins. XII. Sulphated Steroid Glycoside from the Starfish *Hacelia attenuata.* Gazz. Chim. It., **114**, 143 (1984)

128. ZOLLO, F., E. FINAMORE, and L. MINALE: Starfish Saponins. XXIV. Two Novel Steroidal GlycosideSulphates from the Starfish *Echinaster sepositus.* Gazz. Chim. It., **115**, 303 (1985).

129. KICHA, A. A., A. I. KALINOWSKY, E. V. LEVINA, Ya. V. RASHKES, and V. A. STONIK: Khim Prir Soedin, **21**, 356 (1987): Quoted in Ref. 13.

130. MINALE, L., C. PIZZA, F. ZOLLO, and R. RICCIO: Trace Polyhydroxylated Steroids from the Starfish *Hacelia attenuata.* J. Nat. Prod., **46**, 736 (1983).

131. RICCIO, R., M. V. D' AURIA, M. IORIZZI, L. MINALE, D. LAURENT, and D. DUHET: Starfish Saponins. XXV. Steroidal Glycosides from the Starfish *Gomophia watsoni.* Gazz. Chim. It., **115**, 405 (1985).

132. RICCIO, R., L. MINALE, S. BANO, N. BANO, and V. UDDIN AHMAD: Starfish Saponins. Part XXXIII. Two Novel Steroidal Xylofuranosides from the Starfish *Astropecten indicus.* Gazz. Chim. It., **117**, 755 (1987).

133. RICCIO, R., O. SQUILLACE GRECO, and L. MINALE: Starfish Saponins, Part 35. Two Novel Steroidal Glycoside Sulphates from the Starfish *Marthasterias glacialis.* J. Nat. Prod., **51**, 989 (1988).

134. ZOLLO, F., E. FINAMORE, and L. MINALE: Starfish Saponins, XXXI. Novel polyhydroxysteroids and Steroidal Glycosides from the Starfish *Sphaerodiscus placenta.* J. Nat. Prod., **50**, 794 (1987).

135. ANDERSSON, L., A. NASIR, L. BOHLIN, and L. KENNE: Studies of Swedish Marine Organisms. IX: Polyhydroxylated Steroidal Glycoside from the Starfish *Porania pulvillus.* J. Nat. Prod., **50**, 994 (1987).

136. KAPUSTINA, I. I., A. I. KALINOVSKY, S. G. POLONIK, and V. A. STONIK: New Asterosaponins from the Starfish *Dismolasterias nipon.* Khim. Prir. Soedin, **23**, 250 (1987); Index Chemicus **106**, 404500 (1987).

137. BRIDGEMAN, J. E., P. C. CHERRY, A. S. CLEGGY, T. M. EVANS, Sir E. R. M. JONES, A. KASAL, V. KUMAR, G. D. MEAKINS, Y. MORISAWA, E. E. RICHARDS, and P. D. WOODGATE: Microbial Hydroxylation of Steroids. part I. Proton Magnetic Resonance Spectra of Ketones, Alcohols, and Acetates in the Androstane, Pregnane, and Oestrane Series. J. Chem. Soc. Perkin Trans I, 250 (1970).

138. KICHA, A. A., A. I. KALINOVSKY, E. V. LEVINA, and P. V. ANDRIYASHCHENKO: Culcitoside C1 from Starfishes *Culcita novaeguineae* and *Linckia guildingi.* Khim. Prir. Soedin., **21**, 801 (1985); C. A. **104**, 204151 (1986).

139. BRUNO, I., L. MINALE, and R. RICCIO: Starfish Saponins, Part 43. Structures of Two New Sulphated Steroidal Fucofuranosides (Imbricatosides A and B) and Six New Polyhydroxysteroids from the Starfish *Dermasterias imbricata.* J. Nat. Prod., **53**, 366 (1990).

140. D' AURIA, M. V., F. DE RICCARDIS, L. MINALE, and R. RICCIO Synthesis of 24-Methyl-26-hydroxysteroid Side Chains: Models for Stereochemical Assignments in Poly-hydroxylayted Marine Steroids. J. Chem. Soc. Perkin Trans I, 2889 (1990).

141. KICHA, A. A., A. I. KALINOVSKY, P. V. ANDRIYASHCHENKO, and E. V. LEVINA: Culcitosides C2 and C3 from the Starfish *Culcita novaeguineae.* Khim. Prir. Soedin., **22**, 592 (1986); C. A. **106**, 116759 (1987).

142. RICCIO, R., L. MINALE, S. BANO, and V. UDDIN AHMAD: Starfish Saponins Part 32. Structure of a Novel Steroidal 5-O-Methyl Galactofuranoside from the Starfish *Astropecten indicus.* Tetrahedron Lett., **28**, 2291 (1987).

143. FINAMORE, E., F. ZOLLO, L. MINALE, and T. YASUMOTO: Starfish Saponins Part 47. Steroidal Glycoside Sulfates and Polyhydroxysteroids from *Aphelasterias japonica.* J. Nat. Prod., **55**, 767 (1992).

144. PIZZA, C., L. MINALE, D. LAURENT, and J. L. MENOU: Starfish Saponins: XXVII. Steroidal Glycosides from the Starfish *Choriaster granulatus.* Gazz. Chim. It., **115**, 585 (1985).

145. FINDLAY, J. A., and ZHENG-QUAN HE: Polyhydroxylated Steroidal Glycosides from the Starfish *Asterias forbesi.* J. Nat. Prod., **54**, 428 (1991).

146. MINALE, L., C. PIZZA, R. RICCIO, F. ZOLLO, J. PUSSET, and P. LABOUTE: Starfish Saponins, XIII. Occurrence of Nodososide in the Starfish *Acanthaster planci* and *Linkia laevigata.* J. Nat. prod, **47**, 558 (1984).

147. D' AURIA, M. V., L. MINALE, C. PIZZA, R. RICCIO, and F. ZOLLO: Starfish Saponins XV. Stereochemistry at C(24) of Nodososide (24-O-Glycosidated Steroid) and at C(25) of 26-Hydroxysteroids. Gazz. Chim. It., **114**, 469 (1984).

148. LEVINA, E. V., A. I. KALINOVSKII, P. V. ANDRIYASHCHENKO, and A. A. KICHA: Steroid Glycosides from the Starfish *Echinaster sepositus.* Khim. Prir. Soedin. **23**, 246 (1987); Index Chemicus **106**, 404499 (1987).

149. RICCIO, R., M. IORIZZI, O. SQUILLACE GRECO, L. MINALE, D. LAURENT, and Y. BARBIN: Starfish Saponins. XXVI. Steroidal Glycosides from the Starfish *Poraster superbus.* Gazz. Chim, It., **115**, 505 (1985).

150. KICHA, A. A., A. I. KALINOVSKII, and V. A. STONIK: Steroidal Glycosides from the Starfish *Crossaster papposus.* Chem. Nat. Compd. **25**, 569 (1989).

151. RASHKES, YA. V., A. A. KICHA, E. V. LEVINA, and V. A. STONIK: Mass Spectra of Polyhydroxysteroids of the Starfish *Patiria pectinifera.* Chem. Nat. Compd., 337 (1985).

152. BRUNO, I., L. MINALE, R. RICCIO, S. LA BARRE, and D. LAURENT: Isolation and Structure of New Polyhydroxylated Sterols from a Deep-Water Starfish of the Genus *Rosaster.* Gazz. Chim. It., **120**, 449 (1990).

153. MINALE, L., C. PIZZA, R. RICCIO, C. SORRENTINO, F. ZOLLO, J. PUSSET, and G. BARGIBANT: Minor Polyhydroxylated Sterols from the Starfish *Protoreaster nodosus.* J. Nat. Prod., **47**, 790 (1984).

154. MINALE, L., C. PIZZA, F. ZOLLO, and R. RICCIO: 5α-Cholestane-3β,6β,15α,16β,26-pentol: A Polyhydroxylated Sterol from the Starfish *Hacelia attenuata.* Tetrahedron Lett., **23**, 1841 (1982).

155. MINALE, L., C. PIZZA, R. RICCIO, O. SQUILLACE GRECO, F. ZOLLO, J. PUSSET, and J. L. MENOU: New Polyhydroxylated Sterols from the Starfish *Luidia maculata.* J. Nat. Prod. **47**, 784 (1984).

156. DE RICCARDIS, F., L. MINALE, R. RICCIO, B. GIOVANNITTI, M. IORIZZI, and C.

DEBITUS: Phosphated and Sulphated Marine Polyhydroxylated Steroids from the Starfish *Tremaster novaecaledoniae*. Gazz. Chim. It., **122**, 0000 (1992).

157. KICHA, A. A., A. I. KALINOVSKII, and V. A. STONIK: Steroid Hexaol from *Crossaster papposus*, Chem. Nat. Compd., **25**, 376 (1989).

158. RICCIO, R., O. SQUILLACE GRECO, L. MINALE, D. LAURENT, and D. DUHET: Highly Hydroxylated Marine Steroids from the Starfish *Archaster typicus*. J. Chem. Soc. Perkin Trans I, 665 (1986).

159. RICCIO, R., M. SANTANIELLO, O. SQUILLACE GRECO, and L. MINALE: Structure Elucidation of (22E,24R,25R)-24-Methyl-5α-cholest-22-ene-3β,4β,5,6α,8,14,15α,25,26-nonaol and (22E,24S)-24-Methyl-5α-cholest-22-ene-3β,4β,5,6α,8,14,15α,25,28-nonaol, Minor Marine Polyhydroxysteroids Isolated from the Starfish *Archaster typicus*. J. Chem. Soc. Perkin Trans I, 823 (1989).

160. MATTIA, C. A., L. MAZZARELLA, R. PULITI, R. RICCIO, and L. MINALE: Structure and Stereochemistry of (24R)-27-nor-5α-Cholestane-3β,4β,5,6α,7β,8,14,15α,24-nonaol: a Highly Hydroxylated Marine Steroid from the Starfish *Archaster typicus*. Acta Crystallographica, C**44**, 2170 (1988).

161. KOIZUMI, N., Y. FUJIMOTO, T. TAKESHITA, and N. IKEKAWA: Carbon-13 Nuclear Magnetic Resonance of 24-Substituted Steroids. Chem. Pharm. Bull., **27**, 38 (1979).

162. LETOURNEUX, Y. Q., KHUONG-HUU, M. GUT, and G. LUKACS: Identification of C-22 Epimers in Steroids by Carbon-13 Nuclear Magnetic Resonance Spectroscopy. J. Org. Chem., **40**, 1674 (1975).

163. KLUGE, A. F., M. L. MADDOX, and L. G. PARTRIDGE: Synthesis of (20R, 25R)-Cholest-5-ene-3β,26-diol and the Occurrence of Base-Catalyzed 1,5-Hydride Shift in a Steroidal 1,5-ketol. J. Org. Chem., **50**, 2359 (1985).

164. HIRANO, Y., T. EGUCHI, M. ISHIGURO, and N. IKEKAWA: Configuration at the C-23 Position of 23-Hydroxy- and 23,25-Dihydroxycholesterols. Chem. Pharm. Bull., **31**, 394 (1983).

165. RICCIO, R., M. V. D' AURIA, and L. MINALE: Unusual Sulfated Marine Steroids from the Ophiuroid *Ophioderma longicaudum*. Tetrahedron, **41**, 6041 (1985).

166. D' AURIA, M. V., R. RICCIO, L. MINALE, S. LA BARRE, and J. PUSSET: Novel Marine Steroid Sulfates from Pacific Ophiuroids. J. Org. Chem., **52**, 3947 (1987).

167. D' AURIA, M. V., R. RICCIO, E. URIARTE, L. MINALE, J. TANAKA, and T. HIGA: Isolation and Structure Elucidation of Seven New Polyhydroxylated Sulfated Sterols from the Ophiuroid *Ophiolepis superba*. J. Org. Chem., **54**, 234 (1989).

168. D'AURIA, M. V., L. GOMEZ PALOMA, L. MINALE, R. RICCIO, A. ZAMPELLA, T. HIGA, and J. TANAKA: Isolation, Structure Characterization and Conformational Analysis of a Unique 4α,9α-Epoxysteroid Sulphate from the Ophiuroid *Ophiomastix annulosa*. Tetrahedron Lett., **33**, 4641 (1992).

169. FUSETANI, N., S. MATSUNAGA, and S. KONOSU: Bioactive Marine Metabolites II. Halistanol Sulfate, an Antimicrobial Novel Steroid Sulfate from the Marine Sponge *Halicondria* CF. *Moorei* Bergquist. Tetrahedron Lett., **22**, 1985 (1981).

170. ARNOLDS, W., W. MEISTER, and G. ENGLERT, Substituent Increments for the ¹H-NMR. Chemical Shifts of the 18- and 19-Methyl Protons of Steroids. Part II. 9α,10β(Normal)-steroids. Helv. Chim. Acta, **57**, 1559 (1974),

171. D' AURIA, M. V., L. GOMEZ PALOMA, L. MINALE, and R. RICCIO: Stereochemical Assignment at C-24 and C-25 of Marine 24-Ethyl-26-hydroxy Steroids through Comparison with Synthetic (24S, 25S)-, (24S, 25R)-, (24R, 25R)-, and (24R, 25S)-Models. J. Chem. Soc. Perkin. Trans I, 2895 (1990).

172. RICCIO, R., E. FINAMORE, M. SANTANIELLO, and F. ZOLLO: Stereoselective Synthesis of (24S)- and (24R)-24-(Hydroxymethyl)-cholesta-5,22(E)-dien-3β-ol: Model Compounds for Stereochemical Assignments of Polyhydroxylated Marine Steroids. J. Org. Chem., **55**, 2548 (1990).

173. LEVINA, E. V., S. N. FEDOROV, V. A. STONIK, P. V. ADRIYASHCHENKO, A. I. KALINOVSKII, and V. V. ISAKOV: Two New Sulphated Steroid Polyols from *Ophiura sarsi*. Chem. Nat. Compd. **26**, 408 (1990).

174. D' AURIA M. V., L. GOMEZ PALOMA, L. MINALE, and R. RICCIO: unpublished results.

175. KOIZUMI, N., M. MORISAKI and N. IKEKAWA: Absolute Configuration of 24-Hydroxycholesterol and Related Compounds. Tetrahedron Lett., **26**, 2203 (1975).

176. KOIZUMI, N., Y. FUJMOTO, T. TAKESHITA and N. IKEKAWA: Carbon-13 Nuclear Magnetic Resonance of 24-Substituted Steroids. Chem. Pharm. Bull., **27**, 38 (1979).

177. DALE, J. A., and H. S. MOSHER: Nuclear Magnetic Resonance Enantiomer Reagents, Configurational Correlations *via* Nuclear Magnetic Resonance Chemical shifts of Diastereomeric Mandelate, *O*-Methylmandelate, and α-Methoxy-α-trifluoromethylphenylacetate (MTPA). J. Am. Chem. Soc., **95**, 512 (1973).

178. YAMAGUCHI, S.,: Nuclear Magnetic Resonance Analysis Using Chiral Derivatives. In: Asymmetric Synthesis (MORRISON, J. D., ed.), Vol I, pp. 125–152. New York: Academic Press. 1983.

179. GONNELLA, N. C. and K. NAKANISHI: General Method for Determining Absolute Configurations of Acyclic Allylic Alcohols. J. Am. Chem. Soc., **104**, 3775 (1982).

180. SEO, S., Y. YOSHIMURA, T. SATOH, A. UOMORI, and K. TAKEDA: Synthesis of (25S)-[26-$^2H_1$]-Cholesterol and ^{1}H-NMR Signal Assignments of the *pro-R* and *pro-S* Methyl Groups at C-25. J. Chem. Soc. Perkin Trans I, 411 (1986).

181. YASUHARA, F., and S. YAMAGUCHI: Use of Shift Reagents with MTPA Derivatives in ^{1}H NMR Spectroscopy. III. Determination of Absolute Configuration and Enantiomeric Purity of Primary carbinols with Chiral Center at the C-2 Position. Tetrahedron Lett., **18B**, 4085 (1977).

182. SUCROW, W., B. SHUBERT, W. RICHTER, and M. SLOPIANKA: Stereochemie einer Claisen-Umlagerung in der Sterin-Seitenkette. Chem. Ber., **104**, 3689 (1971).

183. WIERSIG, J. R., N. WAESPE-SARCEVIC, and C. DJERASSI: Stereospecific Synthesis of the Side Chain of the Steroidal Plant Sex Hormone Oogoniol. J. Org. Chem., **44**, 3374 (1979).

184. ANASTASIA, M., A. FIECCHI, and A. SCALA: Synthesis of (24R)- and (24S)-3β,29-Dihydroxystigmast-7-ene, a Model for the Side Chain of Oogoniol. J. Chem. Soc. Chem. Comm., 858 (1979).

185. PREUS, M. W., and T. C. McMORRIS: The Configuration at C-24 in Oogoniol [(24R)-3β,11α,15β,29-tetrahydroxystigmast-5-en-7-one] and Identification of 24(28)-Dehydroogoniols as Hormones in Achlya. J. Am. Chem. Soc., **101**, 3066 (1979).

186. SUCROW, W., M. SLOPIANKA, and H. W. KIRCHER: The Occurrence of C_{29} Sterols with Different Configurations at C-24 in *Cucurbita pepo* as shown by 270 MHz NMR. Phytochemistry, **15**, 1533 (1976).

187. HORIBE, J., H. NAKAI, T. STATO, S. SEO, and K. TAKEDA: Stereoselective Synthesis of the C-24 and C-25 Stereoisomeric Pairs of 24-Ethyl-26-hydroxy- and 24-Ethyl-[26-^{2}H] sterols and their Δ^{22}-Derivatives: Reassignment of ^{13}C N.m.r. Signals at C-25 of 24-Ethylsterols. J. Chem. Soc. Perkin Trans I, 1957 (1989).

188. ANASTASIA, M., P. ALLEVI, P. CIUFFREDA, and R. RICCIO: Configuration Assignment of 24R- and 24Ŝ-Isomers of 29-Oxygenated Steroids by ^{1}H and ^{13}C NMR Spectroscopy. Tetrahedron, **42**, 4843 (1986).

189. IRELAND, R. E., R. H. MUELLER, and A. K. WILLARD: The Ester Enolate Claisen Rearrangement. Stereochemical Control through Stereoselective Enolate Formation. J. Am. Chem. Soc., **98**, 2868 (1976).

190. IRELAND, R. E., and A. K. WILLARD: The Stereoselective Generation of Ester Enolates. Tetrahedron Lett., **45**, 3975 (1975).

191. KATSUKI, T., and K. B. SHARPLESS: The First Practical Method for Asymmetric Epoxidation. J. Am. Chem. Soc., **102**, 5974 (1980).

192. GAO, Y., R. M. HANSON, J. M. KLUNDER, S. Y. KO, H. MASAMUNE, and K. B. SHARPLESS: Catalytic Asymmetric Epoxidation and Kinetic Resolution: Modified Procedures Including *in situ* Derivatization. J. Am. Chem. Soc., **109**, 5765 (1987).

193. CHO YUKO, R. HIQUCHI, N. MARUBAYASHI, I. VEDA, and T. KOMORI: X-Ray Crystallographic Analysis of Desulphated Asterosaponin P-1 Isolated from the Starfish *Asterina pectinifera*. Liebigs Ann. Chem., 79 (1992).

(Received June 8, 1992)

Author Index

Subject Index

Fortschritte der Chemie organischer Naturstoffe

Progress in the Chemistry of Organic Natural Products

Founded by L. Zechmeister.
Edited by W. Herz, G.W. Kirby, R.E. Moore, W. Steglich, and C. Tamm

Volume 61:
1993. 4 figures. IX, 206 pages. Cloth DM 220,–, öS 1540,–
ISBN 3-211-82388-3
Contents: D.G.I. Kingston, A.A. Molinero, and J.M. Rimoldi: The Taxane Diterpenoids. • Author Index. • Subject Index.

Volume 60:
1992. 59 figures. VIII, 243 pages. Cloth. DM 184,–, öS 1290,–
ISBN 3-211-82374-3
Contents: I. Wahlberg and A.-M. Eklund: Cyclized Cembranoids of Natural Occurrence. M. Petitou and C.A.A. van Boeckel: Chemical Synthesis of Heparin Fragments and Analogues.

Volume 59:
1992. 1 figure. IX, 328 pages. Cloth DM 260,–, öS 1820,–.
ISBN 3-211-82278-X
Contents: Shin-Ichi Hatanaka: Amino Acids from Mushrooms. • I. Wahlberg and A.-M. Eklund: Cembranoids, Pseudopteranoids, and Cubitanoids of Natural Occurrence.

Prices are subject to change without notice.

Springer-Verlag Wien New York

Volume 58:
1991. 64 figures. VII, 343 pages. Cloth DM 280,–, öS 1960,–.
ISBN 3-211-82265-8
Contents: J.A. Robinson: Chemical and Biochemical Aspects of Polyether-Ionophore Antibiotic Biosynthesis. • R.D.H. Murray: Naturally Occurring Plant Coumarins.

Volume 57:
1991. 26 figures and 2 plates. X, 212 pages. DM 210,–, öS 1470,–.
ISBN 3-211-82245-3
Contents: P. Metzger, C. Largeau, E. Casadevall: Lipids and Macromolecular Lipids of the Hydrocarbon-rich Microalga *Botryococcus braunii*. Chemical Structure and Biosynthesis. Geochemical and Biotechnological Importance. • D.P. Chakraborty and S. Roy: Carbazole Alkaloids III. • G. R. Pettit: The Bryostatins.

Volume 56:
1991. 8 figures. X, 188 pages. Cloth DM 220,–, öS 1540,–.
ISBN 3-211-82188-0
Contents: J. Asselineau: Bacterial Lipids Containing Amino Acids or Peptides Linked by Amine Bonds. • J.Kagan: Naturally Occurring Di- and Trithiophenes.

All Volumes and Cumulative Index 1-20 available
Price reduction for subscribers: 10%

Special reduced price (20% reduction) for the complete Series Vols. 1-60 incl. the Cumulative Index to Vols. 1-20

Prices are subject to change without notice.

Springer-Verlag Wien New York